DX-Centres and other Metastable Defects
in Semiconductors

9780750301534

DX-Centres and other Metastable Defects in Semiconductors

Proceedings of the International Symposium
Mauterndorf, Austria, 18–22 February 1991

Edited by

W Jantsch
Institut für Experimentalphysik
Johannes Kepler Universität
Linz, Austria

and

R A Stradling
Imperial College of Science, Technology and Medicine
London, England

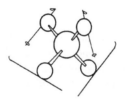

Adam Hilger
Bristol, Philadelphia and New York

British Library Cataloguing in Publication Data

International Symposium on DX-Centres and Other
Metastable Defects in Semiconductors
(1991 Mauterndorf, Austria)
    DX-centres and other metastable defects in
    semiconductors.
    I. Title   II. Stradling, R.A.   III. Jantsch, W.
    621.3815

    ISBN 0-7503-0153-8

Library of Congress Cataloging-in-Publication Data are available

Citation

When citing an item published in this book, authors are requested to cite the original journal of publication. Within *Semiconductor Science and Technology*, the requisite information is given in the 'catchline' at the top of the title page of each article. If required, such a citation may be supplemented by the expansion 'Reprinted in:' followed by the *Semiconductor Science and Technology* reference.

Published under the Adam Hilger imprint by IOP Publishing Ltd
Techno House, Redcliffe Way, Bristol BS1 6NX, England
335 East 45th Street, New York, NY 10017-3483, USA

US Editorial Office: 1411 Walnut Street, Philadelphia, PA 19102

Printed in Great Britain by William Gibbons & Sons Ltd, Wolverhampton

Contents

Contents

(Continued)

Foreword

The first reports on persistent photoconductivity due to metastable defects in III–V and II–VI compound semiconductors appeared in the late 1960s. After cooling a sample in darkness to temperatures of typically less than 100K, illumination with photon energies below the band gap causes a strong increase in the electrical conductivity, which persists after switching off the light source for periods exceeding hours or days. This phenomenon was explained by Lang and Logan in 1977 (*Phys. Rev. Lett.* **39** 635) in terms of a large rearrangement of the defect and its neighbourhood due to a change in the charge state of the defect. The rearrangement of the lattice is a thermally activated process which needs some thermal energy in order to overcome a barrier. This barrier prevents relaxation at low temperature.

Since this early work, the number of reports on defects with metastable excited states has been growing at an ever increasing rate. This topic promises to become one of the main topics of contemporary defect physics. A large number of problems remain unsolved—the microscopic structure and the mechanisms responsible for metastability have been identified only in very few cases, as reviewed here by George Watkins.

Apart from stirring scientific curiousity, the phenomenon of defect metastability is also of some technical importance: in many devices, memory effects may result from defect metastability, which in most cases is intolerable. This problem applies particularly to the so-called DX centres. DX centres apparently are a universal, unavoidable feature of n-doped III–V semiconductors and their alloys, which are employed in ultra-high frequency and electro-optical devices. The name 'DX centre' was created by Dave Lang, who suggested that the deep DX state was caused by a complex of a shallow donor D and some unidentified intrinsic defect X. In the meantime, following the theoretical results of Chadi and Chang, it has been widely agreed that the DX centre is just a distorted state of the isolated substitutional donor which is displaced along a $\langle 111 \rangle$ direction towards the interstitial T_d site.

DX centres cause a rich variety of stunning effects and many of them have been collected in the comprehensive review article by Pat Mooney in the February 1990 issue of *Journal of Applied Physics*, together with their implications for theory. DX centres have been also the subject of a session at the International Conference on the Physics of Semiconductors, held in Thessaloniki in August 1990, where a large number of new experimental findings and also new theoretical and controversial results were presented. There was not enough time, however, to discuss these results in detail and we had the feeling that we should come together again in order to clarify some of the open questions and to try to learn also from knowledge of other metastable defects—the idea of the 'International Symposium on DX Centres and other Metastable Defects in Semiconductors' was created by Janusz Dmochowski, Tony Stradling and myself.

This was the first workshop-like conference in Mauterndorf, a small town in the Austrian Alps with an old castle adapted for small conferences, but not the first meeting on the semiconductor topic. There is already a traditional and very successful biannual winter school, which has already taken place six times, initiated and organized by H Heinrich, G Bauer and F Kuchar, and with our organization we could rely on their experience and know-how.

Here I would like to take the opportunity to thank in particular Gerhard Brunthaler, who took part in the organization of this conference at all stages. Without his diligent and expert support it would not have been possible to arrange the meeting in the relatively short time. Our thanks are also due to Ulla Hannesschläger, Gernot Ostermayer and Alexander Falk, who developed the enchanting enthusiasm in the local organization. We are indebted to Tony Stradling and Janusz Dmochowski for their advice in establishing the scientific programme and to Tony in particular for the editorial work on these proceedings.

All of us would like to thank the invited speakers and all participants for their valuable contributions to make this conference a success. All of us regret deeply that Pat Mooney and Masahi Mizuta were not able to come because of the Gulf War.

Last but not least it is a pleasure to acknowledge the generous financial support received from:

Bundesministerium für Wissenschaft und Forschung, Austria
IOPP—Institute of Physics Publishing Ltd., Bristol England

Bomem Inc., represented by Haider GmbH, Kottingbrunn, Austria
Bruker Analytische Meßtechnik GmbH, Karlsruhe, Germany
Instruments SA (Riber, Jobin–Yvon), Grasbrunn, Germany
GME—Gesellschaft für Mikroelektronik, Vienna, Austria
Hainzl Industriesysteme, Linz, Austria
IBM Österreich GmbH, Vienna, Austria
Multicon Electronic GmbH, Enns, Austria
Österreichische Forschungsgemeinschaft, Vienna, Austria
ÖPG—Österreichische Physikalische Gesellschaft, Austria
Österreichisches Verkehrsbüro, Linz, Austria
Philips Österreich GmbH, Vienna, Austria
Salzburger Landesregierung, Salzburg, Austria
Spectroscopy Instruments GmbH, Gilching, Germany

Wolfgang Jantsch

Semicond. Sci. Technol. **6** (1991) B1–B8. Printed in the UK

Donor-related levels in GaAs and Al$_x$Ga$_{1-x}$As

P M Mooney

IBM Research Division, T J Watson Research Center, PO Box 218, Yorktown Heights, NY 10598 USA

Abstract. It is now widely accepted that DX levels in n-type Al$_x$Ga$_{1-x}$As are ground states of isolated substitutional donors in distorted configurations which are stabilized by trapping two electrons. The leading model is that the distortion occurs when the group IV donor substituted on a group III lattice site, moves along a ⟨111⟩ axis towards an interstitial site. In the case of a group VI donor which substitutes on a group V lattice site, one of the group III neighbours moves toward the interstitial site. Recent work suggests that localized excited states of both donor configurations play an important role in electron transitions between DX levels and the conduction band. The research leading to this view of substitutional donors is reviewed. Some implications of this model for technological applications of Al$_x$Ga$_{1-x}$As and related alloys are discussed.

1. Introduction

The first paper on donor-related deep levels in III–V semiconductor alloys appeared in 1968 [1]. Since then an enormous amount of work has been done to understand both the fundamental nature of the donors in these materials and the effects of the deep levels on heterostructure devices fabricated with them [2]. In the 1970s, studies of deep levels in GaAs$_y$P$_{1-y}$ and Al$_x$Ga$_{1-x}$As were motivated by the usefulness of these alloys for light emitting diodes (LEDs) and heterojunction lasers. The presence of localized electronic states in these materials, when they are doped with any of the group IV or group VI n-type dopants, reduces the conductivity compared with that of GaAs with equivalent dopant concentrations and also results in persistent photoconductivity (PPC) at low temperature. Lang and co-workers, whose comprehensive experiments on n-type Al$_x$Ga$_{1-x}$As revealed a deep state having a thermally activated capture cross section, no observable radiative capture and a large optical ionization energy, concluded that electron capture at these deep states was accompanied by a distortion of the crystal lattice near the donor atoms [3, 4]. They inferred from this 'large lattice relaxation' that the origin of the deep levels must be a defect complex involving the donor (D) and an unknown defect (X); thus these states are known as DX centres.

Research on III–V semiconductor alloys exploded in the 1980s when it was shown that III–V heterojunction structures were also potentially useful for high-speed digital applications [5]. Demonstration that the presence of DX levels in Al$_x$Ga$_{1-x}$As caused instabilities in modu-lation-doped field effect transistors (MODFETs) [6] led to attempts to eliminate DX levels from device structures. One approach, based on the defect complex model of DX centres, was to replace the Si-doped Al$_x$Ga$_{1-x}$As layers with GaAs/AlAs superlattices having the same effective conduction band offset [7]. It was thought that DX levels would be eliminated if the dopant atoms were incorporated only in the GaAs wells of the superlattice, since donors in GaAs are well behaved. The failure of attempts to eliminate instabilities in MODFETs led to an increased effort to understand the fundamental properties of donors in this alloy. Many experiments indicated that DX levels originate from isolated donors rather than defect complexes and several new models for the microscopic structure of the DX levels were proposed during the 1980s. Controversial issues were the magnitude of the lattice distortion and the charge state of the DX centres. A coherent picture of donors in Al$_x$Ga$_{1-x}$As has emerged recently. In this picture isolated donor atoms exist in two different configurations: substitutional configurations usually having shallow hydrogenic ground states, or distorted configurations with the DX level as the ground state. Here the work leading to this picture is reviewed. Some unresolved questions and new problems requiring further work are also examined.

2. Origin of DX levels

Strong evidence that DX levels are states of isolated donor atoms rather than defect complexes comes from hydrostatic pressure experiments. The application of hydrostatic pressure to GaAs compresses the materials, thus reducing the distance between atoms. This modifies the energy position of the conduction band valleys in a way roughly similar to the addition of Al, but without changing the chemical nature of the atoms around the donors. When sufficient pressure is applied to GaAs, PPC and a deep-level transient spectroscopy (DLTS) peak with characteristics similar to the DLTS spectrum of DX levels in $Al_xGa_{1-x}As$ are observed [8, 9]. Infrared absorption measurements of the local vibrational modes of Si in GaAs indicate that 90 % of the Si is substitutional on the Ga site in samples which exhibit large PPC under hydrostatic pressure [10]. Therefore the DX level exists in GaAs where isolated donors are known to occupy substitutional sites. It lies about 260 meV above the bottom of the conduction band in GaAs [11], however, and is occupied only when the material is very heavily doped or when pressure is applied. Additional evidence that DX levels are states of isolated donor atoms is the observation of shallow hydrogenic states associated with the Γ and X minima of the conduction band in $Al_xGa_{1-x}As$ after photoionization of the DX levels at low temperature [12–15]. The hydrogenic levels have been seen in samples with several different n-type dopants by infrared absorption [12, 13], Hall effect [14, 15] and electron paramagnetic resonance [16–18] measurements. The X-valley hydrogenic levels have been studied extensively using spin resonance techniques [19–21]. Hydrogenic levels are states of isolated group IV donors substituted on group III lattice sites or group VI donors substituted on group IV lattice sites, not of defect complexes. Theis *et al* showed that both the DX levels and the hydrogenic levels arise from the same chemical species [12].

Figure 1 shows the alloy composition dependence of the conduction band minima and the important donor-related levels observed in $Al_xGa_{1-x}As$. The reference energy is the top of the valence band in GaAs. The alloy composition dependence of the Γ, L, and X valleys is taken from [22], where the large uncertainty in the position of the L valley was pointed out. The open circles are the Γ-valley hydrogenic levels in Si-doped $Al_xGa_{1-x}As$ [12] and the open squares are the X-valley hydrogenic levels in Te-doped $Al_xGa_{1-x}As$ [13]. The dotted lines represent the energies of the DX levels, as is discussed in detail later on. The conductivity of $Al_xGa_{1-x}As$ is strongly temperature dependent and PPC is observed at low temperature in the composition range where the DX levels are the lowest states.

Up to four different DX levels have been resolved in Si-doped $Al_xGa_{1-x}As$ in experiments which measure both electron emission [23–25] and electron capture [15, 26], whereas there is only a single DX level in GaAs [11, 23, 24, 27]. Figure 2 shows DLTS spectra for GaAs and dilute $Al_xGa_{1-x}As$. The dominant DLTS peak in the alloy appears at a higher temperature than the peak in GaAs. Two peaks, one the same as that in GaAs, are seen in the x = 0.04 sample and a third peak appears as a shoulder on the high-temperature side of the dominant peak in the x = 0.08 sample. The three emission processes giving rise to these DLTS peaks are independent of alloy composition and thus cannot represent transitions from a single deep state to different conduction band minima. This conclusion is supported by DLTS measurements which demonstrate that there is no change in the emission rates when the band structure is modified by hydrostatic pressure [27–29]. Only a single emission rate is seen in GaAs [27], suggesting that the presence of multiple emission rates in the alloy is because Si donors

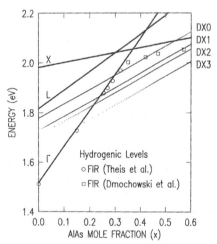

Figure 1. Band edges and important deep and shallow levels in $Al_xGa_{1-x}As$ as a function of alloy composition. The shallow levels are obtained from [12] and [13]. The four DX levels are deduced from various pressure experiments as described in the text. The density of dots is proportional to the relative abundance of the configuration in the random alloy [30].

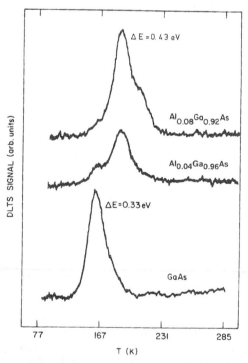

Figure 2. DLTS spectra for Si-doped GaAs and dilute $Al_xGa_{1-x}As$ alloys.

have different configurations of Ga and Al neighbours [23-30]. This interpretation is supported by recent DLTS measurements on Si-doped $In_y(Al_{0.3}Ga_{0.7})_{1-y}As$ which show a new DLTS peak corresponding to electron emission from Si atoms having In close neighbours (figure 3) [31].

Perturbations of the electron emission and capture rates due to the substitution of various group III atoms in alloys give important information about the atomic configurations of the DX centres. Substitutional donors are bonded to four As atoms and have 12 group III atoms as second nearest neighbours; thus one would expect to see many more than four configurations if the DX levels are states of substitutional donors. The DLTS results are, therefore, consistent only with DX levels being states of distorted donor configurations. As was first pointed out by Morgan [30], these data rule out the possibility of DX levels being states of substitutional donors. Numerous other experiments support this conclusion. Among them are measurements of the energy dependence of the photoionization cross section of DX levels in $Al_xGa_{1-x}As$ [4, 32] and in GaAs under pressure [33]. A model-independent analysis of the data yields an optical ionization energy of $\simeq 1.4$ eV [32], much larger than the DX level thermal ionization energy of $\simeq 160$ meV [2] at the Γ-X cross-over composition. The temperature dependence of the photoionization cross section is well described by a large lattice relaxation model [4, 34]. Other characteristics of DX levels that are only easily explained by a large lattice relaxation model are the temperature-dependent capture cross section which results in PPC at low temperature [2-4, 35, 36], the low radiative capture rate [2-4], the large capture cross section for holes [2, 37] and the large vibrational entropy term in the thermal activation energy [38]. The narrowing of the x-ray lattice-diffraction peak of Te-doped $Al_xGa_{1-x}As$ upon photoionization of the DX centre supports a large lattice relaxation model [39].

Figure 4 shows the leading microscopic model for the distorted DX configuration based on pseudopotential calculations by Chadi and Chang [40, 41]. In this model a bond between the Si donor and one of its As neighbours is broken and the Si atom moves along a $\langle 111 \rangle$ axis towards the interstitial site, where it lies very close to three group III atoms (figure 4(a)). The distorted configuration for group VI donors is shown in figure 4(b). In this case it is not the donor but rather one of its group III neighbours which moves toward the interstitial site. The calculations show that this configuration is stable provided that the donor traps two electrons [40, 41]. In figure 4(a) the three group III atoms which are close

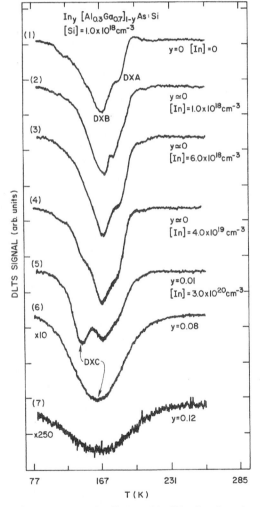

Figure 3. DLTS spectra for Si-doped $In_y(Al_{0.3}Ga_{0.7})_{1-y}As$ alloys. DXA and DXB, both observed when $y = 0$, are identified as configurations having one or two Al close neighbours. DXC is identified with configurations having In close neighbours.

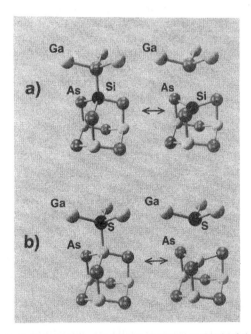

Figure 4. Substitutional (left) and distorted or DX (right) configurations of substitution donors as proposed in [40] and [41] for (a) a group III donor and (b) a group VI donor.

neighbours to the Si donor in the distorted configuration are all Ga atoms. Thus a single DX level in GaAs is predicted by this model. In the alloy, on the other hand, there are four non-equivalent substitutions of Al for Ga atoms on these sites, resulting in configurations having zero, one, two or three Al atoms. This is consistent with the observation of four different DX levels in GaAs/AlAs superlattices [24], where all possible combinations of Ga and Al neighbours occur. In the dilute alloys of figure 2, on the other hand, the probability of a donor atom having three Al close neighbours is negligible [23, 30]. DLTS measurements performed on samples under hydrostatic pressure permit the assignment of the four observed levels to particular configurations [28, 29]. The labels on the four DX levels in figure 1 refer to the number of Al neighbours close to the Si donor in each distorted configuration. The spacing between the dots on the figure is proportional to the probability that each configuration occurs in the random alloy based on the structural model of figure 4 as discussed in [30].

In constructing figure 1, the energy of DX0 in GaAs was taken from [11], the variation of the energy of DX0 with alloy composition was taken from [41] and the energy separation of the four DX levels was taken from [24]. It was assumed that the variation of energy with alloy composition is the same for all four DX levels. A similar value for the energy difference between DX0 and DX1 was determined by a different method in [30], but no additional data are yet available for the other two DX levels. The energy of DX0 in GaAs reported in [24] is about 30 meV higher than that reported in [11]. The lower value for the energy of DX0 in GaAs was used in figure 1 so that DX2 would intersect the Γ valley at $x = 0.22$. DLTS measurements [23] show that DX2 dominates the emission spectrum for alloy compositions of $0.14 < x < 0.27$, suggesting that, in this alloy composition range, the concentration of DX3 is too small to account for a significant portion of the observed PPC. At the Γ–X cross-over composition nearly all electrons are trapped at DX levels, even at room temperature. Assuming that each DX level traps two electrons and using the statistics published in [30], electrons are trapped primarily at DX2 and DX3. The ionization energy determined from Hall effect measurements will therefore reflect a combination of these two DX levels.

As is pointed out in [42], the model appears to be consistent with recent extended x-ray absorption fine structure (EXAFS) experiments, even though these measurements have been interpreted otherwise. In this model the group VI donors remain on the substitutional site which is consistent with EXAFS data [43]. For both group IV and group VI donors the model predicts a decrease in the number of nearest-neighbour bonds from four to three but little change in the length of the three reconstructed bonds. This appears to be consistent with the available data [43, 44], which have not been analysed to determine if the number of bonds is changing. Much more sensitive experiments will be required to rule out or confirm this prediction of the Chadi and Chang model.

The experiments reviewed here clearly support the model shown in figure 4. Donors are bistable, existing either in substitutional configurations having shallow hydrogenic ground states or in distorted configurations with DX levels as ground states. In thermal equilibrium the occupation of the DX levels is determined by their energy position relative to that of the Fermi energy in a given sample. DX configurations are stable states at alloy compositions near the Γ–X cross-over and metastable states in GaAs and dilute $Al_xGa_{1-x}As$. The energy barrier which separates the DX levels from the states of the conduction band is a result of the predicted lattice distortion.

3. Charge state of DX centres

The Chadi and Chang calculations [40, 41] show that the distorted configurations are stabilized by the capture of *two* electrons. In alloys where the DX levels lie below the hydrogenic levels, donors are therefore characterized by a negative effective Hubbard correlation energy (negative U). These calculations stimulated a number of experiments to determine the charge state of the DX centres. The absence of an EPR signal from the DX centres suggests that they are occupied by two electrons [16–18]. Two different experiments showed a persistent increase in the magnetic susceptibility after direct-gap $Al_xGa_{1-x}As$ samples were exposed to light at low temperature [45, 46]. However, the authors arrive at conflicting conclusions as to whether this susceptibility can be associated with DX centres, and thus arrive at conflicting conclusions about their spin states. As pointed out in [15], Hall effect measurements, for practical reasons, cannot distinguish between a negative (two-electron) or neutral (one-electron) charge state. These difficulties can be avoided by calculating the DX ionization enthalpies using the activation energies (enthalpies) for capture and emission determined from kinetic studies [15]. Figures 5(a) and (b) show the activation energies for the emission and capture of electrons in Si-doped $Al_xGa_{1-x}As$ from [35]. In each sample the kinetics are determined by a weighted average of contributions of each of the DX levels existing in the alloy. If DX centres are neutral, a single electron is emitted or captured and the ionization enthalpy of the DX level is simply the difference between the emission and capture activation energies. This is shown by the open circles in figure 5(c). On the other hand, if DX centres are negatively charged, two electrons are captured or emitted. Thus the average ionization enthalpy per electron is half of the difference between the capture and emission activation enthalpies as indicated by the full circles in figure 5(c). Note that for neutral DX centres the data do not fall on a single straight line but have a cusp near the cross-over composition which is inconsistent with experiment. Furthermore, the value of the ionization enthalpy at the Γ–X cross-over composition and the composition at which the DX level intersects the Γ valley calculated for negatively charged DX centres agree much better with experiment than those determined for neutral DX

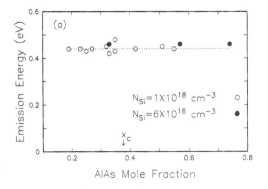

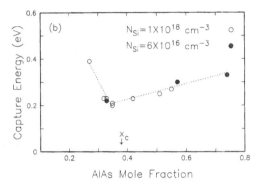

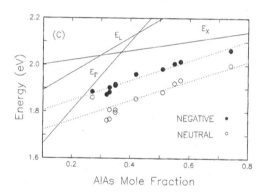

Figure 5. Emission (a) and capture (b) enthalpies for DX levels in Si-doped Al$_x$Ga$_{1-x}$As as a function of alloy composition. The ionization enthalpy of DX levels (c) determined for both neutral and negative DX levels.

on an analysis of the capture and emission kinetics are reported in [48].

A number of other experiments reported during the last year seem to show conclusively that DX centres must be negatively charged. DLTS measurements were performed with hydrostatic pressure applied to samples co-doped with Si and Ge [49]. Since the number of electrons trapped at the DX levels of Ge donors, which lie below the DX levels of Si donor and therefore are preferentially occupied, exceeded the number of Ge donors, it was concluded that DX levels must trap two electrons. An analysis of Mössbauer spectra on Sn-doped GaAs under hydrostatic pressure found that these spectra are consistent with the trapping of either two or three electrons and are thus incompatible with a positive-U model of DX centres [50]. New photoconductance measurements show that the photoionization of DX levels occurs by the successive removal of two electrons [51]. And finally a recent report of local vibrational mode spectra from infrared absorption measurements in Si-doped GaAs under pressure concluded that each DX level traps two electrons [52]. These last three experiments are described in other papers in this volume.

Of all the experiments performed in many different laboratories worldwide, only the magnetic susceptibility measurement of Khachaturyan *et al* seems to contradict this two-electron model. A resolution of the conflict between the two magnetic susceptibility measurements is clearly desirable. Nevertheless, the weight of the evidence overwhelmingly supports a negative charge state for the DX levels. Even if the particular microscopic configuration proposed by Chadi and Chang were to be found inconsistent with the magnetic susceptibility experiment, the negative-U nature of donors in Al$_x$Ga$_{1-x}$As having $0.2 \lesssim x \lesssim 0.7$ has been established.

4. Transitions between DX levels and conduction band states

The trapping of two electrons at DX levels has important implications for electron transitions between DX levels and the conduction band. It has been inferred from studies of the kinetics that the thermal capture of electrons occurs via an intermediate state, a localized state of the substitutional configuration whose energy roughly tracks that of the DX levels [53]. This sequential capture mechanism is readily understood if two electrons are trapped, since the simultaneous capture of two electrons is highly unlikely. In the recently reported two-step photoionization process, the first photon takes an electron from DX$^-$ to the conduction band leaving the donor in a metastable excited state DX^0 [51]. The neutral DX centre may then either be further ionized by a second photon, may be thermally ionized or may capture an electron from the conduction band, thus returning to DX$^-$. A localized state of the distorted configuration is inferred as the intermediate state in the photoionization process. In contrast to photoconductance experiments, photocapacitance experiments probe DX levels in the

centres. The relevant comparison here is with the ionization energy extracted from temperature-dependent Hall effect measurements, most of which have been analysed assuming a single DX level, and with PPC by either Hall effect or capacitance–voltage measurements [2, 47]. The value of the DX ionization enthalpy at the Γ–X crossover is $\simeq 160$ meV and PPC is seen at alloy compositions with $x > 0.22$. A comparison of the data in figure 5(c) with figure 1 is not easy because the parameters which characterize the four individual DX levels are not very precisely known. Keeping this in mind, it is clear that the level structure plotted in figure 1 agrees better with the ionization energy for DX levels determined assuming negatively charged DX centres. Similar conclusions based also space charge region of a Schottky diode or p-n junction where the high electric field sweeps out electrons emitted to the conduction band and prevents their recapture by DX^0 states. Since a second photoionization process was not observed in photocapacitance studies [4, 32–34], it can be inferred that the thermal emission of an electron from DX^0 in the space charge region at these measurement temperatures is very rapid compared with optical ionization, suggesting a very small thermal emission barrier for DX^0. Magneto-optical experiments also imply the existence of an intermediate state, but cannot specify its configuration [54].

As discussed in [42], theory predicts higher-lying localized states. Tight binding theory predicts a spectrum of deep levels, the lowest lying of which should have A_1 symmetry [55]. Magneto-optic and photoluminescence measurements on GaAs under pressure have shown that a donor-related level which does not exhibit PPC lies in the gap, apparently at a somewhat higher energy than the DX levels [56]. Large chemical shifts of this level were identified with Ge, S and Si, indicating that it is a localized state. The absence of PPC indicates at most a small lattice distortion, and thus this state appears to be a state of the substitutional donor configuration. Photoluminescence in Si-doped GaAs under pressure reveals such a state in the gap near the Γ–X cross-over [57]. Pseudopotential calculations predict the existence of both a localized state of the substitutional donor configuration in the gap and also possibly a metastable neutral state of the distorted DX configuration, DX^0, lying about 1 eV above the DX^- level [58]. The thermal emission barrier of this state is at most about 100 meV, consistent with the fast thermal emission inferred from photocapacitance experiments.

Figure 6 shows a suggested configuration coordinate diagram for donors in direct-gap $Al_xGa_{1-x}As$ [42], consistent with the intermediate states inferred from the above-discussed experiments [51, 53] and calculation [58]. The parabolas with energy minima near Q_0 represent the substitutional configuration, with the lower parabola indicating the total energy with two electrons in the conduction band and the upper one the total energy with one electron in the conduction band and one in a neutral excited state. The parabolas having minima near Q_T represent the distorted (DX) configuration, the lower one indicating the energy with two electrons in the DX level. E_{DX^-} is defined as the average energy per electron of the negatively charged state. Thermal transitions of electrons between the conduction band and DX^- are assumed to occur via the excited one-electron state, D^0, as indicated by the curved arrow [42, 53]. An excited one-electron state of the distorted configuration is included to account for the two-step photoionization process [51]. Removal of the first electron (represented by a vertical transition on the diagram) leaves the donor in the excited DX^0 state. Electrons in this state can be thermally excited to the conduction band (upper curved arrow), optically excited to the conduction band, or captured radiatively thus returning the donor to the DX^- state. In indirect-gap material the existence of

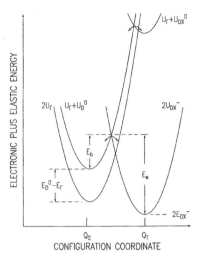

Figure 6. Configuration coordinate diagram appropriate for a negative-U model of the DX centre in direct-gap $Al_xGa_{1-x}As$. See text for a complete explanation.

X-valley hydrogenic states will strongly influence the statistics of population of conduction states and will alter the radiative capture rate.

5. Applications of the model

As suggested in the introduction to this paper, the microscopic model of DX levels has implications outside the physics community. While physicists have the task of making models, engineers have the job of using them, in this case to design and fabricate electronic and optoelectronic devices using III–V compound semiconductors. As pointed out earlier, attempts to eliminate instabilities in MODFETs using the defect complex model failed. It is therefore interesting to ask what the implications of the current model are for technology.

A few years ago it was suggested that DX centres are eliminated from Sn-doped $Al_xGa_{1-x}As$ grown by metal-organic chemical vapour deposition (MOCVD) when the material is grown at extremely high temperature [59–60]. Hall measurements showed reduced carrier freeze-out and little PPC in such material. Growing the material at high temperature might affect the presence of point defects and thus this result might be explained in terms of a defect complex model. However, this explanation is in conflict with our present understanding of the nature of DX levels. Recent studies of Sn-doped GaAs found that Sn diffuses rapidly at the temperatures where 'DX-free' Sn-doped AlGaAs was grown [61]. These experiments suggested that Sn atoms deposited in the AlGaAs layer diffuse into the underlying GaAs during growth at high temperature. If this were happening, parallel conduction in the GaAs would mask the observation of DX levels in the AlGaAs layer by Hall effect measurements. Recent DLTS measurements, which probed only the AlGaAs layer, revealed the presence of DX levels, and SIMS measurements confirmed the diffusion of Sn into the underlying GaAs in these samples [62]. Thus the MOCVD growth of Sn-doped AlGaAs layers at high temperature does not contradict the present model.

Another approach is to modify the characteristics of DX levels by the addition of other group III atoms such as B or In during crystal growth. In order for n-type AlGaAs to be useful for MODFETs or similar structures, the capture barrier separating the DX levels from the conduction band states must be higher than that shown in figure 5(b), or alternatively the DX levels must lie at energies high enough above the bottom of the conduction band for their occupation to be negligible. The preceding discussion shows that the presence of Al close neighbours lowers the energy of the DX levels with respect to its energy when there are only Ga neighbours [28, 29]. On the other hand, the presence of In close neighbours significantly reduces the height of the capture barrier but has little effect on the energies of the DX levels [31]. Thus the addition of In has an effect opposite to that desired. The effect of Boron close neighbours is less clear. It has been reported that the DLTS peak characteristic of DX levels was not seen in samples of LEC-grown Si-doped GaAs containing B as a contaminant [63]. DLTS peaks attributed to unidentified deep levels believed to be related to the presence of B were observed instead. The capture cross sections of these B-related traps are much larger than those of DX levels. The proposed explanation is that B forms a complex with Si, thus modifying the characteristics of the donor levels. Experiments on Si-doped Al$_x$Ga$_{1-x}$As where controlled amounts of B are added during growth are needed to test this explanation and to learn if such an approach could be useful for technology.

An alternative to using Al$_{0.35}$Ga$_{0.65}$As/GaAs for fabricating MODFETs and other devices is Al$_{0.48}$In$_{0.52}$As/Ga$_{0.47}$In$_{0.53}$As, which is lattice matched to InP. A very recent study of such modulation-doped structures grown by molecular beam epitaxy (MBE) found little PPC compared with that in Al$_x$Ga$_{1-x}$As/GaAs having the same conduction band offset [64]. Preliminary DLTS measurements in Al$_{0.48}$In$_{0.52}$As layers grown under the same conditions show no indication of DX-like levels [65]. The only deep trap observed was one occurring in very low concentration compared with that of the Si-donor. Unlike DX levels, this level has a capture cross section which is independent of temperature. Thus this material appears to be a promising alternative. However, earlier papers reported a deep level with DX-like characteristics in Al$_{0.48}$In$_{0.52}$As [66, 67], suggesting that the properties of this material may vary with alloy composition or strain or other unknown parameters. In the model of Chadi and Chang, DX levels are expected to lie below the bottom of the conduction band whenever the average density of states in the conduction band lies near the band edge. A study of Al$_y$In$_{1-y}$As at alloy compositions near the direct–indirect gap cross-over composition would be useful to determine both the pervasiveness of DX-like deep levels in III–V semiconductor alloys and the generality of the leading DX model.

6. Conclusions

It is now widely accepted that DX levels in Al$_x$Ga$_{1-x}$As are ground states of distorted configurations of isolated substitutional donors which are stable (i.e. are the lowest donor states) at alloy compositions near the direct to indirect-gap cross-over composition. Furthermore, recent experiments have conclusively shown that DX levels trap two electrons and thus are negatively charged. In this alloy composition range, therefore, donors have the property of negative U. Alloy peturbation experiments on Si-doped Al$_x$Ga$_{1-x}$As support the model for the distorted configuration shown in figure 4(a). Other experiments which measure atom positions have not yet been performed with sufficient sensitivity to confirm or contradict this model. Kinetic studies suggest that thermal capture occurs via an intermediate state, possibly a localized neutral state of the substitutional configuration, whereas photoionization occurs via the metastable DX^0 state. Recent calculations predict the existence of both these states. Our current understanding is that DX levels are intrinsic states of isolated donor atoms. Efforts to eliminate these levels from Al$_x$Ga$_{1-x}$As or to significantly modify their properties have been unsuccessful.

References

[1] Craford M G, Stillman G E, Rossi J A and Holonyak N Jr 1968 *Phys. Rev.* **168** 867
[2] Mooney P M 1990 *Appl. Phys. Rev. J. Appl. Phys.* **67** R1
[3] Lang D V and Logan R A 1977 *Phys. Rev. Lett.* **39** 635
[4] Lang D V, Logan R A and Jaros M 1979 *Phys. Rev. B* **19** 1015
[5] Solomon P M and Morkoç H 1984 *IEEE Trans. Electron Devices* **31** 1015
[6] Kastalsky A and Kiehl R A 1986 *IEEE Trans. Electron Devices* **33** 414
[7] Baba T, Mizutani T, Ogawa M and Ohata K 1983 *Japan. J. Appl. Phys.* **23** L654; Hueken M, Prost W, Kugler S and Heime K 1987 *Gallium Arsenide and Related Compounds 1986 (Inst. Phys. Conf. Ser. 83)* ed W T Lindley (Bristol: Institute of Physics) p 563
[8] Tachikawa M, Fujisawa T, Kukimoto H, Oomi G and Minomura S 1985 *Japan. J. Appl. Phys.* **24** L893
[9] Mizuta M, Tachikawa H, Kukimoto H and Minomura S 1985 *Japan. J. Appl. Phys.* **24** L143
[10] Eaves L *et al* 1988 *Gallium Arsenide and Related Compounds 1987 (Inst. Phys. Conf. Ser. 91)* ed A Christov and H S Rupprecht (Bristol: Institute of Physics) p355; 1989 *Shallow Impurities in Semiconductors 1988 (Inst. Phys. Conf. Ser. 95)* ed B Monemar (Bristol: Institute of Physics) p 315
[11] Theis T N, Mooney P M and Wright S L 1988 *Phys. Rev. Lett.* **60** 361
[12] Theis T N, Kuech T F, Palmateer L and Mooney P M 1984 *Gallium Arsenide and Related Compounds 1984 (Inst. Phys. Conf. Ser. 74)* ed B de Cremoux (Bristol: Institute of Physics) p 241
[13] Dmochowski J E, Langer J, Raczynska J and Jantsch W 1988 *Phys. Rev. B* **38** 3276; Dmochowski J E, Dobaczewski J E, Dobaczewski L, Langer J M and Jantsch W 1989 *Phys. Rev. B* **40** 9671
[14] Mizuta M and Mori K 1988 *Phys. Rev. B* **37** 1043
[15] Theis T N, Mooney P M and Parker B D 1991 *J. Electron. Mater.* **20** 35

[16] Mooney P M, Wilkening W, Kaufmann U and Kuech T F 1989 *Phys. Rev.* B **39** 5554

[17] von Bardeleben H J, Bourgoin J C, Basmasji P and Gibart P 1989 *Phys. Rev.* B **40** 5892

[18] Khachaturyan K, Weber E R, Crawford M G and Stillman G E 1991 *J. Electron. Mater.* **20** 59

[19] Glaser E, Kennedy T A and Molnar B 1989 *Shallow Impurities in Semiconductors 1988 (Inst. Phys. Conf. Ser. 95)* ed B Monemar (Bristol: Institute of Physics) p 233

[20] Glaser E, Kennedy T A, Sillmon R S and Spenser M G 1989 *Phys. Rev.* B **40** 3447

[21] Kaufmann U, Wilkening W, Mooney P M and Kuech T F 1990 *Phys. Rev.* B **41** 10206

[22] Theis T N 1989 *Shallow Impurities in Semiconductors 1988 (Inst. Phys. Conf. Ser. 95)* ed B Monemar (Bristol: Institute of Physics) p 307

[23] Mooney P M, Theis T N and Wright S L 1988 *Appl. Phys. Lett.* **53** 2546

[24] Baba T, Mizuta M, Fujisawa T, Yoshino J and Kukimoto H 1989 *Japan. J. Appl. Phys.* **28** L891

[25] Calleja E, Gomez A, Criado J and Muñoz E 1989 *Mater. Sci. Forum* **38-41** 1115

[26] Brunthaler G and Köhler K 1990 *Appl. Phys. Lett.* **57** 2225

[27] Calleja E, Mooney P M, Theis T N and Wright S L 1990 *Appl. Phys. Lett.* **56** 2102

[28] Calleja E, Garcia F, Gomez A, Muñoz E, Mooney P M, Morgan T N and Wright S L 1990 *Appl. Phys. Lett.* **56** 934

[29] Mooney P M, Theis T N and Calleja E 1991 *J. Electron. Mater.* **20** 23

[30] Morgan T N 1991 *J. Electron. Mater.* **20** 63

[31] Pann L S, Tischler M A, Mooney P M and Neumark G F 1990 *J. Appl. Phys.* **68** 1674

[32] Mooney P M, Northrop G A, Morgan T N and Grimmeiss H G 1988 *Phys. Rev.* B **37** 8298

[33] Li M F, Yu P Y, Weber E R and Hansen W 1987 *Phys. Rev.* B **36** 4531

[34] Legros R, Mooney P M and Wright S L 1987 *Phys. Rev.* B **35** 7505

[35] Mooney P M, Caswell N S and Wright S L 1987 *J. Appl. Phys.* **62** 4786

[36] Mooney P M, Calleja E, Wright S L and Heiblum M 1986 *Mater. Sci. Forum* **10-12** 417

[37] Brunthaler G, Ploog K and Jantsch W 1989 *Phys. Rev. Lett.* **63** 2278

[38] Theis T N, Morgan T N, Parker B D and Wright S L 1989 *Mater. Sci. Forum* **38-41** 1073

[39] Leszcnski M, Suski T and Kowalski G 1991 *Semicond. Sci. Technol.* **6** 59

[40] Chadi D J and Chang K J 1988 *Phys. Rev. Lett.* **61** 873

[41] Chadi D J and Chang K J 1989 *Phys. Rev.* B **39** 10063

[42] Theis T N 1991 *Proc. Symp. on Defects in Materials, Materials Research Society Fall Meeting, Boston, Nov 26–Dec 1, 1990* (in press)

[43] Mizuta M and Kitano T 1987 *Appl. Phys. Lett.* **52** 126

[44] Hayes T M, Williamson D L, Outsourhit A, Small P, Gibart P and Rudra A 1989 *J. Electron. Mater.* **18** 207

[45] Khachaturyan K A, Awschalom D D, Rosen J R and Weber E R 1989 *Phys. Rev. Lett.* **63** 1311

[46] Katsumoto S, Matsunaga N, Yishida Y, Sugiyama K and Kobayashi S 1990 *Japan. J. Appl. Phys.* **29** L1572

[47] Chand N, Henderson T, Clem J, Masselink W T, Fischer R, Chang Y C and Morkoç 1984 *Phys. Rev.* B **30** 4481

[48] Mosser V, Contreras S, Piotrzkowski R, Lorenzini Ph, Robert J L, Rochette J F and Marty A 1991 *Semicond. Sci. Technol.* **6** 505

[49] Fujisawa T, Yoshino J and Kukimoto H 1990 *Japan. J. Appl. Phys.* **29** L388

[50] Gibart P, Williamson D L, Moser J and Basmaji P 1990 *Phys. Rev. Lett.* **65** 1144

[51] Dobaczewski L and Kaczor P 1991 *Phys. Rev. Lett.* **66** 68

[52] Wolk J A, Kruger M B, Heyman J N, Walukiewicz W, Jeanloz R and Haller E E 1991 *Phys. Rev. Lett.* **66** 774

[53] Theis T N and Mooney P M 1990 *Mater. Res. Soc. Symp. Proc.* **163** 729

[54] Fockele M, Spaeth J-M and Gibart P 1990 *Proc. 20th Int. Conf. on the Physics of Semiconductors* ed E M Anastassakis and J D Joannopoulos (Singapore: World Scientific) p 517

[55] Hjalmarson H P, Vogel P, Wolford D J and Dow J D 1980 *Phys. Rev. Lett.* **44** 810

[56] Dmochowski J E, Wang P D and Stradling R A 1990 *Proc. 20th Int. Conf. on the Physics of Semiconductors* ed E M Anastassakis and J D Joannopoulos (Singapore: World Scientific) p 658

[57] Liu X, Samuelson L, Pistol M-E, Gerling M and Nilsson S 1990 *Phys. Rev.* B **41** 11791

[58] Dabrowski J, Scheffler M and Strehlow R 1990 *Proc. 20th Int. Conf. on the Physics of Semiconductors* ed E M Anastassakis and J D Joannopoulos (Singapore: World Scientific) p 489

[59] Basmaji P, Guittard M and Gibart P 1987 *Phys. Status Solidi* a **100** K41

[60] Basmaji P, Zazouk A, Gibart P, Gauthier D and Portal J C 1989 *Appl. Phys. Lett.* **54** 1121

[61] Kuech T F, Tischler M A, Potemski R, Cardone F and Scilla G 1989 *J. Crystal Growth* **98** 174

[62] Mooney P M, Parker B D, Basmaji P and Gibart P unpublished

[63] Li M F, Yu P Y and Shan W 1989 *Appl. Phys. Lett.* **54** 1344

[64] Tischler M A and Parker B D 1991 *Appl. Phys. Lett.* **58** 1614

[65] Tischler M A, Parker B D, Mooney P M and Goorsky M S unpublished

[66] Nakashima K, Nijima S, Kawamura Y and Asahi H 1987 *Phys. Status Solidi* a **103** 511

[67] Hong W-P, Dahr S, Bharracharya P K and Chin A 1987 *J. Electron. Mater.* **16** 271

Semicond. Sci. Technol. **6** (1991) B9–B15. Printed in the UK

Comparison of three DX structural calculations presented at Thessaloniki

G A Baraff

AT & T Bell Laboratories, Murray Hill, NJ 07974, USA

Abstract. Three structural calculations of the DX centre presented at Thessaloniki are of especial interest. Although they were carried out using nominally the same calculational apparatus, namely first-principles pseudopotentials, local density approximation, large unit cells etc, the three reached vastly different conclusions about the nature of the DX centre. In this paper we study these calculations in terms of their internal consistency and find that only one is essentially error-free in the picture it presents. Our analysis suggests that the size and possibly the sign of the U of the DX centre may well depend on alloy concentration or pressure. The negative-U property may therefore not be as universal as heretofore assumed.

1. Background

At the recent International Conference on the Physics of Semiconductors [1], there were three state-of-the-art calculations of the DX centre. The question studied in each calculation was: how does pressure and/or alloy concentration affect electronic levels, equilibrium positions of the atoms and the charge states? In spite of the similarities of the calculational methods used, the conclusions of the three calculations were quite different. The differences go to the heart of the question of whether the DX is a large lattice relaxation (LLR) system or a small lattice relaxation (SLR) one. They also differ as to which states, if any, are present in addition to the well known deep state and effective mass states.

We have been asked to look at these three papers (sections 1.1–1.3) through the eyes of a theorist to explain the differences and help choose between the competing pictures these papers present. Clearly, the last word on this subject will be had by the experiments: one or another of the models will turn out to be closest to reality. However, the fact that some of the papers reach conclusions that are not justified, even within their own calculational framework, will be useful in narrowing the choice of models the experimentalist should consider.

1.1. The Chadi paper

The first of these calculations was by Zhang and Chadi [2]. Preceding their calculations were earlier papers by Chadi and Chang [3, 4] in which the well known large lattice relaxation model was first presented. Those calculations, one recalls, were carried out for an Si defect in a GaAs host at normal pressure. Under these conditions, the DX centre is not observed. However, by reasoning about how the energy balance would change if either hydrostatic pressure were applied or an AlGaAs alloy were substituted for the GaAs, Chadi and Chang arrived at the model which has been discussed so often at this conference.

The Zhang and Chadi calculation was an attempt to deal with criticism that the earlier papers were incomplete because they did not calculate the DX centre in GaAs under pressure or in the AlGaAs alloy. Not surprisingly, the authors carried our the new calculation using the same calculational apparatus as the old, and, not surprisingly, they confirmed the earlier model in its entirety.

1.2. Yamaguchi, Shiraishi and Ohno

The second of these calculations was presented by Yamaguchi, Shiraishi and Ohno [5]. Yamaguchi had earlier proposed a small-lattice relaxation model for the DX centre [6, 7]. Using a Green function technique to calculate the electronic structure of substitutional Si in $Al_x Ga_{1-x} As$ alloys, he had previously found that a compact state of A1 symmetry emerged from the conduction band and that its binding energy varied with x in the same way as did the DX level. A similar result in GaAs under hydrostatic pressure had been found in a Green function calculation performed by Oshiyama [8]. Thus, Yamaguchi had reason to challenge the LLR model, and the calculation [5] started with just this challenge.

In the first of their calculations [5], Yamaguchi *et al* chose a lattice constant for GaAs that corresponded to a hydrostatic pressure of 30 kbar, and substituted two Si atoms for two gallium atoms. These corresponded to the two d^0 defects of the Chadi model. One of these two was moved along the (111) antibonding direction, just as in the Chadi model, and the total energy was calculated as a function of the position of the displaced atom. They found two minima, one at the undisplaced position and the other close to the position identified by Chadi and Chang [3, 4] as the configuration for DX. However, Yamaguchi et al found that the Chadi minimum energy configuration had a higher total energy than did the undisplaced configuration. This served to justify their earlier calculations and model for DX.

At this point, we must ask why the Chadi minimum was, in this calculation, the higher-lying minimum, while the opposite was so in Chadi's calculation. Frankly, this is one question that cannot be answered definitively from what we can find out about the two calculations. Certainly, there are technical differences between the two. The sizes of unit cells were different, the cut-off for the plane waves differed, the exchange correlation functionals might have been different, etc. On a number of technical points, the Yamaguchi calculation seems to be better converged, but that better convergence may have actually been a source of the problem, for the following reason.

Local density functional calculations used with first-principle pseudopotentials lead to the so-called band gap problem: the calculated gap is invariably too small. Chadi's less well converged calculation had a larger gap than did Yamaguchi's better converged calculation. In the calculations being discussed here, two electrons from the bottom of the conduction band fall into a state which drops into the gap as the Si atom is displaced. Therefore a band gap that is too small robs the displaced silicon of some of the energy which should drive the displacement. This is only a speculation as to the cause of the difference. We cannot really answer the question in a satisfactory way yet. Alternatively, as we shall discuss in the next section, there is a possibility that, in some ranges of alloy composition or hydrostatic pressure, the finding of Yamaguchi et al is indeed correct.

To return now to a description of the Yamaguchi paper [5]: having found that the undisplaced configuration gave the minimum energy, Yamaguchi et al investigated its properties as a function of hydrostatic pressure. They confirmed their earlier finding of an A1 level which tracked the pressure as did the DX level. In addition, they found a T2 level about 1 eV higher in energy, and calculated that the optical transition strength to it would be so great as to provide an explanation for the well known large optical absorption threshold energy.

At this point, two questions arise. First, why did Chadi and Chang not find the A1 state? Second, why was there not appreciable optical absorption between the A1 state and the X and L conduction band minima which lie about 0.1 to 0.3 eV above the A1 level? We shall answer both these questions in the next section.

1.3. Dabrowski, Scheffler and Strehlow

The third calculation was by Dabrowski, Scheffler, and Strehlow [9]. They used a 54-atom unit cell and investigated both the diplaced (Chadi) configuration and the undisplaced (Yamaguchi) configuration. They found a $(0/+)$ level which confirmed the work of Yamaguchi et al [5-7]. They found the $(-/+)$ level which confirmed this aspect of Chadi's work [2-4]. They also found a T2 level at about 0.3 eV above the A1 level, contradicting Yamaguchi's placement of the T2 level at 1 eV above the A1. They commented that, due to the band gap problem, they were unable to convincingly study which of the two configurations was of lower energy. They also commented that the calculated optical threshold from the A1 state is far too small to accord with experiment. Whether they meant that the optical absorption would be from the A1 level to the X and L bands or to the T2 state is not clear and does not matter: all these features lie at about the same energy in their calculation and would give an optical threshold of about 0.2 eV.

Their finding that the $(-/+)$ level is below the $(0/+)$ level does not imply that the LLR configuration is the more stable. Levels are a measure of the energy to add electrons to the system: stability is a matter of the smallest total energy, and these issues are quite independent of each other. The authors were careful to state that, within their own ability to calculate, they could not decide which configuration had the lower energy. It is clear that they felt that they were confirming the basic validity of the Chadi model, albeit with an extra A1 state associated with the SLR configuration.

The question raised by this calculation [9] compared with that of Yamaguchi [5] is why the energy of the T2 state is so much lower. The question raised by this calculation [9] compared with that of Chadi [2-4] is why the A1 state is absent from Chadi's work.

2. Analysis

2.1. The missing A1 state

We shall now answer some of the questions we have raised. Why did Chadi fail to find the A1 state for the SLR configuration that both Yamaguchi et al and Dabrowski et al had found? The answer is that Chadi had not expected to find it: if (or more likely, when) it showed up in his calculation, he was not sufficiently interested in it to study it further, perhaps because his main interest was in justifying his earlier calculations. Evidence for this assertion appears in the text of the *Physical Review* article by Zhang and Chadi [10]. This is the full paper corresponding to the Thessaloniki paper [2]. There, we find: 'For an eighteen atom cell, the calculated pressure dependence of E_0 for the effective mass state is substantially smaller than the value in $Al_xGa_{1-x}As$ alloys. This discrepancy is caused by the fact that the shallow d^0 donor state is forced to be overly confined by the small size of the supercell.'

This is true, but if we accept the results of Yamaguchi et al and of Dabrowski et al that there is a genuinely confined A1 state waiting to come down, it is likely that the confinement mentioned by Zhang and Chadi was not only caused by the small size of the supercell, but was genuine in its own right. In that case, they had correctly calculated the existence of the confined A1 state but, not expecting it, were confused by its appearance. At any rate, Zhang and Chadi did not further investigate this state. The text tells what was done instead: 'To avoid errors caused by this problem', they continue 'we avoid the supercell calculations for E^0 and instead make use of the following expression:

$$E^0 = E^+ + \varepsilon_g^{LDA}$$

where ε_g^{LDA} is the corresponding band gap obtained from an LDA calculation'.

The meaning of this mathematical step is that the A1 level is completely ignored and instead replaced by an effective mass level tied to the bottom of the conduction bands. In other words, at this point, they simply throw the A1 level away. In retrospect, this may seem careless, but it must be viewed in the context of the problem that the authors were most concerned with, namely the effect of pressure or alloying on the properties of the LLR situation they identified with the relaxed DX defect.

2.2. The thermal ionization energy

There is another idea, brilliantly put forth by Chadi and Chang, that is useful both with regard to the DX centre and also to the EL2 centre which is discussed elsewhere in this issue. The idea is that if a localized state is made up only from the lowest conduction bands, then its variation in energy with respect to some external parameter, e.g. pressure or alloy concentration, will be the same as the average variation of the lowest conduction band with respect to the same parameter. The average conduction band energy can be evaluated using a special points scheme

$$E_{cb} = [E(\Gamma) + 3E(X) + 4E(L)]/8.$$

One finds that, to within 20% for alloy composition as the external parameter, or to within 30% for pressure as the external parameter, the energy of the localized state calculated by this formula varies like the L-point minimum. Chadi and Chang had originally put forth this argument to explain why a localized DX state could appear to be connected with the L-band minimum alone, even though it is clear that a localized state has to involve the entire Brillouin zone. The argument, however, implies that any localized state made up from the lower conduction bands will vary in this way. Therefore, having a localized state give rise to a level which appears to follow the DX level under pressure or alloy concentration does not mean that the state *is* the DX level. Any localized state will give rise to the same behaviour as long as no large lattice relaxation is involved in determining the level position.

As an example of the validity of this insight, we now use it to calculate the pressure dependence of the EL2 thermal ionization energy. Calculations by Bachelet et al [11] showed that the midgap A1 state of As_{Ga} is made up of 1/6 valence bands and 5/6 conduction bands. This is apparent in figure 1, which is reproduced from their

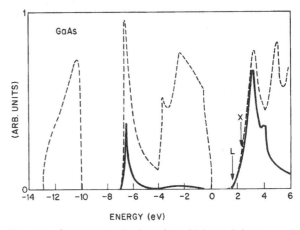

Figure 1. Spectral distribution of the A1 level defect wavefunction when projected onto crystal Bloch functions (full curve). The position of the L and X conduction band edges are indicated as well as the perfect crystal density of states (broken curve).

paper. This A1 state is the state from which electrons are ionized out of EL2 into the conduction band. The pressure variation of this level should be given by 5/6 the variation of Chadi's special point expression for the average conduction band energy, plus 1/6 the variation of the energy of the valence bands. Using accepted pressure coefficients for the Γ, X and L points [12], the pressure coefficient of the $\Gamma - A1$ difference is 9 meV $kbar^{-1}$, which is virtually the same as the measured 8.7 meV $kbar^{-1}$ [13]. The validity of the insight gives the reason why the A1 state of Yamaguchi et al is able to explain the pressure dependence of the DX thermal level.

2.3. The optical ionization threshold

Now we investigate the claim of Yamaguchi et al that the optical threshold of 1 eV arises because of intense transitions between the A1 and the T2 states. They are not the first to have asserted this [14], although they may have been the first to calculate that the T2 state is up that high.

Yamaguchi et al have widely circulated a preprint of an article submitted to the *Physical Review* [15]. This article corresponds completely to the Thessaloniki paper, and it contains enough details of the optical threshold calculation to make it possible to understand what they have done.

In the preprint, they present the band structure calculated for their 64-atom cell containing one substitutional silicon atom. The size and shape of this unit cell is such that both the L point in the band structure of the perfect crystal and the X point fold back onto the Γ point of the large unit cell. The energy bands of the perfect crystal are modified by the presence of the defect. In addition, new bands, identified as being the A1 and T2 states caused by the defect, also appear.

The authors label the energy of the modified conduction bands at Γ (the $k = 0$ point of their large unit cell) as Γ^*, L^* and X^*. The asterisk indicates that the band has been slightly shifted by the presence of the defect. The label also indicates the point in the standard Brillouin zone from which the state at Γ originated. They also present charge contour plots for the A1 and T2 states they identified in that calculation. They note correctly that the A1 state is clearly localized, and they point out that the T2 state has much more delocalized character, extending with appreciable amplitude through the entire cell. In spite of its delocalization, however, they identify it as the 'localized' T2 state.

To calculate the optical transitions in this system, they evaluate the optical matrix element between their A1 state and all states whose energy lies at, or below their T2 state. This evaluation is carried out only at the Γ point in their Brillouin zone. In a supercell calculation, this evaluation should have been carried out at all values of k within the Brillouin zone and the result summed by integration. This would have broadened out the otherwise discrete transition energies into bands, but no matter: the authors here are interested in showing only which transitions are strong and which are weak, and, aside from symmetry selection rules which would operate at Γ and at selected points on the cell boundary, this is a legitimate way to make the point.

For each state, there is a definite transition energy and a definite transition probability. This probability is equal to the square of the optical matrix element. There are five such energies and transition probabilities in all. Taking the transition probability between the A1 and T2 states as 1.00, the energies and transition probabilities they find are shown in table 1.

The transition probability to the Γ* state should have been rigorously zero because of a symmetry selection rule, so the quoted value of $\approx 10^{-23}$ indicates a fairly careful numerical integration. In the same way, we can infer that the small transition probability to the unnamed state was also caused by a symmetry selection rule. Thus there are three transitions in this 64-atom unit cell which have appreciable probability, and the threshold for these is at 0.2 eV.

At this point, Yamaguchi et al argue that, if they had calculated using a cell which was larger by a factor M than their 64-atom unit cell, any transition between the localized A1 state and a delocalized band state would have been reduced in intensity by a factor M, due to the normalization of the band state which still had to have unit integrated density in the enlarged unit cell. On the other hand, the transition to the T2 state (which being a localized state, does not change its amplitude as the unit cell is enlarged) would have retained its strength. On this

basis, they argue that if M were made large enough, all the transitions they had calculated above, save only that to the T2 state, would become unobservably small. This would leave only the A1-T2 transition at 1.0 eV as important.

This argument is clearly suspect on physical grounds: what is the mechanism by which spacing the Si atoms at greater and greater distances causes a threshold which was originally at 0.2 eV to increase to 1.0 eV? The region of space sampled by the optical matrix element is confined to the neighbourhood of each individual silicon atom because of the localization of the A1 initial state. Therefore an optical transition at one Si atom is, in this sense, independent of what happens at its neighbours. The potential perturbation from one silicon atom to the next is negligible even at the spacing corresponding to the 64-atom unit cell. Therefore, there is no direct influence from this cause on the transitions at one atom to the next. The only thing that links the atoms is the spatial extent of the final state, and this influence is via phase factors, interferences. These usually suppress transitions in certain energy regions. Constructing a larger unit cell, placing the atoms further apart, should act to destroy the phase correlation and allow the optical intensity to appear in energy regions from which it had been previously excluded. Yet the Yamaguchi et al argument is that just the opposite occurs: on separating the atoms, the optical intensity now is excluded from the low energy region below the T2 level.

In terms of the calculation presented, the error in the argument is that it ignores the fact that, as the unit cell is made larger, the corresponding Brillouin zone is made smaller, and therefore many more k points from the unit cell fold back onto the Γ point of the reduced Brillouin zone. For a unit cell M times larger, there are M times as many bands in the reduced Brillouin zone that have to be considered. Although each individual transition to an extended state drops in intensity by a factor M, there are now M times as many transitions to be considered. Yamaguchi et al failed to recognize that the new transitions could have appreciable amplitude.

Although we have argued that it is reasonable that the optical threshold stays close to the band edge (where it lies in the 64-atom cell calculation), we have not proved that it does so. Only a calculation can supply the proof. The way to carry out such a calculation in the limit of an infinitely large unit cell, i.e. in the limit of a single defect in an otherwise perfect infinite crystal, has already been given by Petit et al [16]. These authors showed how to express the optical cross section in terms of the perturbed Green function. The Green function is a sum over all the states of the crystal containing a single defect. That sum contains a factor that resolves the states according to energy. The Green function is the natural quantity to use to calculate an optical cross section.

Figure 2 is taken from [16]. It shows the optical cross section for the undistorted vacancy in silicon caused by transitions between the gap state and the valence band. As we show in the diagram which we have added to the figure, the undistorted vacancy in silicon gives rise to a

Table 1. Energies and transition probabilities between A1 state of Yamaguchi et al [15] and final states below T2.

Energy (eV)	Final state	Probability
0.2	X*	6.21×10^{-1}
0.25	Γ*	8.26×10^{-24}
0.35	L*	9.14×10^{-2}
0.6	unnamed	$\approx 10^{-23}$
1.0	T2	1.00

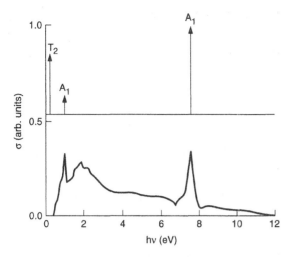

Figure 2. Optical cross section for transitions between the T2 bound state of the undistorted Si vacancy and the valence band (from [16]). Also indicated are the positions of the T2 bound state, the higher-lying A1 resonance and the strong low-lying deep A1 resonance.

T2 state low in the gap, an A1 resonance about 1 eV below it and another A1 resonance about 7 eV below that. The T2 state and upper A1 resonance form the A1-T2 pair which is mentioned so prominently in the articles by Watkins and by Lannoo in this conference. The only reason that the A1 resonance has any discernible identity this far (1 eV) down into the valence bands is that the very top of the valence band contains no state transforming as A1: therefore, there is a low background density of states of the proper symmetry to smear out this A1 resonance.

The other stronger A1 resonance, at ≈ -8 eV, arises from the quasi gap. In silicon, this is a region where the overall density of states is anomalously low. In a heteropolar semiconductor there would have been a genuine gap here, the heteropolar gap, with zero density of states, and one would have had a genuine A1 state in this gap. As it is, in silicon, the density of states is merely very low, and so one has a well defined resonance.

There are two points to be made from figure 2. The first is that a resonance can appear only if the background density of states is low. The second is that, even when the resonance does appear in the optical cross section, it does not dominate that cross section. The optical threshold associated with the upper A1 resonance is clearly at the band edge: the resonance itself is not orders of magnitude stronger than the background cross section one would attribute to absorption by the extended valence band states.

The optical cross section here is for a system which has a resonance about 1 eV from the band edge, the same situation as considered by Yamaguchi *et al*. In silicon, the resonance is still recognizable as a distinct feature because of the low background density of A1 states. In the case of DX, the resonance will have T2 symmetry, but the conduction band in which it is is embedded has (except in the Γ valley) a large background density of T2 states.

One would expect that the resonance would be even less prominent in this situation. One would expect the optical threshold to be at the L or X minimum, whichever is lower. That is, the situation that Yamaguchi *et al* calculated for the 64-atom cell, namely, a threshold at the lowest X band state, would be expected to remain even in the limit of the infinitely large unit cell.

Fortunately that expectation can be tested. Several years ago, in collaboration with Michael Schluter, we were studying the optical transitions of the isolated antisite, the electronic structure of which is shown in figure 3. There is a midgap state of A1 symmetry, and a T2 state or resonance which lies low in the conduction band, fairly close to the L-valley minimum. We were interested in how the optical absorption intensity would be distributed between the T2 state and the background density of states in the L valleys and X valleys. Since our electronic structure calculations were carried out in the Green function formalism, we did have available the Green function needed to evaluate the cross section according to the formula of Petit *et al* [16].

Because these calculations were preliminary to other studies, we carried them out in a very rough way, and never published these results. They have relevance here because, as in Yamaguchi's work, the initial state was of A1 symmetry, the final states were of T2 symmetry and the band structure was that of GaAs. The defect potential was not a self-consistent one, but was one that had approximately the correct spatial form and extent, and was adjusted so that we could move the T2 resonance up and down through the lower parts of the conduction band and into the gap. This was ideal for what we wanted to know: how quickly would the T2 state in the gap lose its individuality as it is forced up into the conduction band, first through the Γ valley, then through the L valley, then through the X?

Figure 4 shows the results. The final state of T2 symmetry gave rise to a delta function in the optical cross section when it was in the gap (not shown). It was broadened imperceptibly as it passed through the Γ

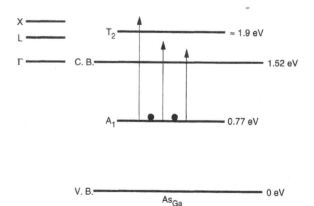

Figure 3. Schematic representation of the level structure of the isolated As_{Ga} defect in GaAs. The energies of the various conduction band minima are depicted, but the exact position of the T2 resonance relative to them is only approximately known.

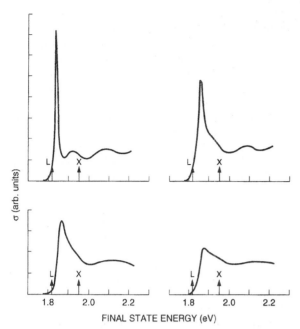

Figure 4. Optical cross section between the midgap A1 level and the lower part of the conduction bands, displayed as a function of the final state energy. Four possible locations of the final T2 resonance are considered: at 25, 50, 75 and 100 meV above the L-band minimum.

valley because there was almost no background of T2 states to hybridize with in the Γ valley (also not shown). The four panels in figure 4 do show the calculated optical cross section for the resonance placed successively at 25, 50, 75 and 100 meV above the L-valley edge. By the time the resonance is 0.1 eV above the edge, it is barely identifiable, and for all practical purposes the optical threshold is at the edge itself.

Is there any chance that the resonance will strengthen again in the region about 1 eV higher in energy? Although we did not study this situation, the background density of states increases continuously through this region (see figure 1) and there is no reason why a resonance should re-emerge.

Having established that the T2 resonance loses its individuality at about 0.1 eV above the lowest relevant minimum, we can return to the question of why Yamaguchi et al found the T2 state at about 1.0 eV above the A1 while Dabrowski et al found it about 0.2 eV above. If the state has no real independent identity in an infinite crystal calculation, then there is a certain amount of arbitrariness in what is identified to be 'the T2 state' in a large unit cell calculation. The criterion to differentiate between the new localized state and the alteration of pre-existing energy bands may not have been the same in the two calculations. Even if it was the same criterion, its application to two different unit cells (differing in both shape and volume) could well have selected two different states. In any case, the question has no importance since the T2 state as such has no relevance to the physical question of the threshold energy for optical absorption.

2.4. A composition-dependent negative U?

The final matter to be discussed in this section is a possible effect of the deep A1 state on the energy balance of the Chadi and Chang reaction

$$2d^0 \rightarrow d^+ + DX^-.$$

The amount of energy released by the reaction proceeding in the direction shown is $-U$. When the reaction proceeds in the direction indicated and the energy released is positive, this is a 'negative-U' reaction. If each electron associated with the neutral d^0 defect resides in a deep A1 state (instead of in the effective mass state tied to the bottom of the conduction bands as Chang and Chadi have assumed), the total energy of the initial $2d^0$ situation will be reduced by twice the binding energy of the A1 state. This will reduce the energy driving the Chadi and Chang reaction, and could even render it energetically unfavourable, depending both on the amount of the reduction and the (previously calculated) size of U. If the sign of U were really reversed, then a calculation such as that carried out by Yamaguchi et al [5, 15] would find that the minimum energy at the Chadi configuration is greater than that of the unrelaxed configuration. This is the situation we had in mind when we said that the relative ordering of the minima as found by Yamaguchi et al might, under certain conditions, be correct.

Clearly, the binding energy of the A1 state depends on pressure or on alloy concentration, so the amount by which the driving energy is reduced will depend on these. Equally clearly, the size (and possibly the sign) of the actual 'negative U' will also depend on these. This possibility may have relevance in understanding those experimental situations where a negative U was sought and found either inconclusively, or not found at all. In interpreting the experiments, it should be kept in mind that the question of whether or not the distorted configuration is stabilized by two electrons and the question of the sign of U are not related. Depending on the details of the experiment, one could still have some fraction of the DX centres containing and stabilized by the two electrons, even if the Chadi reaction were not energetically favourable.

3. Summary

Our analysis has shown that the situation in regard to the three claims of the Yamaguchi paper at Thessaloniki is the following: the claim that the LLR configuration lies at a higher energy than the SLR configuration can not be adequately decided on the basis of any information available about the three calculations. Clearly, the experimentalists have very stong ideas about the validity of this claim, but from a theorist's study of the calculations themselves, nothing can be said. The claim that the SLR configuration and its associated A1 state explain the thermal ionization energy of the DX is true, but not conclusive, since any localized state drawn down close to the bottom of the conduction band would behave in the same way. The claim that the SLR configuration and its associated A1–T2 pair explain the optical threshold at about 1 eV is simply wrong.

We also concluded that the situation with regard to the existence of the A1 localized state associated with the SLR configuration is the following: Zhang and Chadi missed finding it because they were not expecting it. Dabrowski *et al* found it but were not able, on the basis of what they could reliably calculate within LDA, to say whether the energy of the SLR configuration was higher or lower than the LLR configuration. They called attention to this limitation of their work. Yamaguchi *et al* provided a valuable service in dramatizing its existence to the DX community. That community had not paid much attention to the earlier papers which had calculated the existence of the A1 state, certainly not after the appearance of the Chadi model and its multiple successes.

Clearly, the SLR configuration, and the fact that it can support both an A1 localized state as well as effective-mass-like states, will play an important role in interpreting DX experiments. This is true even if its existence is irrelevant to the experimental finding that the optical absorption threshold of DX is much greater than the thermal ionization energy. Our analysis suggested that the deep A1 state will reduce the size of the negative U and perhaps, under some conditions, even render it positive. Certainly, it makes the size (and possibly the sign) of the negative U depend on alloy composition and pressure. We speculated that this might provide an explanation for those experiments where a negative U was sought and found inconclusively, or even not found at all.

Acknowledgments

I should like to thank D J Chadi for the frank and open discussions about the calculations with which he was involved. I should also like to thank Matthias Scheffler for providing me with both a copy of the paper and the view graphs which were used for the presentation of the talk at Thessaloniki, and his subsequent thoughts about their meaning.

References

[1] Anastassakis E M and Joannopoulos J D (ed) 1990 *Proc. 20th Int. Conf. on the Physics of Semiconductors* (Singapore: World Scientific)
[2] Chadi D J 1990 *Proc. 20th Int. Conf. on the Physics of Semiconductors* ed E M Anastassakis and J D Joannopoulos (Singapore: World Scientific) p 493
[3] Chadi D J and Chang K J 1988 *Phys. Rev. Lett.* **61** 873
[4] Chadi D J and Chang K J 1989 *Phys. Rev.* B **39** 10063
[5] Yamaguchi E, Shiraishi K and Ohno T 1990 *Proc. 20th Int. Conf. on the Physics of Semiconductors* ed E M Anastassakis and J D Joannopoulos (Singapore: World Scientific) p 501
[6] Yamaguchi E 1986 *Japan. J. Appl. Phys.* **25** L643
[7] Yamaguchi E 1987 *J. Phys. Soc. Japan* **56** 2835
[8] Oshiyama A 1988 *Proc. 19th Int. Conf. on the Physics of Semiconductors* ed W Zawadski (Warsaw: Institute of Physics, Polish Academy of Science) p 1089
[9] Dabrowski J, Scheffler M and Strehlow R 1990 *Proc. 20th Int. Conf. on the Physics of Semiconductors* ed E M Anastassakis and J D Joannopoulos (Singapore: World Scientific) p 489
[10] Zhang S B and Chadi D J 1990 *Phys. Rev.* B **42** 7174
[11] Bachelet G B, Schluter M and Baraff G A 1983 *Phys. Rev.* B **27** 2545
[12] von Bardeleben H J 1989 *Phys. Rev.* B **40** 12546
[13] Dreszer P and Baj M 1988 *Acta Phys. Polon.* A **73** 219
[14] Bourgoin J C, Feng S L and von Bardeleben H J 1989 *Phys. Rev.* B **40** 7663
[15] Yamaguchi E, Shiraishi K and Ohno T 1991 *Preprint* submitted to *Phys. Rev.* B 14 January 1991
[16] Petit J, Allen G and Lannoo M 1986 *Phys. Rev.* B **33** 8595

Semicond. Sci. Technol. **6** (1991) B16–B22. Printed in the UK

Theoretical treatments of DX and EL2

M Lannoo

Laboratoire d'Etude des Surfaces et Interfaces, URA CNRS N° 253 ISEN, 41 boulevard Vauban, 59046 Lille Cédex, France

Abstract. The available theoretical calculations concerning deep donors in GaAlAs compounds are discussed. The duality of the deep and shallow behaviours is analysed and shown to result in an inverted A_1–T_2 splitting for the L-derived states. This is applied to the interpretation of the stress splitting of the optical absorption of EL2. The electron–lattice interaction is then considered, with a comparison of the different possible ground state configurations for DX. Finally the information provided by recent magnetic resonance data is taken into consideration and the conclusions concerning the possible level scheme are drawn.

1. Introduction

This paper will mainly discuss the properties of the DX centre but some analogies with EL2 will also be investigated.

There is strong evidence that DX is a single donor (for recent reviews see [1, 2]). However, it seems to exhibit a dual nature since, in addition to a relatively deep ground state (which exists for $x > 0.22$ in $Ga_{1-x}Al_xAs$ alloys) one can also observe states following the predictions of effective mass theory (EMT). It also exhibits peculiar properties such as persistent photoconductivity and absence of EPR signal, which can be taken as evidence of strong lattice relaxation such that one gets negative-U behaviour [3, 4]. As regards EL2 in GaAs it is connected with the As_{Ga} antisite and should behave as a double donor. Again this defect is found to have conflicting properties so that two current models are proposed: the isolated antisite [5] or the As_{Ga}-X complex, X being possibly the arsenic interstitial As_i [6, 7].

The similarities between the two defects concern their dual behaviour (deep or shallow, strongly relaxed or not). Here we analyse such a situation, first in the absence of lattice relaxation, by trying to make a synthesis of the localized and EMT descriptions of donor defects. This is applied to the interpretation of the stress splitting of the zero-phonon line of EL2 [5]. We then consider the influence of the electron–lattice coupling and discuss the accuracy of Chadi and Chang's prediction [4] for the DX ground state. We finally consider recent results for the paramagnetic state D^0 and discuss their possible interpretations.

2. The two extreme theoretical descriptions of donors

The dual nature of donors is usually treated from two different points of view, a localized description or effec-

tive-mass theory. We discuss both of them and try to synthesize their predictions.

2.1. Localized description

This is in principle valid for deep states whose spatial extension is fairly limited so that only the short-range part U_{SR} of the defect potential has to be known accurately. This allows to make use of Green functions or supercell techniques, either in the local density (LDA) or in the tight binding (TB) approximations.

To understand the physical nature of the deep donor states let us start with the simplest TB description, i.e. the defect molecule model [8], which we illustrate for the Si_{Ga} case. For this we first remove the Ga atom, creating a vacancy V_{Ga}. We then replace it by the Si impurity. In this view the s and p atomic states (respectively of A_1 and T_2 symmetry) interact with the A_1 and T_2 symmetry combinations of the V_{Ga} dangling bonds states. Each pair of states of the same symmetry will lead to one bonding and one antibonding combination. This leads to the level structure of figure 1. For attractive donor potentials the bonding $A_1(b)$ and $T_2(b)$ levels fall in the valence band and are filled. Among the antibonding combinations $A_1(ab)$ can possibly be lowered into the gap while $T_2(ab)$ will be resonant in the conduction band. The single donor will be neutral when $A_1(ab)$ is occupied by one electron.

This simple picture is fully confirmed by numerical calculations. Let us begin with the results of a charge-dependent TB calculation [8]. Considering first the arsenic antisite As_{Ga} as a test one finds the level $A_1(ab)$ at $E_c - 0.2$ eV in the neutral charge state and $E_c - 0.65$ eV in the positive state, indicating the existence of a deep donor state in agreement with other calculations. The case of the single donor Si in GaAs or AlAs is more difficult since $A_1(ab)$ falls in the vicinity of E_c and the

0268-1242/91/100B16+07 $03.50 © 1991 IOP Publishing Ltd

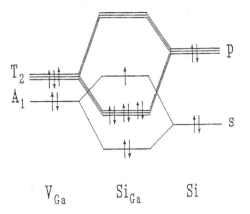

Figure 1. The molecular model of Si_{Ga} showing the interaction of the s and p states of Si with the A_1 and T_2 dangling bond states of the Ga vacancy.

method is not accurate enough. Anyway, from the results of [8], one can draw some general conclusions:

(i) A_1(ab) is probably resonant for Si_{Ga} in GaAs but certainly within the gap for Si_{Al} in AlAs;

(ii) the A_1(ab) → T_2(ab) distance varies in energy from ∼0.7 eV in GaAs to ∼1 eV in AlAs;

(iii) the effective Coulomb interaction U is of order 0.4–0.5 eV.

Let us now come to the results of the first-principles LDA calculations. For the unrelaxed situation Chadi and Chang [4] simply indicate that A_1(ab) is resonant in the Γ band for GaAs. This is confirmed by the more recent work of Yamaguchi et al [9] whose results can be summarized as follows:

(i) the A_1(ab) state is at $E_c + 0.17$ eV for GaAs but can be stabilized within the gap by a hydrostatic pressure (figure 2);

(ii) the A_1(ab) → T_2(ab) transition lies between 0.8 and 0.9 eV;

(iii) A_1(ab) depends on the local environment in $Ga_{1-x}Al_xAs$ and can be found in the gap for $x > 0.3$.

These results confirm the TB trends and will be discussed later.

2.2. EMT calculations

These are opposite in spirit since EMT applies to cases of a smooth long-range potential U_{LR} usually taken as $-Ze^2/\varepsilon r$. Each conduction band minimum, when treated independently, gives rise to a series of hydrogenic states. Those connected with the lowest conduction band will be truly localized, the others will form resonant states with a lifetime of order $10^{-8} - 10^{-9}$ s. Numerical values of the ground state binding energy E_B in this independent minimum approximation are, for GaAs:

(i) Γ minimum: $E_{Br} = 5.7$ meV since $m_\Gamma = 0.065$.

(ii) L minimum: one has anisotropy with $m_l = 1.9$ and $m_t = 0.075$. From the numerical calculations of [10] one gets $E_{BL} = 14$ meV.

(iii) X minimum: here $m_l = 1.9$ and $m_t = 0.19$ so that [10] gives $E_{BX} \sim 30$ meV.

The case of N equivalent minima ($N = 4$ for L, 3 for X) leads to hydrogenic states with N-fold degeneracy which is lifted by intervalley coupling, mainly due to the central cell correction. An interesting point is that this will not occur for the X minima when considering group IV donors on the Ga site. In this case, due to symmetry, the X minima will remain uncoupled leading to a triply degenerate $T_2(X)$ state [11]. This will not be true of the L minima ($N = 4$) for which the intervalley coupling gives, for the 1s states, an $A_{1L}(1s)$-$T_{2L}(1s)$ splitting.

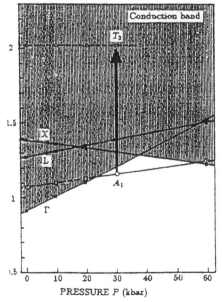

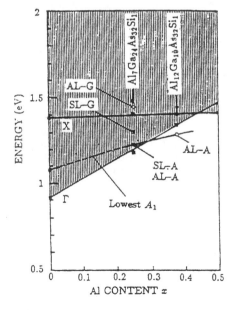

Figure 2. Results from the LDA calculation of [9]: (a) versus hydrostatic pressure, (b) versus composition.

Bourgoin and Mauger [12] have argued that the central cell correction is strong enough for the $A_{1L}(1s)$ to become moderately deep while retaining its pure L character. They identify this level with the DX state which would explain why it follows the L band when x is varied. In their model, at the limit of applicability of EMT, the DX level is expected to be weakly coupled to the lattice.

2.3. Synthesis of the two points of view

In an 'exact' calculation of the donor states the full potential seen by the electron reduces to U_{SR} in the vicinity of the defect and to U_{LR} in the outer region. The first consequence is that any deep state (i.e. having a compact wavefunction) only experiences U_{SR} and is correctly predicted by a localized description ignoring U_{LR}. For the deep donors considered here this can occur for the $A_1(ab)$ state and not for $T_2(ab)$ which is always resonant in the conduction band.

If this is the case, once $A_1(ab)$ is determined, one should calculate the other states which are shallow. For this the best procedure is to orthogonalize the hydrogenic trial wavefunctions to the deep state wavefunction. This is equivalent to replacing the true hydrogenic potential by a pseudopotential (see [13] for a review) whose short-range attractive part is suppressed, leading now to a small repulsive central cell correction which only concerns the ns hydrogenic states. The net conclusion is that the deep $A_1(ab)$ state and the shallow hydrogenic states will coexist but with some modified features which are discussed in the next section.

3. The shallow states of deep donors: Se in Ge and EL2 [14]

An interesting test case of the above discussion is provided by donors in Ge [15] for which the conduction band minimum is at the L points. In cases where a deep donor $A_1(ab)$ state is stabilized in the gap then, as we have seen, the hydrogenic ns states will experience a small repulsive central cell correction. This will lead to an inverted splitting of the L-derived ns states, with $A_{1L}(ns) > T_{2L}(ns)$. This important result can be viewed in another way: as the central cell potential is made more attractive, the $A_{1L}(1s)$ state becomes deeper and, by coupling with other bands, becomes $A_1(ab)$. In the same way all $A_1(ns)$ states with $n > 1$ are lowered with respect to their ns EMT value but tend towards the $(n-1)s$ EMT value as a limiting case.

This is fully confirmed by table 1 which reproduces the results of [15] with changes in notation corresponding to our relabelling of the $A_{1L}(ns)$ states. The first column represents the results of EMT theory for independent minima. They quantitatively agree with experiment for the single donor P with a small $A_{1L}(1s)$–$T_{2L}(1s)$ splitting of ~ 3 meV, the two levels being in normal order. On the other hand, the double donor Se in Ge leads to a deep state which must correspond to $A_1(ab)$. The lowest two excited states of Se^0 are close to the EMT 1s value and correspond to $A_1(1s)$ and $T_2(1s)$ in inverted order as discussed above. For Se^+ the $A_1(ab)$ state is still deeper but the observed states correspond to EMT with an impurity charge $+2e$.

This discussion can be directly applied to the optical absorption spectrum of EL2 in which the stress splitting of the zero-phonon line observed by Kaminska et al [5] provides a strong argument in favour of the identification of EL2 with the isolated antisite As_{Ga}. Recently von Bardeleben [16] has convincingly argued that it could be due to an $A_1(ab) \to T_{2L}$ transition, an attribution that has been given further support by Spaeth et al [17] for different antisite-related defects in GaAs. Let us then examine the isolated antisite As_{Ga} in GaAs and the possible transitions from its deep $A_1(ab)$ state to the L-derived hydrogenic states. This situation is comparable to the case of Se in Ge since the effective masses in the L band are of comparable magnitude. The L-derived 1s states will thus be in inverted order $T_{2L}(1s) < A_{1L}(1s)$ and, if the zero-phonon line is due to the $A_1(ab) \to T_{2L}(1s)$ transition, its stress splitting can be determined from the behaviour of the $T_{2L}(1s)$–$A_{1L}(1s)$ system under stress. To lowest order this can be done simply by considering that the independent valley 1s states rigidly follow their corresponding minimum. The description

Table 1. Comparison of experimental and theoretical energy levels for donors in germanium (taken from [15] with changes in notation as described in the text). For Se^+ the values $A_1(1s)$, $T_2(1s)$, $T_2(2s)$ in parentheses are simply scaled from Se^0 by a factor of four, the others being experimental values. The $\Delta E(Se^+)$ correspond to energy differences.

	EMT	P	Se^0	Se^+	$\Delta E\ Se^+$
$3p_\pm$	1.03	1.05	1.04	4×0.98	
$2p_\pm$	1.73	1.73	1.73	4×1.73	
$3p_0$	2.56	2.56	2.57	4×2.55	4.1
$2s$	3.52		$T_2(2s): 3.58$	$(T_2(2s): 4 \times 3.6)$	4.9
$2p_s$	4.74	4.74	4.75	4×4.80	10.4
$1s$	9.81	$T_2(1s): 9.93$	$A_1(1s): 7.4$	$(A_1(1s): 4 \times 7.4)$	10.2
		$A_1(1s): 12.76$	$T_2(1s): 9.95$	$(T_2(1s): 4 \times 9.95)$	
$A_1(ab)$			268.2	512.4	

then contains two parameters: d which describes the stress splitting of the L conduction bands in the bulk crystal and Δ the $T_{2L}(1s)$–$A_{1L}(1s)$ distance in energy at zero stress. As shown in [14] this leads to the same matrix as the one used by Kaminska et al [5]. The advantage is that d is known from bulk properties ($d = -77$ meV GPa^{-1} [18]) while Δ is known from Se in Ge, i.e. from table 1, $\Delta = 2.55$ meV or 10.2 meV for Se^0 or Se$^+$. From [19] a fit to experiment leads to two sets of values $d = -76$ meV GPa^{-1} or -90 meV GPa^{-1} and $\Delta = 7.5$ meV or 10 meV. Thus our description leads directly to the correct value for d, but also for Δ if the charge state is considered to be the $+$ state. This value of the charge state is confirmed by the positions of the higher hydrogenic states which explain in a natural way the first four peaks in the absorption spectrum of [5] (the so-called replica).

From this one can draw important consequences:

(i) the zero-phonon line indeed corresponds to the $A_1(ab) \rightarrow T_{2L}(1s)$ transition;

(ii) the corresponding charge state (EL2$^+$) rules out the isolated antisite which would be paramagnetic and gives support to the As$_{Ga}^0$–As$_i^+$ complex proposed in [6, 7];

(iii) the apparent symmetry will be T_d since the short-ranged axial potential of the pair will not split the extremely shallow $T_{2L}(1s)$ state.

4. The electron–lattice interaction

As there is experimental evidence of large lattice relaxation several models and calculations have been devoted to the study of the electron–lattice interaction.

Let us begin with the model of Toyozawa [20] in the EMT limit. In this model, the Hamiltonian is obtained by adding to the usual hydrogenic part a short-range contribution U_{SR} and an electron–lattice coupling term U_{eL} corresponding to a position-dependent dilation. Starting from a trial wavefunction

$$\psi(r) = \frac{1}{\sqrt{\pi a^3}} \exp\left(-\frac{r}{a}\right) \tag{1}$$

where a should in principle be much larger than the lattice constant, one can easily show [21] that the kinetic energy scales like a^{-2}, the Coulomb potential like a^{-1} while both U_{SR} and U_{eL} scale like a^{-3}. The energy $E(a)$ is thus a cubic function which has to be minimized with respect to the variational parameter a, noting that a lower bound for a is a_0, the atomic cell radius, and that in this limit EMT is no longer valid. The behaviour of the possible solutions is discussed in [20] and here we would like to mention two interesting situations: (i) a metastable one with a normal minimum and a second one at $a = a_0$ and (ii) a continuously decreasing $E(a)$ curve ending on a deep state at $a = a_0$. However, in both cases, the minimum at $a = a_0$ cannot be treated correctly in EMT and should be handled by localized descriptions.

A model of the DX centre has been proposed by Morgan [11] which can be viewed as an extension of Toyozawa's model. In this the $T_{2L}(1s)$ state, being triply degenerate is considered to be subject to a Jahn–Teller splitting. The most natural motion to consider first is a motion of the impurity itself, forming a lattice mode of T_2 symmetry. It is known that such a case leads to (100) or (111) distortions. Morgan argues that the second category of displacements is favoured, the displacement occurring in a (111) antibonding direction. Such a problem could be treated in a similar way as Toyozawa's model in the context of EMT. Again, if a shallow–deep instability occurs then EMT fails and a localized description is needed.

The lack of EPR observations for the DX ground state has led to the proposal that DX should behave as a negative-U system. This was first discussed on the basis of Toyozawa's model [3]. Strong support for this possibility was then provided by the LDA calculation of Chadi and Chang [4]. From an 18-atom supercell calculation they found that D$^-$ has a metastable configuration corresponding to the large atomic displacement pictured in figure 3. For the Si$_{Ga}$ impurity this situation can be viewed as a distorted V$_{Ga}$–Si$_i$ pair in which the Si atom forms three bonds and has two lone-pair electrons, leaving one As dangling bond. For the Si$_{Ga}$ case in GaAs the energy of this metastable state is only slightly unfavourable (by ~ 0.22 eV). Later on similar calculations were performed for GaAs under pressure and for the Ga$_{1-x}$Al$_x$As compounds as a function of x [22]. The results, reproduced in figures 4 (a, b), are in striking agreement with experiment. They thus seem to provide extremely strong support in favour of DX$^-$ (i.e. this strongly distorted configuration) being the observed DX ground state. However, we shall see in the next section that the situation is perhaps not so simple.

5. Large or small relaxation for DX?

Here we discuss the available theoretical arguments on this subject.

One of the main arguments in favour of the large lattice relaxation (LLR) model is the difference between the optical ionization energy (centred on 1 eV), and the thermal ionization energy ($\lesssim 200$ meV). In a simple

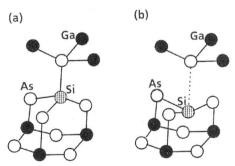

Figure 3. Chadi and Chang's DX$^-$ stable configuration [4]: (a) normal Si$_{Ga}$ configuration, (b) distorted configuration.

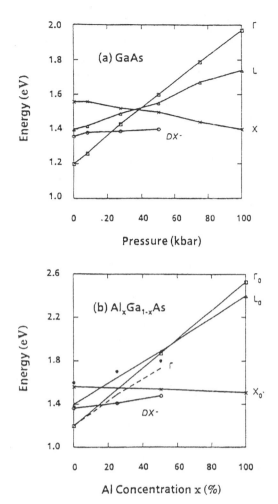

Figure 4. Pressure and concentration dependence of the DX⁻ level $\varepsilon(+, -)$.

configuration coordination diagram this implies a large Franck–Condon shift (~ 0.9 eV) which would thus support the strongly distorted DX⁻ state. However, we have seen before that TB theories as well as the LDA calculation of Yamaguchi show that it is likely that the $A_1(ab)$ state of the neutral donor lies in the gap (see figure 2) for some range of x values or hydrostatic pressures. From such a state the dominant optical transition would be towards the $T_2(ab)$ resonant state, located ~ 1 eV above. This directly corresponds to experiment, so that there is no need to invoke a LLR at least for this property. However, other properties like the almost infinite lifetime of the resonant DX state in GaAs or the lack of EPR signal for the ground state have yet to be explained in a small lattice relaxation (SLR) model.

The fact that the DX state practically follows the L band has been taken by Bourgoin and Mauger [12] as evidence that DX corresponds to the A_{1L} (1s) state. However, considering the large binding energy of DX with respect to the L band (~ 200 meV), this state is necessarily deep, so that EMT is no longer valid and this state should be given by the localized descriptions, i.e. should be $A_1(ab)$.

Thus the relevant information must come from such descriptions, in particular from the first-principles LDA calculations. However, the two of them we have discussed lead to opposite conclusions. First, as we have seen, the main result of Chadi and Chang [4] is that DX corresponds to the strongly distorted DX⁻. On the other hand Yamaguchi *et al* [9] find that this distortion only leads to a secondary minimum for the energy, 0.65 eV higher than the absolute one which keeps T_d symmetry and corresponds to the $A_1(ab)$ state with no negative-U properties. Furthermore, as shown on figure 2, the $A_1(ab)$ exhibits the same behaviour versus pressure or composition as the experimentally observed DX state. The comparison between these calculations is thus not decisive at all and one can wonder what the differences are between the two treatments. The following points emerge.

(i) Both calculations have different forms for the exchange-correlation potential V_{XC}.

(ii) Yamaguchi *et al* consider large supercells and directly include pairs of Si atoms in the supercell.

It is hard to deduce which of the two predictions is the more accurate. The difference might well be due to the sensitivity to V_{XC} but could also be attributed to technical inaccuracies. At least must one have in mind that the main drawback of LDA treatments is the band-gap problem related to the local form of V_{XC}: with the 'best' available V_{XC} the predicted gaps for Si and Ge are 0.6 eV and 0 eV compared with the experimental values 1.2 eV and 0.7 eV. This could be reflected in substantial inaccuracies concerning the energies of deep defects. Thus the most reasonable position is to consider that there are two likely possibilities for the DX ground state: the distorted DX⁻ configuration and the undistorted $D^0 A_1(ab)$ state. Both of them will then have to be confronted with the whole body of experimental information. At this point we should mention a third LDA calculation by Dabrowski *et al* [23] finding that hydrostatic pressure can stabilize both states, DX⁻ being the more stable situation.

More interesting information comes from the DLTS measurements which give, for a given impurity, an activation energy E_A independent of x. As E_A is the sum of the thermal ionization energy $\Delta(x)$ and of the activation energy for the capture cross section $E_B(x)$ (the capture barrier) one gets for this

$$E_B(x) = E_{B0} - \Delta(x). \tag{2}$$

Such a relation can be accidental but it is more convincing if one can derive it from a physical argument. The simplest one comes from the theory of multiphonon capture (see [24]) in the strong-coupling limit. Then the capture cross section is thermally activated with

$$E_B(x) = \frac{(\Delta(x) - d_{FC})^2}{4 d_{FC}}. \tag{3}$$

If the Franck–Condon shift d_{FC} is much larger than $\Delta(x)$ one can write approximately

$$E_B(x) \simeq \frac{d_{FC}}{4} - \frac{\Delta(x)}{2} \tag{4}$$

which is similar to (2) but with a slope of only half the experimental value. However, if one goes beyond this oversimplified theory it is easy to get the correct order of magnitude. For instance if the force constant k in the filled state is assumed to differ from k_0 in the ionized state then the same simple configuration coordinate diagram leads for $k_0/k = 0.5$ to a slope of -0.81, much closer to the experimental one. One can thus conclude that the LLR model is capable of giving the correct dependence of the barrier for capture versus the thermal ionization energy.

6. Magnetic resonance on the paramagnetic state

There have been a lot of recent results on the paramagnetic state D_0 of the donors. We discuss them shortly since they will be analysed in detail in other contributions.

For clarity we restrict ourselves to the group IV donors Si and Sn which seem to exhibit fairly different behaviour. Published studies on Si_{Ga} have been performed for indirect-gap materials ($x > 0.4$) by ODMR or conventional EPR [25] mainly for epitaxial layers $Ga_{1-x}Al_xAs/GaAs$. The spectrum is anisotropic and corresponds to near tetragonal symmetry. It has been explained as resulting from the $T_2(X)$ state derived from the X minimum. However, in bulk material this should give rise to an isotropic spectrum as confirmed by experiment. The anisotropy in epitaxial layers is due to the lattice mismatch between the layer and substrate which, although small (maximum value 0.18%), leads to a strained layer, with symmetry reduced to tetragonal. The X minima are no longer equivalent and one can show [26, 27] that X_z is raised in energy while X_x and X_y are lowered. The splitting should be linear with x with a maximum value of ~ 15 meV so that X_z can be assumed to be depopulated. The spectra can only be interpreted by assuming that the coupling between the equivalent minima X_x and X_y is completely quenched. This is thought to be due to random strains which split these minima by an amount larger than their coupling, due to the spin–orbit interaction. Finally the anisotropy decreases with x, which could indicate an increased interaction between minima.

On the other hand studies of Sn_{Ga} have been performed in direct-gap material. They result in an isotropic line observed either by EPR or by MCDA and ODEPR [28–31]. The major feature of such studies is that one observes a strong hyperfine interaction whose value indicates an unpaired 's' electron density at the Sn nucleus of order 20%. This is typical of the deep $A_1(ab)$ state. This result is thus in contradiction with the Si_{Ga} case for which the $T_2(X)$ state was observed.

How can we try to reconcile these apparently different results? A possible answer comes from figure 2 where the lowest $A_1(ab)$ level found by Yamaguchi et al [9] is seen to cross the Γ band at $x \sim 0.3$. If we extrapolate it a crossing with the X level would occur at $x \sim 0.5$. This would perfectly explain the Sn_{Ga} results in this interval

for which the stable D^0 state would be $A_1(ab)$. On the other hand, for $x > 0.5$ the $T_2(X)$ level would be the lowest, explaining the Si_{Ga} results. However, a problem would arise in the overlap region where the Sn and Si situations are the same (the predicted $A_1(ab)$ positions are identical [9]). In T_d symmetry $A_1(ab)$ and $T_2(X)$ would not mix. However, a spontaneous distortion such as a displacement of the impurity, e.g. along 100 axes, would lower the symmetry to C_{2v} and have several effects:

(i) It would induce a linear Jahn–Teller splitting of $T_2(X)$ resulting in a moderate shallow–deep instability in the spirit of Toyozawa's model (note that the experimental $T_2(X)$ binding energy can amount to 70 to 80 meV, about twice the EMT value). This would lead to equivalent (100) minima in configurational space, the wavefunction at the (100) minimum, for instance, being the product of the one-valley X_x state times an harmonic oscillator wavefunction centred on the corresponding stable atomic configuration. An advantage of this possibility is that it leaves unchanged the usual description where the minima are split by the strain in the epitaxial layer but it also gives rise to quenching of the spin–valley coupling by vibrational overlaps, avoiding the need for important random strains.

(ii) It would induce an electron-lattice coupling between the $A_1(ab)$ and $T_2(X)$ states, i.e. a pseudo Jahn–Teller effect. For decreasing x the ground state would have an increasing admixture of the $A_1(ab)$ state contributing to the reduced anisotropy.

(iii) The symmetry, at least for $x \gtrsim 0.4$, would be C_{2v}, i.e. lower than tetragonal. This would naturally explain the observation of three distinct principal components for the g tensor (see, in particular figure 8 of [32]).

7. Conclusion

We have investigated the possible nature of the states characterizing deep donors like DX or EL2. We have shown that there is coexistence of the deep donor state and of the hydrogenic states. We have pointed out the existence of an inverted splitting of the L-derived hydrogenic states which provides a complete explanation for the stress splitting of the zero-phonon line of EL2. We have also examined the different predicted possibilities for DX. The LDA calculations differ in their conclusions leading either to a DX^- or to a $D^0A_1(ab)$ ground state. There is strong experimental support for the DX^- charge state to be the stable situation. However, we have pointed out that magnetic resonance studies of D^0 show that $A_1(ab)$ is observed for Sn in direct-gap materials. It also seems likely that the hydrogenic $T_2(X)$ state is subject to a Jahn–Teller distortion reducing its symmetry to C_{2v} which also induces a pseudo Jahn–Teller effect between the $A_1(ab)$ and $T_2(X)$ states.

References

[1] Bourgoin J C (ed) 1990 *Physics of DX Centers in GaAs Alloys* (Vaduz: Sci. Tech. Publications)
[2] Mooney P M 1990 *J. Appl. Phys.* **67** 3 R1
[3] Kachaturyan K, Weber E R and Kaminska M 1989 *Mater. Sci. Forum* **38-41** 1067
[4] Chadi D J and Chang K J 1988 *Phys. Rev. Lett.* **61** 873; 1989 *Phys. Rev.* B **39** 10063
[5] Kaminska M, Skowrowski M and Kuszko W 1985 *Phys. Rev. Lett.* **55** 2204
[6] von Bardeleben H J, Stiévenard D and Bourgoin J C 1985 *Appl. Phys. Lett.* **47** 970
[7] Meyer B K, Hofmann D M, Niklas J R and Spaeth J M 1987 *Phys. Rev.* B **36** 1332
[8] Foulon Y, Lannoo M and Allan G 1990 *Physics of DX Centers in GaAs Alloys* ed J C Bourgoin (Vaduz: Sci. Tech. Publications) 195
[9] Yamaguchi E, Shiraishi K and Ohno T 1990 *Proc. 20th Int. Conf. on the Physics of Semiconductors* ed E M Anastassakis and J D Joannopoulos (Singapore: World Scientific) p 501
[10] Faulkner R A 1969 *Phys. Rev.* **184** 713
[11] Morgan T 1986 *Phys. Rev.* B **34** 2664
[12] Bourgoin J and Mauger A 1988 *Appl. Phys. Lett.* **53** 749
[13] Cohen M L, Heine V and Weaire D 1970 *Solid State Phys.* **24** (New York: Academic)
[14] Lannoo M 1991 *Phys. Rev.* B to be published
[15] Grimmeiss H G, Montelius L and Larsson K 1988 *Phys. Rev.* B **37** 6916
[16] von Bardeleben H J 1989 *Phys. Rev.* B **40** 12546
[17] Spaeth J M, Kambrock K and Hofmann D M 1990 *Proc. 20th. Int. Conf. on the Physics of Semiconductors* ed E M Anastassakis and J D Joannopoulos (Singapore: World Scientific) p 441
[18] Mirlin D N, Sapega V F, Karlik I Ya and Katilius R 1987 *Solid State Commun.* **61** 799
[19] Davies G 1990 *Phys. Rev.* B **41** 12303
[20] Toyozawa Y 1983 *Physica* **116B** 7
[21] Lannoo M 1990 *Physics of DX Centers in GaAs Alloys* ed J C Bourgoin (Vaduz: Sci. Tech. Publications) p 209
[22] Zhang S B and Chadi D J 1990 *Phys. Rev.* B **42** 7174
[23] Dabrowski J, Scheffler M and Strehlow R 1990 *Proc. 20th Int. Conf. on the Physics of Semiconductors* ed E M Anastassakis and J D Joannopoulos (Singapore: World Scientific) p 489
[24] Bourgoin J and Lannoo M 1983 *Point Defects in Semiconductors II* (*Springer Series in Solid State Science 35*) (Berlin: Springer)
[25] See review papers by T A Kennedy and E Glaser and by H J von Bardeleben 1990 *Physics of DX Centers in GaAs Alloys* ed J C Bourgoin (Vaduz: Sci. Tech. Publications) pp 53 and 181, respectively
[26] Glaser E, Kennedy T A, Sillmon R S and Spencer M G 1989 *Phys. Rev.* B **40** 3447
[27] Kaufmann E, Wilkening W, Mooney P M and Kuech T F 1990 *Phys. Rev.* B **41** 10206
[28] von Bardeleben H J, Bourgoin J C, Basmaji P and Gibart P 1989 *Phys. Rev.* B **40** 5892
[29] Fockele M, Spaeth J M and Gibart P 1990 *Proc. 20th Int. Conf. on the Physics of Semiconductors* ed E M Anastassakis and J D Joannopoulos (Singapore: World Scientific) p 517
[30] von Bardeleben H J, Bourgoin J C, Delerue C and Lannoo M to be published
[31] Fockele M, Spaeth J M, Overhof H and Gibart P to be published
[32] Montie E A, Henning J C M and Cosman E C 1990 *Phys. Rev.* B **42** 11808

Semicond. Sci. Technol. **6** (1991) B23–B26. Printed in the UK

The DX centre

T N Morgan

IBM T J Watson Research Center, Yorktown Heights, NY 10598, USA

Abstract. A new model for DX centres is proposed and shown to be the only model which agrees with DLTS, metastability and ballistic phonon scattering experiments. In this model an As antisite complex, formed from a substitutional donor by the exchange of two adjacent atoms, binds two electrons. Furthermore, if these electrons are assumed to occupy a triplet spin state, the model can explain the paramagnetic susceptibility reported for filled DX centres. The stability of such a configuration could be due to the unusual properties of the antisite complex. Extension of this analysis to the related EL2 defect in GaAs suggests that the normal EL2 state may be $^3T_2(A_2)$ and the metastability of the excited state due, in part, to spin conservation.

1. Introduction

No completely satisfactory microscopic explanation has yet been found for persistent photoconductivity (PPC) and other unique properties of DX centres in III-V crystals [1]. Several key experiments have therefore been examined and found to lead to a unique explanation. In this paper the evidence is presented; the new model which this explanation requires is described and its consequences discussed. It is assumed initially that in its formation, each DX centre undergoes a large lattice relaxation [1] and binds two electrons to become DX⁻ [2], two important properties which now seem well established.

The appearance of only three or four peaks in the DLTS spectra of Mooney et al [3] and of Baba et al [4] provides strong evidence that the Si donor in $Al_xGa_{1-x}As$ or GaAs moves, when it forms a DX centre, nearer to three of its surrounding group III atoms. This conclusion led to the proposal of the vacancy-interstitial (V-I) model [5, 6], in which the Si moves away from a nearest neighbour As into an interstitial site, as the simplest and, therefore, most probable distortion. There is, however, an alternative relaxation of the Si, which also leaves it with three group III neighbours. This occurs in the exchanged site (X-S) model, in which the Si exchanges sites with a neighbouring As. It is this model which, as shown below, agrees best with experiment.

The search for a new model was initially prompted by the expectation that the V-I model could not explain the large enthalpy difference of 100 meV found by DLTS between configurations containing zero and one Al atom [3]. Although a very recent calculation† suggests that this energy difference may be consistent with the V-I model, more serious contradictions are found in other properties.

2. The new model

In the X-S model an Si atom exchanges sites with one of its neighbouring As atoms to form a donor-acceptor complex, as shown in figure 1(a). Thus, after interchange, both atoms remain substitutional, so that all tetrahedral bonds are complete and the lattice is stable for small displacements. Also, the Si becomes a Si_{As}^- acceptor which is bonded to three group III atoms, and the As becomes an antisite double donor, As_{Ga}^{+2}, binding one or two electrons in a deep state. The formation energy of such a defect is known to be small [7], and the capture barrier is reduced by the binding of the two captured electrons. The statistics of occupancy of the various configurations in Si-doped crystals containing both Ga and Al atoms are exactly the same as for the V-I model [6, 8].

The As_{Ga} antisite, which is known to generate the deep EL2 defect in GaAs [9], stabilizes DX by binding two electrons, so that the occupied DX has a negative charge [2],

$$Si_{Ga}^+ + As_{As} + 2e^- \rightarrow Si_{As}^- + As_{Ga}^0 \equiv DX^-.$$

The repulsive potential of the neighboring negative ion, however, in the simplest approximation reduces the binding energies of these electrons by $\simeq 0.25$ eV from those of EL2, and lowers their symmetry.

Let us consider the energy difference found between centres with and without Al. In the process of capturing two electrons to form the deep DX state, three Ga site atoms which were bonded to As rebond to a Si_{As}^- ion.

---

†A recent calculation (S B Zhang, private communication) claims a similar energy difference for the V-I model. If confirmed, this result would agree with experiment.

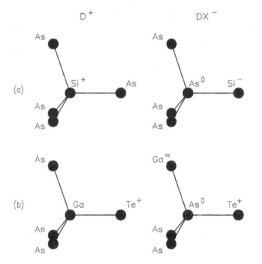

Figure 1. Atomic rearrangements in the X–S model of DX centres in GaAs: (a) group IV donors (Si); (b) group VI donors (Te). Each atom on the right is bonded to three Ga atoms.

is metastable in its exchanged position and contains a *dipole* oriented along ⟨111⟩ only *after* optical excitation. In contrast, the defect in the V–I model, even if metastable, would be *neutral* after photoionization with an anisotropy *reduced* from that of DX$^-$. The orthorhombic symmetry [11] found for Te suggests a more complex centre.

The most probable model for group VI donors is suggested by the phonon results and shown in figure 1(b). Here, two adjacent host atoms interchange sites to produce an antisite pair, which binds two electrons on the As and is adjacent to the group VI donor ion. We note that this complex is stabilized relative to an isolated antisite pair by the positive donor ion. Again, the stabilizing energy of DX comes from the two electrons on the As antisite, the neighbouring ions have a net charge of -1, and the binding energy and symmetry are little different from those of group IV donors.

Two DX centre energy level diagrams are shown in figure 2. Figure 2(a), is valid for any model invoking a large lattice distortion and capture of two electrons. It shows that the total energies associated with the defect when zero (DX$^+$), one (DX^0) or two (DX$^-$) electrons are bound by the centre. All curves are shown with local

Hence, the gain, ΔE, in binding energy when one Al replaces one Ga equals the difference in bond energy in the final state minus the difference in the initial state

$$\Delta E = (E^{\mathrm{B}}_{\mathrm{Al-Si}} - E^{\mathrm{B}}_{\mathrm{Ga-Si}}) - (E^{\mathrm{B}}_{\mathrm{Al-As}} - E^{\mathrm{B}}_{\mathrm{Ga-As}})$$

where E^{B}_i is the energy per bond in bond i. The AlAs and GaAs energies are known, and to approximate the energies of the bonds with $\mathrm{Si}^-_{\mathrm{As}}$ we can use the values for AlP and GaP [10]. Hence,

$$\Delta E \simeq (2.13 - 1.78) - (1.89 - 1.63)\,\mathrm{eV} = 90\,\mathrm{meV}$$

in good agreement with the measured value of 100 meV.

Thermal capture and emission processes involve simultaneous electronic transitions and atomic rearrangements. In an optical transition, however, atomic rearrangements need not occur if both configurations are stable with respect to small lattice distortions. Many experiments confirm that interconversion between D and DX occurs above a threshold temperature [1]. At lower temperatures, however, removal of one or two electrons from DX$^-$ usually leaves a metastable DX^0 or DX$^+$ configuration which can recapture electrons [11, 12]. Thus, after photoexcitation of one electron, the system shows properties of both the DX^0 defect and the released electron, and the latter may be trapped on the Ga- or As-site shallow donors.

This model neatly explains the puzzling phonon scattering results of Narayanamurti *et al* [11] where ballistic phonons in Al$_x$Ga$_{1-x}$As at low temperature were found to be scattered anisotropically by the DX centres *after* they had been photoionized and were expected to have returned to their stable configurations. The trigonal symmetry deduced from the scattering by Sn centres in this experiment is in agreement with the X–S model as described above (figure 1(a)), as the centre

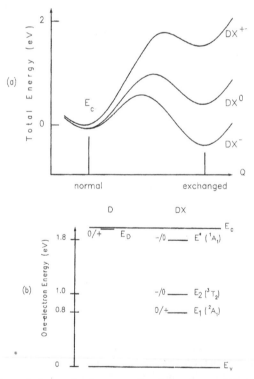

Figure 2. Approximate energy levels for donors (D) and DX centres under conditions making DX$^-$ lowest. (a) Total energy of two electrons at high temperature under a distortion Q which interchanges the atoms. Unbound electrons are assumed to be at the conduction band edge. (b) One-electron energies at low temperature, where the D and DX centres act independently. The symmetry assignments refer to the spin triplet form of the model.

minima near the exchanged position because of the metastability found experimentally (and expected for the X-S model). For DX$^-$ to be stable in the distorted configuration, the condition of 'negative U', the sum of the one-electron energies gained by capture of the two electrons, must exceed the energy cost of the distortion, i.e. DX$^-$ on the right must lie below D$^+$ on the left. When this condition is satisfied, capture of two electrons occurs over the barrier in the DX$^-$ curve, whose energy has been lowered by the binding of two electrons.

Thus, as figure 2(b) shows, a DX centre binds one electron in a deep state at an energy E_1 and a second at E_2 with a normal, 'positive U', ordering of levels. The gain in energy from capture of these electrons is $2E_c - (E_1 + E_2)$, where E_c is the conduction band energy. Note that at low temperatures the two types of defect, D and DX, are distinct and act independently, as experiment confirms. They also occur in approximately equal numbers, if there is no compensation.

As figure 2 also shows, we expect two energy thresholds for optical excitation of electrons from DX centres, one at $h\nu = E_c - E_2$ from DX$^-$ and one at $h\nu = E_c - E_1$ from DX^0. Furthermore, one expects emission and recapture to occur in two ways—either within one type of centre, D or DX, or with conversion of one into the other. Examples of the latter, which occurs with a large lattice relaxation and requires a source of thermal or other energy to surmount the barrier, are (i) the thermal conversion of D$^+$ into DX$^-$ with the capture of two electrons and (ii) the photoconversion of DX^0 or DX$^-$ into D$^+$ with the release of one or two electrons. At low temperatures, as process (i) does not occur, process (ii), when it does occur, generates PPC.

The metastability of DX^0 and DX$^+$ is shown in the studies by von Bardeleben et al [12] in Sn-doped Al$_x$Ga$_{1-x}$As. This work also helps locate the energy levels if we identify the first threshold, at which a long-lived ($\tau \simeq 10^4$ s) EPR spectrum appears, with the energy $E_c - E_2 \simeq 0.8$ eV, and the second ($\tau \simeq 10^2$ s) with $E_c - E_1 \simeq 1.0$ eV. We deduce, therefore, for $x \simeq 0.3$, an upper limit near 1.8 eV for the (lattice) energy of formation of DX$^+$ from D$^+$. Fockele et al [13] using magnetic circular dichroism also found the first (long-lived) threshold for Sn near 0.8 eV, and, for Si, Mooney et al, with $x \simeq 0.4$, find, using EPR [14], two thresholds near 0.8 and 1.1 eV and, using photocapacitance [15], a strong threshold near 1 eV†. Further work to determine these thresholds and lifetimes in various materials is needed.

The metastability of these deep states explains the data reported by Samuelson et al [16] and by Dmochowski et al [17] in GaAs under hydrostatic pressure. As figure 2(b) shows, the optical transitions, which lie above 1.5 eV, cannot involve the ground DX$^-$ state, whose second electron level lies only about 1 eV above the valence band, but involve instead higher lying states of A$_1$ symmetry. These may be: (i) the antibonding D^0

† It appears in [15] that the intensity of the lower energy transition has been reduced by the light used to initialize the diodes prior to measurement.

states proposed by Hjalmarson et al [18] or (ii) excited states of DX$^-$ labelled E*. Experiments at low temperatures indicate the presence of both types: (i) D^0 states which appear when the pressure is increased in the dark, so that DX centres are not present, and (ii) DX$^-$* states, which require light exposure. Either might explain the anticrossings observed [17] and the small Franck–Condon shifts [16] reported, but only the latter can account for the narrowness and small energy shifts of the pair lines and the exponential decay of the luminescence [16]. These effects will be explored in detail in a future publication.

3. The model and magnetic susceptibility

I have shown that experiment establishes the structure and energy levels of DX centres and favors the X-S model for group IV and, probably, group VI donors, but have said nothing about the symmetries of the states. Accepted theory predicts that the ground state will contain two electrons with antiparallel spins in a level of symmetry A$_1$, as Chadi and Chang propose [5]. Such a state must be non-magnetic, however, and therefore cannot explain the magnetic susceptibility measurements of Khachaturyan et al [19], who find the occupied DX state to be paramagnetic with a susceptibility consistent with one electron per donor. A conflicting result has recently been reported by Katsumoto et al [20], who find a smaller susceptibility. Both sets of experiments are being repeated.

I shall, in the remainder of this paper, develop the consequences of the magnetic results and show that both may be consistent with the X-S model. First, if [19] is found to be in error, conventional theory can explain these data in either model. If, however, the large susceptibility is confirmed, they can be explained only if the two DX electrons form a triplet spin state, and this requires that they occupy different orbitals, presumably one of A$_1$ and one of T$_2$ symmetry. Thus, the total symmetry would be ^{3}T$_2$ (slightly perturbed by the neighbouring ion), and the magnetic susceptibility could be the same, within experimental uncertainty, as the data reported. There are complications, however, because of the spin-orbit coupling. For example, an EPR signal, which is not found [14], need not be seen in a ^{3}T$_2$ state because of spin-orbit splitting and the broadening of the resonance by random strains [21].

Further, such a state would be split by lattice strain, and this can account for the differences between the two experiments. In [20], the sample was removed from the substrate and unstrained, so that the non-degenerate (and non-magnetic) spin-orbit split component of the ^{3}T$_2$ state would lie lowest. In [19], however, the samples, measured on their GaAs substrates, were strained. The strain, which lifts the orbital degeneracy, may have caused the magnetic, $m_s = \pm 1$, component to lie lowest and reveal the large susceptibility.

An additional consequence of these assumptions is that the DX$^-$* state, of symmetry ^{1}A$_1$, could account for the pair spectra reported in [16]. The existence of a $^3T_2$ two-electron state at a lower energy than $^1A_1$, although not expected, is required if the large paramagnetism is confirmed. It would imply that a one-electron $T_2 - A_1$ energy difference of about 0.25 eV has been reduced by an exchange energy of about 1 eV to put $^3T_2$ 0.75 eV below $^1A_1$. It is possible that the low energy of the T_2 level could be a consequence of the complex formed by the As antisite and the donor.

It is of interest that an extension of this analysis to EL2 defects in GaAs, which also involve As_{Ga}, [9] appears to agree with experiment. The two levels above the valence band at 0.55 and 0.75 eV correspond to the $^2A_1$ and $^3T_2$ levels, which, in DX, lie higher by the Coulomb energy of 0.25 eV. Note that the spin–orbit split A_2 level of $^3T_2$ should lie lowest. By assuming an equal shift from the DX level near 1.8 eV, we predict a $^1A_1$ EL2 level about at the GaAs band edge, where the metastable state is thought to be. Further, transitions from this singlet level to the normal, $^3T_2$ levels are spin forbidden, which could account in part for the metastability. These assignments for the states of EL2 disagree, however, with calculations by Dobrowsky and Scheffler using density functional theory [22] and require further confirmation.

4. Summary

A new 'X–S' model has been introduced. When this and the V–I model are compared with experiments on DX centres, only the former appears able to explain all results. More complete quantitative analyses and comparisons with experiment are in preparation.

Acknowledgments

I should particularly like to thank Pat Mooney, Tom Theis and Don Wolford for assistance and encouragement. This work has been partially supported by the US Office of Naval Research under Contract Nos N00014-85-C-0868 and N00014-90-C-0077.

References

[1] See Mooney P M 1990 *J. Appl. Phys.* **67** R1 and references therein
[2] Many recent experiments establish that DX binds two electrons. See e.g., Fujisawa T, Yoshino J and Kukimoto H 1990 *Proc. 20th Int. Conf. on the Physics of Semiconductors*, ed E M Anastassakis and J D Joannopoulos (Singapore: World Scientific) p 509 and other papers in the same volume
[3] Mooney P M, Theis T N and Wright S L p 1109; *Appl. Phys. Lett.* **53** 2546
[4] Baba T, Mizuta M, Fujisawa T, Yoshino J and Kukimoto 1989 *Japan. J. Appl. Phys.* **28** L891
[5] Chadi J D and Chang K 1988 *Phys. Rev. Lett.* **61** 873
[6] Morgan T N 1989 *Defects in Semiconductors 15* (*Mater. Sci. Forum* **38–41**) ed G Ferenczi (Switzerland: Trans. Tech.) p 1079
[7] The enthalpy of formation of an antisite pair in GaAs is calculated to be about 0.7 eV: Van Vechten J A 1975 *J. Electrochem. Soc.* **122** 423
[8] Morgan T N 1991 *J. Electron. Mater.* **20** 63
[9] See Bourgoin J C, von Bardeleben H J and Stievenard D 1988 *J. Appl. Phys.* **64** R65
[10] Harrison W A 1980 *Electronic Structure and the Properties of Solids* (San Francisco: W H Freeman) table 7-3
[11] Narayanamurti V, Logan R A and Chin M A 1979 *Phys. Rev. Lett.* **43** 1536
[12] von Bardeleben H J, Bourgoin J C, Basmaji P and Gibart P 1989 *Phys. Rev. B* **40** 5892. The second resonance may be due to shallow A_1 states derived from X and bound to DX^+
[13] Fockele M, Spaeth J-M and Gibart P 1990 *Proc. 20th Int. Conf. on the Physics of Semiconductors* ed E M Anastassakis and J D Joannopoulos (Singapore: World Scientific) p 517
[14] Mooney P M, Wilkening W, Kaufmann U and Kuech T F 1989 *Phys. Rev. B* **39** 5554
[15] Mooney P M, Northrop G A, Morgan T N and Grimmeiss H G 1988 *Phys. Rev. B* **37** 8298
[16] Samuelson L, Gerling M, Liu X, Nilsson S, Omling P, Pistol M-E and Silverberg P 1988 *Proc. 19th Int. Conference Physics of Semiconductors* ed W Zawadski (Warsaw: Institute of Physics, Polish Academy of Science) p 967;
Xiao Liu, Samuelson L, Pistol M-E, Gerling M and Nilsson S 1990 *Phys. Rev. B* **42** 791
[17] Dmochowski J E, Wang P D and Stradling R A 1990 *Proc. 20th Int. Conf. on the Physics of Semiconductors* ed E M Anastassakis and J D Joannopoulos (Singapore: World Scientific) p 658
[18] Hjalmarson H P, Vogl P, Wolford D J and Dow J D 1980 *Phys. Rev. Lett.* **44** 810
[19] Khachaturyan K A, Awaschalom D D, Rozen J R and Weber E R 1989 *Phys. Rev. Lett.* **63** 1311
[20] Katsumoto S, Matsunaga N, Yoshida Y, Sugiyama K and Kobayashi S-I 1990 *Japan. J. Appl. Phys.* **29** L1572
[21] Mehran F, Morgan T N, Title R S and Blum S E 1972 *Phys. Rev. B* **6**, 3917
[22] Dabrowski J and Scheffler M 1989 *Phys. Rev. B* **40** 10391

Semicond. Sci. Technol. **6** (1991) B27–B30. Printed in the UK

Hall measurements under weak persistent photoexcitation in Si-doped $Al_xGa_{1-x}As$

A Baraldi†, C Ghezzi†, A Parisini†, A Bosacchi‡ and S Franchi‡

† Dipartimento di Fisica dell'Università, Viale delle Scienze-43100 Parma, Italy
‡ CNR-Istituto MASPEC, Via Chiavari, 18/A-43100 Parma, Italy

Abstract. The low-temperature Hall mobility of photoexcited electrons has been measured in Si-doped MBE AlGaAs samples. Different fractions of occupied DX centres were obtained by selecting different free-electron densities: possible systematic errors in Hall measurements due to the method of photoexcitation are demonstrated and critically analysed. Using suitable values for the acceptor density and the alloy scattering potential a fair fitting of the experimental data was achieved within both negative-U and positive-U models for the DX centre. Discrepancies between calculated and experimental mobility versus temperature curves are observed, which are more evident the lower the free-electron densities. They are tentatively explained as being due to electrons in an impurity band originated by the shallow effective-mass state related to the Γ minimum of the conduction band.

1. Introduction

The DX centre is a relatively deep level related to an isolated substitutional donor impurity which is responsible for unusual electrical properties observed in n-doped $Al_xGa_{1-x}As$ [1, 2]. This level is resonant with the conduction band (CB) up to an $x = 0.22$ AlAs mole fraction, but for $x > 0.22$ it becomes the lowest donor state thus controlling the free-electron concentration [3, 4]. The centre is characterized by a thermally activated capture rate [5] which can explain the low-temperature persistence of photoexcited electrons in the conduction band after the light is switched off (persistent photoconductivity, PPC). The microscopic structure of the centre is the object of controversy [2] with particular interest in the charge of the impurity in the ground state [6].

In this work we analyse the temperature dependence of the Hall mobility of photoexcited electrons. Mobility data are taken under both conditions of saturated (SPPC) and unsaturated (PPC) persistent photoconductivity; in this way different mobility versus T curves refer to different fractions of occupied DX centres and different electron mean energies. The procedure employed to reach these conditions is critically analysed. A theoretical fit of all mobility curves is given and the agreement with the experimental data is discussed by taking into account the possible contribution of a impurity band conduction when the Hall density n_H of photoexcited electrons is small enough ($n_H < 10^{16}$ cm^{-3}). At higher n_H values the conduction is due only to free electrons in the Γ valley of the conduction band. The consistency of both positive- and negative-U approaches with the measured data

under these latter conditions is here briefly underlined, a full detailed analysis being given elsewhere [7, 8].

2. Theory

The mobility is calculated within the relaxation time approximation and Fermi-Dirac statistics used: the Fermi energy was derived from the electron concentration data after correction for the Hall factor $r_H = \langle \tau(E)^2 \rangle / \langle \tau(E) \rangle^2$, which is self-consistently evaluated within a 1 % accuracy. $\langle \tau(E) \rangle$ is the mean on the energy of the total scattering time, which is obtained from $\tau(E) = [\Sigma_i (1/\tau_i(E))]^{-1}$; τ_i indicates the different scattering mechanisms. In order to avoid complications due to multivalley conduction, we limited our calculations to the temperature range in which only the Γ valley is populated ($\approx$ for $T < 100$ K). Non-parabolicity effects of the Γ valley are roughly taken into account by introducing a non-parabolic effective mass averaged on the electron energy.

Ionized impurity scattering turns out to play the most important role in the low-temperature mobility behaviour. The Brooks–Herring relaxation time [9] was considered for this mechanism, which alone is able to explain the temperature dependence of the mobility taken both in the dark and in PPC conditions. Different expressions for the N_i ionized impurity density were assumed within the positive-U or the negative-U approach, according to the proper charge neutrality equation. A possible role of spatial correlations among charged impurities is neglected here. Among the other scattering mechanisms the alloy scattering mechanism

0268-1242/91/100B27+04 $03.50 © 1991 IOP Publishing Ltd

[10] was by far the dominant one, phonon scattering being negligible when $T < 100$ K. The acceptor density N_a and the alloy scattering potential V_{al} were allowed to vary as free parameters. More details on the calculations are given elsewhere [8].

3. Experiment

The $Al_xGa_{1-x}As$ samples were Si-doped layers, grown by molecular beam epitaxy at 600 °C; their compositions were $x = 0.25$ (5.4 μm thick), $x = 0.30$ (5.9 μm) and $x = 0.35$ (5.5 μm). The doping level was $\approx 10^{18}$ cm^{-3}. Between the undoped semi-insulating GaAs substrate and the doped AlGaAs layer were grown, from the bottom, an undoped 0.5 μm thick GaAs layer and an undoped AlGaAs buffer layer (thickness 0.2 μm). The latter was introduced to avoid effects due to two-dimensional electron gas (2DEG) [11]. Au-Ge ohmic contacts were made on the corners of the sample in order to use the van der Pauw method. The experiment was made in the 7–300 K temperature range. A GaAs LED was used as a light source for the photo-Hall measurements and a set of curves at low temperature were taken corresponding to different densities of photoexcited electrons.

4. Results and discussion

An example of the n_H Hall electron density as a function of the temperature is reported in figure 1 for the $x = 0.30$ sample. Above $T \approx 100$ K the freezing of the electrons in the DX centre is observable, whereas below $T \approx 100$ K the electron density remains nearly constant, owing to the vanishing capture rate of the DX centre. In the $x = 0.25$ sample the low-temperature electron density varies in the range 10^{17}–2×10^{18} cm^{-3} and all curves are T independent for $T < 100$ K. Here the important point is that a weak T dependence of n_H has been observed in the $x = 0.30$ and $x = 0.35$ (see insert, figure 1) samples and it is more evident when n_H is below about 10^{16} cm^{-3}.

The experimental mobility data are reported in figure 2. The continuous curves indicate the results of our fits by assuming a neutral charge state for the occupied DX centres (positive-U model). Comparable agreements between experimental and calculated data were also obtained within the frame of the negative-U model, as reported and discussed elsewhere [7, 8]. The same pair of N_a and V_{al} values was obtained by fitting all curves for $n_H > 10^{16}$ cm^{-3} in a given sample: by changing the photoionized electron density, the ionized impurity scattering and the alloy scattering contribute to the total mobility with different T dependence [8], the independence of the two parameters thus being assured. The positive-U approach can explain the increase of the electron mobility after photoexcitation if a sufficiently high acceptor density is assumed. The negative-U model results are consistent with the experimental data irrespective of the presence of compensating acceptors in the material, according to the acceptor-like character of the DX$^-$.

In both the approaches similar fairly good fits have been obtained for the higher-mobility curves of figure 2,

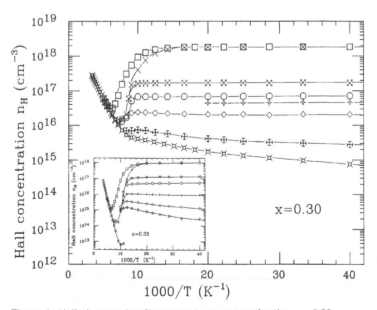

Figure 1. Hall electron density versus temperature for the $x = 0.30$ sample. ⋈, dark curve; □, saturated PPC conditions. The latter curve is taken under continuous photoexcitation at every temperature. The intermediate curves refer to unsaturated PPC conditions. Continuous curves are guides for the eye. In the insert the electron-density data for the $x = 0.35$ sample are reported, with a similar meaning of the symbols.

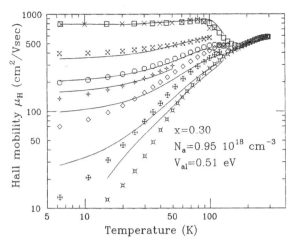

Figure 2. Hall mobility versus temperature for the $x = 0.30$ sample. Corresponding $n_H(T)$ curves of figure 1 and $\mu_H(T)$ data are indicated with the same symbols. The continuous curves give the calculated values within the positive-U model. The estimated values of N_a and V_{al} are indicated in the figure.

corresponding to lower fractions of occupied DX centres, but disagreements appear for the lower curves, which are more pronounced the lower the electron density. In our opinion these disagreements are due to a failure of the hypothesis that the electrical conduction is only due to free electrons in the Γ valley of the CB. In fact, when the capture rate of the DX centre becomes too small, the next lower impurity level, that is the shallow level coupled to Γ or X minimum, can control the freezing of the electrons [12–15]. Moreover, in our case the high doping density makes it reasonable to suppose the formation of an impurity band related to the shallow level. When the band is partially filled, conduction through these states is possible and a two-band model needs to be used to calculate both the effective mobility and carrier density. In this case the apparent $1/eR_H$ and μ_H measurements are no longer simply related to the $n_\Gamma(T)$ electron density and the $\mu_\Gamma(T)$ mobility in the Γ valley, but depend also on the electron fraction and the mobility in the impurity band. On the other hand, when the fraction density increases, the freezing of the electrons in the impurity band is expected to become less effective owing to stronger screening effects. In particular when the condition $n(T)^{-1/3} > 4a_H$ is approximately fulfilled, the occupation of the impurity band is inhibited. Here a_H is the Bohr radius for an hydrogenic impurity level coupled to the Γ minimum: $n_c^{-1/3} = 4a_H$ gives the critical electron density for the Mott transition n_c. Above n_c the measured electron density is then due to free electrons in the Γ minimum and the calculated mobility is expected to agree with the experimental data. This is really observed in our data, the estimated critical electron density being approximately $n_c \approx (4–6) \times 10^{16}$ cm^{-3} for all samples. These values were obtained by considering respectively the formulae $m_\Gamma(x) = (0.067 + 0.083x)m_0$ and $\varepsilon(x) = 12.4 - 2.79x$ for the effective mass relative to the Γ conduction band minimum and the relative dielectric

constant, m_0 being the free-electron mass and x the AlAs mole fraction. Moreover, a slow T dependence in the $n_H(T)$ curves is expected when $n_H < n_c$, as is indeed observed.

Unfortunately a calculation which accounts for the impurity band conduction is unreliable owing to the number of unknowns in the proper equations. Magneto-transport measurements, taken for different photo-excited electron densities could confirm, in principle, this hypothesis.

Let us now briefly comment upon the procedure we employed to select different photoexcited electron densities. We found that the mobility measurements can give different results when the same electron density is obtained through different photoexcitation methods. We have considered two possibilities: a given electron density is reached

(i) after a set of controlled light flashes at the lowest temperature or

(ii) through the following steps: (a) complete photo-ionization of the deep centre at the lowest temperature (SPPC), (b) heating the sample in the dark until the electron density decreases to the given value, (c) cooling again rapidly at the lowest temperature and making the measurement.

The mobility data obtained through the above procedures are reported in figure 3; the comparison was repeated for different electron densities, with similar conclusions. In every case, by following method (i) a lower mobility is observed. In order to investigate the origin of this effect, possible causes of spatial inhomogeneity of the sample must first be considered. Sometimes, due to unintentional geometrical misalignments, a non-uniform illumination of the sample can be realized. Moreover, the regions covered by the metallic contacts are weakly and indirectly illuminated. Only after a long exposure time is the density of photoexcited electrons then expected to reach spatially homogeneous conditions within the sample. This is indeed the case when saturated PPC conditions are first reached and the desired n_H value is then adjusted by heating the sample in the dark and subsequent rapid cooling. Method (ii), in spite of its greater complexity, must then be preferred to achieve spatially homogeneous conditions and mobility data unaffected by systematic experimental errors (see data in figure 2). On the contrary, when method (i) is applied, high-resistivity regions of macroscopic size can remain under the contacts and an apparently lower mobility is measured.

In summary we have analysed Hall mobility data versus temperature in Si-doped MBE Al$_x$Ga$_{1-x}$As samples of different AlAs mole fraction. The measurements were taken both in the dark and for different densities of photoexcited electrons. The mobility appears to be dominated by ionized impurity scattering, the alloy scattering being the most important of the other mechanisms. Using suitable values for the acceptor density and the alloy-scattering potential, a fair fitting of the mobility versus temperature was achieved, no matter whether a positive-

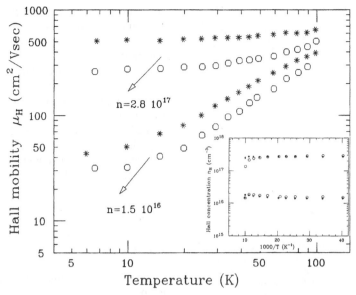

Figure 3. Mobility versus temperature data when a given n_H density of photoexcited electrons has been achieved through different procedures: ○, low-temperature light flashes; ∗, reaching SPPC conditions, heating and rapid cooling of the sample (see text). The n_H values versus T are given in the insert ($x = 0.30$).

or negative-U approach was used for the DX centre. Systematic discrepancies between experimental and calculated mobility data have been observed to become more and more pronounced the lower the electron density. We pointed out that this takes place when the electron density is lower than the critical Mott value evaluated for a shallow effective-mass level coupled to the Γ minimum. We then suggested that under these conditions a fraction of electrons contributes to electrical conduction, that fraction being located in a partially filled impurity band originating from that level. Finally the experimental procedures employed to achieve unsaturated PPC conditions have been critically analysed. Differences have been demonstrated in the experimental mobility results by following two different methods to achieve the same electron density of photoexcited electrons. In one of them spatially inhomogeneous photoionization of the impurity centres is believed to cause systematic errors in the Hall measurements.

Acknowledgment

This work was partially supported by the finalized research project 'Material devices for solid state electronics' of the National Research Center.

References

[1] Mooney P M 1990 *J. Appl. Phys.* **67** R1
[2] Bourgoin J C *Physics of DX Centres in GaAs Alloys (Solid State Phenomena 10)* (Vaduz: Sci. Tech. Publications)
[3] Chand N, Henderson T, Klem J, Masselink T, Fischer R, Chang Y C and Morkoc H 1984 *Phys. Rev.* B **30** 4481
[4] Bhattacharya P 1988 *Semicond. Sci. Technol.* **3** 1145
[5] Mooney P M, Caswell N S and Wright S L 1987 *J. Appl. Phys.* **62** 4786
[6] Chadi D J and Chang K J 1989 *Phys. Rev.* B **39** 10063
[7] Baraldi A, Ghezzi C, Parisini A, Bosacchi A and Franchi S 1991 *Appl. Surf. Sci.* in press
[8] Baraldi A, Ghezzi C, Parisini A, Bosacchi A and Franchi S 1991 *Phys. Rev.* B in press
[9] Chattopadhyay D and Queisser H J 1981 *Rev. Mod. Phys.* **53** 745
[10] Harrison W and Hauser J R 1976 *Phys. Rev.* B **13** 5347
[11] Collins D M, Mars D E, Fischer B and Kocot C 1983 *J. Appl. Phys.* **54** 857
[12] Mizuta M 1990 *Physics of DX Centers in GaAs Alloys (Solid State Phenomena 10)* ed J C Bourgoin (Vaduz: Sci. Tech. Publications) p 65
[13] Theis T N, Kuech T F, Palmateer L F and Mooney P M 1985 *Gallium Arsenide and Related Compounds 1984 (Inst. Phys. Conf. Ser. 74)* ed B de Cremoux (Bristol: Institute of Physics) p 241
[14] Lavielle D, Goutiers B, Kadri A, Ranz E, Dmowski L, Portal J C, Grattepain C, Chand N, Sallese J M and Gibart P 1990 *Proc. 20th Int. Conf on the Physics of Semiconductors* ed E M Anastassakis and J D Joannopoulos (Singapore: World Scientific) p 521
[15] Dmochowski J E, Dobaczewski L, Langer J M and Jantsch W 1989 *Phys. Rev.* B **40** 9671

Semicond. Sci. Technol. **6** (1991) B31–B33. Printed in the UK

The influence of the DX centre C–V and I–V characteristics of Schottky barriers in n-type AlGaAs

C Ghezzi†, E Gombia‡ and R Mosca‡

† Dipartimento di Fisica, Università degli Studi di Parma, Viale delle Scienze,
43100 Parma, Italy
‡ Istituto MASPEC–CNR, Via Chiavari 18/A, 43100 Parma, Italy

Abstract. Due to the vanishing of both electron emission and capture rates, the DX centre shows a non-equilibrium occupancy at low temperature. It is shown that this effect results in a non-uniform free-electron density profile that can be controlled by biasing the junction during cooling. Preliminary results are reported showing that the I–V characteristics of Schottky barriers on AlGaAs ($x = 0.25$) are affected by the DX centre occupation.

1. Introduction

In recent years the DX centre in AlGaAs has attracted much attention due to its technological importance and its intriguing physical properties [1]. It is nowadays agreed that the DX centre is originated by the dopant impurities and that its features are strongly influenced by the structure of the conduction band, depending on the Al content. Capacitance measurements have been frequently used in the study of the DX centre, although the very peculiar properties of the centre require some caution in the analysis of the experimental data. In fact, contrary to the usual situations, the DX centre density is much larger than the density of shallow donors, if any. Furthermore, the DX centre has a relatively small ionization energy, so that in the flat band region there is a large fraction of unoccupied centres, and in the space charge region of a junction device the size of the so-called λ region is vanishingly small. Finally, at low temperature both electron emission and capture rates are unobservably small, so that the charge state of the DX centre is frozen and can only be changed by photoionization. As a matter of fact, as shown by capture-time data reported by Mooney *et al* [2], the presence of a large capture barrier neglects that in practical experiments the DX centre occupancy may reach, at low temperature, values corresponding to thermal equilibrium. This is also supported by the observation of the persistent photoconductivity (PPC) effect [3]. A consequence of this situation is that the electron density in the flat band region depends on the cooling rate of the sample, as recently reported [4].

This paper is mainly concerned with the influence of the DX centre non-equilibrium occupancy on the results of C–V and I–V measurements carried out on Schottky barriers at low temperature.

2. Samples

The samples were grown by MBE and consisted of an n$^+$ substrate, a 0.1 μm thick n$^+$ GaAs buffer layer, a 3 μm thick Si-doped Al$_x$Ga$_{1-x}$As ($x = 0.25$) layer and a 200 Å thick undoped GaAs cap. Ohmic contacts were obtained by evaporating an AuGe alloy on the reverse side of the samples and annealing at 410 °C for 1 min in a forming gas atmosphere. Schottky barriers were fabricated by evaporating Au dots with an area of 0.26 mm^2.

3. Results and discussion

From quasistatic C–V measurements on Schottky diodes it is possible to get a density N as usually defined through the slope of a linear $1/C^2$ versus V plot (C = diode capacitance per unit area, V = applied voltage) and an apparent built-in potential given by the intercept V_a of the $1/C^2$ straight line on the V axis. Owing to band bending at the barrier, a non-uniform spatial distribution of positively charged centres is expected. Such a distribution shows a step-like profile that is controlled by the bias voltage V_0 used during cooling of the sample, whereas, since the DX centre occupancy is frozen at low temperature, it is independent of the voltage V applied for the low-temperature C–V measurements. The $1/C^2$ versus V plot is thus expected to consist of two regions: the first, at reverse voltage V larger than V_0 and with a larger slope, refers to the low-density region, while the second, at reverse bias less than V_0 or at forward bias and with a smaller slope, refers to the high-density region.

Figure 1 shows the $1/C^2$ versus V plots obtained after cooling the sample with $V_0 = -1.8$ V and with $V_0 = +0.6$ V. The presence of two density regions is evident

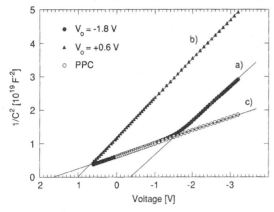

Figure 1. $1/C^2$ versus V plots obtained at 77 K: (*a*) after cooling the sample with $V_0 = -1.8$ V, (*b*) after cooling the sample with $V_0 = +0.6$ V and (*c*) in saturated PPC conditions. The fittings reported in the figure (full lines) give (*a*) $N = 1.69 \times 10^{17}$ cm^{-3} and $V_a = -0.39$ V, (*b*) $N = 1.5 \times 10^{17}$ cm^{-3} and $V_a = 1.02$ V and (*c*) $N = 4.52 \times 10^{17}$ cm^{-3} and $V_a = 1.58$ V, respectively.

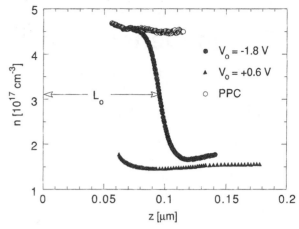

Figure 2. Free-electron density profile as obtained by C–V measurements (●) after cooling the sample with $V_0 = -1.8$ V, (▲) after cooling the sample with $V_0 = +0.6$ V and (○) in saturated PPC conditions. The step width L_0 is defined as the distance from the metal–semiconductor interface where $n = (N_{DX}^+(\infty) + N_{DX})/2$.

when $V_0 = -1.8$ V: the values of N and V_a achieved at bias less than V_0 are 4.52×10^{17} cm^{-3} and 1.58 V respectively, in agreement with the results of the measurements carried out under saturated PPC conditions (figure 1). The results obtained for reverse bias V larger than V_0 show that $N = 1.69 \times 10^{17}$ cm^{-3} and $V_a = -0.39$ V when $V_0 = -1.8$ V, whereas $N = 1.5 \times 10^{17}$ cm^{-3} and $V_a = 1.02$ V when $V_0 = +0.6$ V.

In the considered experimental conditions, the significance of N and V_a can be understood by integrating the Poisson equation under the depletion approximation [5]. It can be shown that, in agreement with results obtained under PPC conditions, when reverse voltages V less than V_0 are considered, V_a gives the built-in potential $\Phi_0 (V_a = \Phi_0)$ and $N = N_{DX} - N_a$ gives the net DX centre concentration ($N_{DX} = $ DX centre density, $N_a = $ ionized acceptor density). In contrast, for reverse bias larger than V_0, V_a gives an apparent built-in potential ($V_a = \Phi_0 - V_1$), differing from the built-in potential Φ_0 by an amount V_1 which is larger the larger V_0. On the other hand, no matter what the value V_0, N gives the free-electron density n in the flat band region and can be influenced by the different rates used in cooling the sample [4]. If a negative-U character [6] for the DX centre is assumed, $n = 2N_{DX}^+(\infty) - N_{DX} - N_a$ ($N_{DX}^+(\infty) = $ density of positively charged DX centres in the asymptotic flat band region). It is worth noting that equivalent conclusions can be reached if, for example, a positive-U character is assumed for the DX centre or if the presence of hypothetical shallow donors not related to DX centres [7] is accounted for.

Due to the presence of knee-shaped $1/C^2$ versus V plots, non-uniform free-electron densities $n(z)$ can be observed by C-V profiling. Figure 2 shows the $n(z)$ profile obtained under the experimental conditions considered in figure 1 ($V_0 = -1.8$ V, $V_0 = +0.6$ V and PPC conditions). It is worthwhile to note that the choice of V_0

actually affects the step width L_0, that can be strongly reduced by forward biasing the junction when cooling.

Some consequences of the non-uniform distribution of charged centres on the derivation of density data are discussed elsewhere [5]. Here we will focus our attention on the role that this profile can have on the I-V curves of the device. The influence of the DX centre on the device characteristics has been previously considered in HBT [8] and in AlGaAs/GaAs laser structures [9]: in the former case the I-V characteristics have been found to be unaffected by the presence of the DX centre, whereas hysteretic behaviours, attributed to changes in the DX centre occupation, have been observed in the latter case. To the best of our knowledge, no similar study on Schottky barriers has been reported. Nevertheless the possibility to change the density of positively charged DX centres by photoexcitation appears rather attractive. In fact, if for example tunnelling contributions are present in the I-V characteristics, the currents flowing through the barrier should be affected by the density of positively charged DX centres in the semiconductor, since different densities result in different barrier widths. To check this point, I-V measurements have been carried out in the dark at 77 K after cooling the sample with $V_0 = -1.8$ V, with $V_0 = +0.6$ V and in persistent PPC conditions. The free-electron profiles corresponding to the three experimental situations are shown in figure 2. The forward branches (figure 3) are affected to only a minor extent by the occupation of DX centres, although the data taken after photoionization appear to be shifted towards larger currents. In contrast figure 4 clearly shows that the reverse branches are influenced by the occupation of the DX centre: at fixed bias the largest current values are observed when the sample is photoexcited and, as expected, the effect of the photoionization is larger when $V_0 = +0.6$ V than when $V_0 = -1.8$ V. This seems to suggest that a tunnelling contribution to

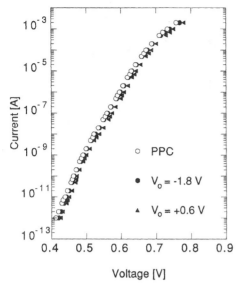

Figure 3. Forward branches of the *I–V* characteristics obtained (○) in saturated PPC conditions, (●) after cooling the sample with $V_0 = -1.8$ V and (▲) after cooling the sample with $V_0 = +0.6$ V.

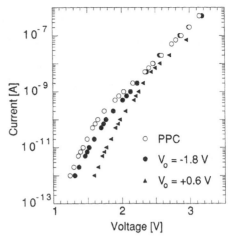

Figure 4. Reverse branches of the *I–V* characteristics obtained (○) in saturated PPC conditions, (●) after cooling the sample with $V_0 = -1.8$ V and (▲) after cooling the sample with $V_0 = +0.6$ V.

the reverse current is present, although other mechanisms should be taken into account to quantitatively correlate the *I-V* characteristics with the occupation profile of the DX centre. It is worth noting that, although in the present case the net DX centre concentration is only three times larger than the dark free-electron density *n* in the flat band region, the *I-V* characteristics are observed to be influenced by changes in the occupation of the DX centre. Nevertheless, in order to achieve some insights into the transport mechanisms in Schottky barriers prepared on AlGaAs, samples with larger AlAs mole fraction ($x \sim 0.3$) should be used. In fact, the effects reported in the present paper should be amplified for this Al content, since larger changes in the density of positively charged DX centres can be obtained by photoexcitation.

4. Summary

Summarizing, it has been shown that in Schottky barriers the low-temperature non-equilibrium occupancy of the DX centre results in a non-uniform free-electron density profile that can be controlled by biasing the junction during cooling. Preliminary results have been reported, showing that the *I-V* characteristics of Schottky barriers on AlGaAs ($x = 0.25$) are affected by the DX centre occupation.

Acknowledgments

The authors wish to thank Dr A Bosacchi, Dr S Franchi and Mr P Allegri for providing the samples.

References

[1] Mooney P M 1990 *J. Appl. Phys.* **67** R1
[2] Mooney P M, Caswell N S and Wright S L 1987 *J. Appl. Phys.* **62** 4786
[3] Lang D V, Logan R A and Jaros M 1979 *Phys. Rev. B* **19** 1015
[4] Ghezzi C, Mosca R, Bosacchi A, Franchi S and Gombia E 1991 *Solid State Commun.* **78** 159
[5] Ghezzi C, Mosca R, Bosacchi A, Franchi S and Gombia E *Appl. Surf. Sci.* to be published
[6] Langer J M, Dmochowski J E, Dobaczewski L, Jantsch W and Brunthaler G 1990 *Physics of DX Centers in GaAs Alloys (Solid State Phenomena 10)* ed J C Bourgoin (Vaduz: Sci. Tech. Publications) p 233
[7] Li M F, Jia Y B, Yu P Y, Zhou J and Gao J L 1989 *Phys. Rev. B* **40** 1430
[8] Nathan M I, Tiwari S, Mooney P M and Wright S L 1987 *J. Appl. Phys.* **62** 3234
[9] Hubik P, Smid V, Hulicius E, Kristofik J, Mares J J, Hlinomaz P and Zeman J 1990 *Gallium Arsenide and Related Compounds 1989 (Inst. Phys. Conf. Ser. 106)* ed T Ikoma and H Watanabe (Bristol: Institute of Physics) p 303

Semicond. Sci. Technol. **6** (1991) B34–B37. Printed in the UK

Appearance and destruction of spatial correlation of DX charges in GaAs

Z Wilamowski†, J Kossut†, T Suski‡, P Wiśniewski‡ and L Dmowski‡

† Institute of Physics, Polish Academy of Sciences, Al. Lotników 32/46, 02–668 Warsaw, Poland
‡ High Pressure Research Center, Polish Academy of Sciences, Warsaw, Poland

Abstract. Measurements of the Hall coefficient and the conductivity in n-GaAs heavily doped with silicon were performed at 4.2 K under hydrostatic pressure applied at an elevated temperature. The procedure ensures that a metastable occupation of the DX centres by electrons is induced. As shown earlier, electric charges on DX centres form, under these conditions, a spatially correlated system of charges. The correlation marks its presence as an enhancement of the electron mobility. Here we study the destruction of the spatial correlation of donor charges by illumination with an LED. A calculation within the short-range correlation model accounts well for the decrease of the mobility seen after illumination.

1. Introduction

An anomalous increase of the low-temperature electron mobility in n-GaAs with DX centres subject to hydrostatic pressure applied at an elevated temperature (typically at T above ~ 100 K) was first reported by Maude *et al* (1987). This took place at pressure values that brought the DX level (which at ambient pressure is a resonant donor level degenerate with the conduction band continuum) down to energetic coincidence with the Fermi level. The pinning of the Fermi level to the DX state is marked as an onset of the decrease of the conduction-electron concentration (measured by the Hall voltage). The full symbols in figure 1 show an example of such behaviour observed by ourselves in an MBE-grown GaAs:Si sample. Similar behaviour was also found in many GaAs samples doped with Si, Sn and Te. While the decrease of the concentration of electrons is qualitatively understandable in terms of the resonant DX level, whose energy, with respect to the conduction band minimum, decreases with applied pressure, the behaviour of the mobility is not so straightforward.

It was suggested by Dietl *et al* (1990) and, independently, by O'Reilly (1989) (see also Kossut *et al* 1990b, 1991, Suski *et al* 1990) that such mobility behaviour can be understood in both competing models of the DX centre (singly (DX^0) and doubly (DX^-) occupied, i.e. characterized by positive ($U > 0$) and, respectively, negative ($U < 0$) Hubbard correlation energy of electrons on the impurity site) provided that the correlation of the spatial positions of the impurity charges are taken into consideration. The correlation arises because of the Coulomb interaction between the charges localized on donors. In the case of a DX centre model with positive correlation energy ($U > 0$), when the impurity is either occupied and neutral with respect to the lattice or ionized and positively charged, the interaction that drives the

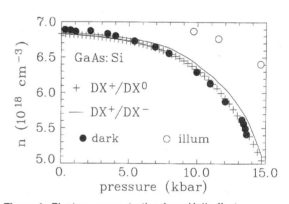

Figure 1. Electron concentration from Hall effect measurements at 4.2 K in GaAs sample doped with Si as a function of hydrostatic pressure applied at a temperature exceeding 100 K. The full circles represent experimental results in the sample kept in darkness ($n = N_e^i$), while the open circles are obtained after a prolonged illumination with an LED. The full curve and crosses show the results of the calculation within the short-range correlation model for negative and positive electron correlation energy on the DX centre, respectively, for the case of the dark sample with DX level energy E_{DX}, and its pressure coefficient γ, chosen to be $E_{DX} = 280$ meV and 300 meV and $\gamma = -11$ meV kbar^{-1} and -12 meV kbar^{-1} respectively for $U > 0$ and $U < 0$ (see Kossut *et al* 1990b, 1991 for details of the calculation).

0268-1242/91/100B34+04 $03.50 © 1991 IOP Publishing Ltd

appearance of the correlation is the repulsion between the like charges of the ionized donors. In the case of the $U < 0$ model of DX centres the most important correlation is due to the attraction of positively (empty) and negatively charged (occupied) donors and can be viewed as a formation of dipole-like objects. In both cases, the scattering rate of the conduction electrons by ionized impurities is reduced (i.e. the mobility is enhanced) because of the fact that, contrary to the case of chaotic distribution of scattering potentials, some of the scattering events have a coherent character and do not contribute to momentum relaxation. For similar reasons, they also do not shorten the electron lifetime measured by the Dingle temperature characterizing the amplitude of quantum oscillations in the presence of strong magnetic fields. Possibly an increase of the amplitude observed in the Shubnikov–de Haas effect study of Zrenner et al (1988) in delta-doped GaAs subject to hydrostatic pressure is ascribable to the existence of spatial correlation of DX charges.

Of course, the necessary condition for the correlation to arise is the condition of partial occupation of the donor centres by electrons. When DX centres are completely empty, or when they are fully occupied, the redistribution of charges between various impurity centres is not possible.

The spatial correlation of impurity charges in the case of $U < 0$ and $U > 0$ models of the DX centres in GaAs has its analogues in the situation occurring in HgCdTe with resonant acceptors and HgSe:Fe (where Fe is a resonant donor) (see Wilamowski et al 1990).

The degree of correlation that can be achieved depends on the ratio of the concentration of ionized to occupied donor centres. This ratio can be varied by changing the pressure.

2. Illumination experiments

Here we study the effect of illumination with an LED, which destroys initial correlation induced by application of a hydrostatic pressure, on the mobility of conduction electrons. Experimentally, the illumination causes electron transfer from the DX centres to the conduction band (concentration of electrons grows from N_e^{dark} to N_e^i), where the limiting value of N_e^i for very prolonged illumination is that found in the uncompressed sample (see open symbols in figure 1). Of course, any value of N_e^i can be obtained by illumination of intermediate duration. This increase of the conduction-electron concentration is accompanied by a decrease of the mobility (see experimental points in figure 2). Such behaviour is to be expected since the incident photons excite electrons on DX centres (and cause their transfer to the conduction band) regardless of the position of the donor with respect to other donors. If the sample is kept at a low temperature the correlated state cannot be re-established because of the configurational barriers that inhibit recapture of the electrons by the DX centres.

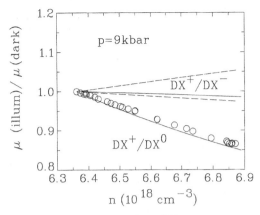

Figure 2. Electron mobility in GaAs:Si kept at pressure $p = 9$ kbar (applied at an elevated temperature) measured at 4.2 K in the illuminated sample (relative to the mobility at $p = 9$ kbar in darkness) as a function of the conduction-electron concentration varied by illumination ($n = N_e^i$). The open circles show the results of experiment; the lines are calculated without (broken lines) and with spatial correlation of DX charges taken into account (full lines). The calculations were done within the short-range correlation model for both positive- and negative-U models (as indicated in the figure) of the DX centre.

3. Destruction of spatial correlation—theory

We analyse the observed mobility dependences on the conduction-electron concentration varied by the illumination (i.e. on the duration of the illumination) in terms of the short-range correlation model which we have found earlier to describe properly the mobility in dark samples. The applicability of this extremely simplified model was confirmed by our numerical simulation studies (see Wilamowski et al 1990, Kossut et al 1990a, 1991).

Since the debate concerning which of the two competing models of the DX state in GaAs is the correct one has not yet been settled, both are considered here in parallel.

In the case of $U > 0$ model, i.e. for DX centres that can accommodate only one electron, each ionized donor is surrounded by a region (assumed to be a sphere of radius r_c) where a probability of finding other ionized donors is depleted owing to the Coulomb repulsion of their like charges. We assume, therefore, that the pair correlation function for the ionized impurities $g_{++}(r)$ has a simple step-like form, and is equal to zero for $r < r_c$, whereas it is equal to 1 for $r > r_c$ (i.e. for $r > r_c$ there are no spatial correlations of the donor charges). Let us remember that the pair correlation function is proportional to the probability of finding, in this case, an ionized donor at a distance r from a site where there is already an ionized donor. The correlation radius r_c depends on the ratio of the concentrations of the ionized, N_+, to occupied, N_0, donors and is given by the solution of the equation (Kossut et al 1990b, 1991)

$$N_e V_c = 1 - \exp(-N V_c) \tag{1}$$

with $V_c = 4\pi r_c^3/3$ and N being the total number of donors, $N = N_e + N_0$. By taking the Fourier transform of the correlation function it is easy to calculate the corresponding structure factor needed in the calculation of the scattering rate from the system of charged donors. It is obvious that the pair correlation function of occupied centres $g_{00}(r)$ is equal to unity for all inter-donor distances. By virtue of the equality

$$N^2 = N_0^2 g_{00}(r) + N_+^2 g_{++}(r) + 2N_0 N_+ g_{0+}(r) \quad (2)$$

which simply states that the impurities are introduced into the host semiconductor in a random fashion, we can determine the pair correlation function g_{0+}. It is equal to $g_{+0} = g_{0+} = (N + N_0)/(2N_0)$ for $r < r_c$ and, of course, is equal to 1 for $r > r_c$.

It is easy to express the pair correlation function of the ionized donors after the illumination in terms of all three of the above introduced pair correlation functions in the dark sample. In general we have for all values of r

$$g_{++}^i(r) = \frac{N_+^{\text{dark}}}{N_+^i} \left(g_{++}^{\text{dark}}(r) \frac{N_+^{\text{dark}}}{N_+^i} + g_{+0}^{\text{dark}}(r) \frac{\Delta N_+}{N_+^i} \right)$$
$$+ \frac{\Delta N_+}{N_+^i} \left(g_{0+}^{\text{dark}}(r) \frac{N_+^{\text{dark}}}{N_+^i} + g_{00}^{\text{dark}}(r) \frac{\Delta N_+}{N_+^i} \right) \quad (3)$$

where the superscript i denotes that the quantity is measured in the illuminated sample and ΔN_+ denotes the change of the concentration of ionized impurities induced by the illumination, $N_e^i = N_e^{\text{dark}} + \Delta N_+$. Substituting our simple step-like correlation functions we obtain for $r < r_c$

$$g_{++}^i(r) = \left(\frac{N_{\text{DX}} + N_0^{\text{dark}}}{N_0^{\text{dark}}} \Delta N_+ N_+^{\text{dark}} + (\Delta N_+)^2 \right) \frac{1}{(N_+^i)^2} \quad (4)$$

while for $r < r_c$ the pair correlation function remains unchanged and equal to 1. Using equation (4) we can calculate the corresponding $S(q)$ and, further, the relaxation time for the scattering by these ionized impurities. The result is shown in figure 2 by the full line labelled DX^+/DX^0.

The DX centre characterized by a negative correlation energy requires consideration of three pair correlation functions $g_{++}(r)$, $g_{+-}(r) = g_{-+}(r)$ and $g_{--}(r)$ where the subscripts specify which of the possible charge states of two impurities are concerned. The most important of the three correlation functions is $g_{+-}(r)$. Since the charges of opposite sign attract each other there is a tendency to formation of closely spaced pairs of occupied and ionized donors, DX^-–DX^+. Therefore, it is not unreasonable to assume a step-like form of $g_{+-}(r)$ with values exceeding unity for $r < r_d$ and equal to 1 (no correlation) for $r > r_d$. The parameter r_d can be interpreted as a maximum dipole length of the DX^-–DX^+ pair. As in the case of $U > 0$ model, one can derive an equation for r_d (Kossut *et al* 1990b, 1991). This has a form resembling equation (1)

$$N_- V_d = 1 - \exp(-NV_d)(1 + NV_d) \quad (5)$$

where $V_d = 4\pi r_d^3/3$. The concentration of ionized and occupied centres can now be expressed by the total concentration of DX centres N and the conduction-electron concentration N_e by

$$N_-^\alpha = (N - N_e^\alpha)/2 \quad \text{and} \quad N_+^\alpha = (N + N_e^\alpha)/2 \quad (6)$$

($\alpha = $ dark, or $\alpha = $ i for illuminated). The changes of the respective concentrations induced by the illumination are given by

$$\Delta N_+ = (N_e^i - N_e^{\text{dark}})/2 \quad \text{and} \quad \Delta N_- = -(N_e^i - N_e^{\text{dark}})/2. \quad (7)$$

Making use of the analogue of equation (3) expressing the pair correlation functions in the illuminated sample by those in the dark sample, for example

$$g_{+-}(r) = \frac{N_+^{\text{dark}}}{N_+^i} \left(g_{+-}^{\text{dark}}(r) \frac{N_-^{\text{dark}}}{N_-^i} + g_{++}^{\text{dark}}(r) \frac{\Delta N_-}{N_-^i} \right)$$
$$+ \frac{\Delta N_+}{N_+^i} \left(g_{--}^{\text{dark}}(r) \frac{N_-^{\text{dark}}}{N_-^i} + g_{-+}^{\text{dark}}(r) \frac{\Delta N_-}{N_-^i} \right) \quad (8)$$

one can again calculate the structure factor and, then, the inverse momentum relaxation time for the scattering from the system of negatively charged occupied DX centres and positively charged ionized DX centres whose positions are intercorrelated. The result of the calculation is shown in figure 2 by the full line labelled DX^+/DX^-.

4. Discussion

Inspection of figure 2 shows clearly that the destruction of the spatial correlation of donor charges by the illumination does form a good basis for understanding the observed mobility decrease in GaAs with DX centres. It is tempting to conclude, after comparing the results presented in figure 2, that the $U > 0$ case provides a better description of the observed values of the mobility decrease. Such a conclusion must, however, be regarded as premature, since our simple approach ignores finite temperature effects as well as the presence of compensating acceptors. That the latter are probably present in the investigated sample is clear from the fact that the absolute values of the electron mobility are considerably smaller than those calculated assuming that the only scattering mechanism is that of charged donors. In fact, in order to bring the calculated absolute values of the mobility into agreement with the observed ones, one has to assume that there are 3.4×10^{18} cm^{-3} acceptors in the $U > 0$ case and 1.7×10^{18} cm^{-3} acceptors in the $U < 0$ case. Such high compensation ratios are suggested by Walukiewicz (1989, 1990) to necessarily occur in heavily doped GaAs. Negatively charged acceptors certainly modify the correlation of the location of donor charges. On the other hand, the discrepancies between the calculated and measured mobility values may mark the failure of the approach based on the first Born approximation and be related to multiple scattering effects. The fact that the correlation is frozen at a finite temperature means that our estimate of the degree of the correlation is too high. We plan to address ourselves to these questions in the future.

Finally, a comment is due concerning the values of

the concentration of the conduction electrons that were recovered after a prolonged illumination. As can be seen in figure 1 at $p = 9$ kbar the electron concentration in the illuminated sample returned nearly completely to its value at $p = 0$. At 11 kbar and at 14 kbar the recovered concentrations are markedly smaller than the concentration at $p = 0$. This fact suggests the existence of some efficient trap for electrons whose energy with respect to the conduction band minimum decreases monotonically with the hydrostatic pressure. However, at 17 kbar and 19 kbar (not shown in figure 1) the concentration of electrons that was measured after prolonged illumination was again very close to that at ambient pressure. This problem requires further experimental investigation.

References

Dietl T, Dmowski L, Kossut J, Litwin-Staszewska E, Piotrzkowski R, Suski T, Świątek K and Wilamowski Z 1990 *Acta Phys. Polon.* A **77** 29

Kossut J, Dobrowolski W, Wilamowski Z, Dietl T and Świątek K 1990a *Semicond. Sci. Technol.* **5** S260

Kossut J, Wilamowski Z, Dietl T and Świątek K 1990b *Proc. 20th Int. Conf. on the Physics of Semiconductors* ed E M Anastassakis and J D Joannopoulos (Singapore: World Scientific) p 613

——1991 *Acta Phys. Polon.* A **79** 49

Maude D K, Eaves L, Foster T J and Portal J C 1989 *Phys. Rev. Lett.* **62** 1922

Maude D K, Portal J C, Dmowski L, Foster T, Eaves L, Heiblum N M, Harris J J and Beal R B 1987 *Phys. Rev. Lett.* **59** 815

O'Reilly E P 1989 *Appl. Phys. Lett.* **55** 1409

Suski T, Wiśniewski P, Litwin-Staszewska E, Kossut J, Wilamowski Z, Dietl T, Świątek K, Ploog K and Knecht J 1990 *Semicond. Sci. Technol.* **5** 261

Walukiewicz W 1989 *J. Vac. Sci. Technol.* B **6** 1256

——1990 *Phys. Rev.* B **41** 10218

Wilamowski Z, Świątek K, Dietl T and Kossut J 1990 *Solid State Commun.* **73** 833

Zrenner A, Koch F, Williams R L, Stradling R A, Ploog K and Weimann G 1988 *Semicond. Sci. Technol.* **3** 1203

Semicond. Sci. Technol. **6** (1991) B38–B46. Printed in the UK

DX centres and Coulomb potential fluctuations

Z Wilamowski†, J Kossut†, W Jantsch‡ and G Ostermayer‡

† Institute of Physics, Polish Academy of Sciences, Al. Lotnikow 32/46, 02–668 Warsaw, Poland
‡ Insitut für Experimentalphysik, Johannes-Kepler-Universität, A-4040 Linz-Auhof, Austria

Abstract. DX centres, as well as other charged impurities with random location in the host crystal, give rise to fluctuations in the local potential. These fluctuations are comparable in amplitude to the level splittings of the DX centres due to different numbers of Al neighbours in $Al_xGa_{1-x}As$:Si and also to the variation in the quasi-Fermi level in experiments involving the kinetics of electron capture and emission processes. Thus, to describe the situation in $Al_xGa_{1-x}As$:Si in a quantitative way we develop a self-consistent model of the fluctuations. The resulting broadening of the DX level is taken into account in evaluating transport experiments. The proposed model yields the ground state energies and the barrier height for the four types of Si DX centre as functions of alloy composition and hydrostatic pressure. We explain also the non-exponential behaviour of the capture kinetics as well as the mobility in terms of the impurity level broadening and self-screening which results from a minimization of the Coulomb energy of interacting charges localized on donors.

1. Introduction

There is growing evidence that the Si DX centre in $Al_xGa_{1-x}As$ is a two-electron system with a negative Hubbard correlation energy U (Mooney 1990, Jantsch et al 1991, Mosser et al 1991). The D^- state is now widely believed to be identical to the two-electron DX state with large lattice relaxation obtained by Chadi and Chang (1988, 1989) and by Dabrowski et al (1990) from ab initio calculations. As a consequence, all defects (ionized and occupied) are charged and thus subject to mutual Coulomb interactions.

The importance of the inter-impurity interaction has been recognized particularly in the context of hopping conductivity and the formation of the Coulomb gap in the impurity density of states (for review see Efros and Shklovskii (1985)). On the other hand, discussing the conduction electron scattering, the effects of the interimpurity interactions are often neglected and the impurities are treated as isolated entities. Only in recent years has the subject of many-body interaction between scattering centres received growing attention. For instance, it was considered by Mycielski (1986) in order to explain the anomalously large values of the mobility of the conduction electrons in the zero-gap semiconductor HgSe:Fe (Dobrowolski et al 1987, Pool et al 1987, Wilamowski et al 1990). Iron as a dopant in HgSe forms a resonant

donor level approximately 200 meV above the conduction band edge. If the Fe concentration exceeds 5×10^{18} cm^{-3} the Fermi level becomes pinned to the donor level. Then the neutral (=occupied) and the positively charged (=ionized) donors coexist. Because of the partial occupation, there is a possibility for the charges to be redistributed among the randomly distributed donors in such a way that the total Coulomb energy is minimized. In other words, the inter-donor Coulomb interactions drive the formation of a spatially correlated system of donor charges.

In his approach to the problem of the mobility in HgSe:Fe, Mycielski (1986) showed that the total Coulomb energy is indeed lower by 10 meV per Fe ion when the electrons are distributed among the donor sites in such a way that the remaining ionized donors form a three-dimensional periodic charge superlattice within the crystal as compared with the value of this energy for a random distribution of charges. As a consequence of this ordering, the electron mobility must sharply increase since the ionized impurity scattering is not effective for a periodic arrangement of the scatterers. Later on, more realistic models were developed (Wilamowski et al 1988, 1990a, Kossut et al 1990a, b) which showed that the degree of ordering that occurs in HgSe:Fe is rather modest and that only a strong short-ranged order appears in the occupancy of the Fe donors in HgSe.

0268-1242/91/100B38+09 $03.50 © 1991 IOP Publishing Ltd

Striking similarities between the behaviour of the mobility in HgSe:Fe and in GaAs with DX centres were noted (Kossut *et al* 1990a, b, Jantsch and Wilamowski 1990). It was observed in GaAs heavily doped with Si and Sn that the mobility increases if the hydrostatic pressure exceeds some value (Maude *et al* 1987, 1989). This effect persists at low temperatures even after releasing the pressure (Suski *et al* 1990). The increase can be explained in terms of a spatially correlated arrangement of D^+ and D^- states (see Kossut *et al* 1990a, b, O'Reilly 1989).

In the case of DX centres in $Al_xGa_{1-x}As$ alloys another complication arises. It is now widely accepted that group IV donors in $Al_xGa_{1-x}As$ experience a splitting of their ground state due to different possible numbers of Al neighbours in the donor environment (Mooney *et al* 1988, Brunthaler and Köhler 1990, Piotrzkowski *et al* 1990). According to Chadi and Chang (1988, 1989), the Si-related DX state is accomplished by means of a large lattice relaxation of the substitutional donor atom in a $\langle 111 \rangle$ direction where it approaches a cluster of three group III host atoms. Of these, $i = 0, 1, 2$ or 3 can be Al. Depending on the actual number i for a given DX centre one of the four possible states is realized. This provides a natural explanation for the fourfold DX level splitting observed in alloys. Of course, the abundance of a given type of the effect is a function of the alloy composition (Morgan 1991).

In the analysis of various transport experiments, broadening of these four DX levels has not been considered so far. In this paper we show that the broadening due to potential fluctuation in the Coulomb potential of the randomly incorporated charged impurities amounts to up to 50 meV. Therefore, they cannot be neglected in the quantitative description of the experimental data.

In this paper, after reviewing briefly the approaches to the problem of correlated impurity charges used previously, we present a new description based on the notion of the self-screening of the impurity potentials by charges localized on neighbouring impurity sites. The proposed approach is free of any adjustable parameters and is self-consistent. It incorporates in a natural way the effects of finite temperature which previously were found to be difficult to account for. This formulation is then used to analyse data for the concentration of conduction electrons in $Al_xGa_{1-x}As$ enabling us to construct a quantitative energy versus x and/or hydrostatic pressure diagram including the ground state energies of the four types of Si DX centres and the barrier heights. Further on, we discuss the kinetics of photoconductivity and show that it is also affected by the broadening of the DX density of states. In particular, we show that it is the capture process that is sensitive to the local potential fluctuations whereas the emission remains unaffected. This effect enables the direct observation of the alloy splitting of the emission barrier. On the other hand, we show that anomalies of the capture kinetics are not associated with any alloy splitting of the capture barrier (as suggested in earlier literature) but, suprisingly, again with the splitting of the emission barrier. We show that there exist two temperature regions with entirely different emission rates leading to a characteristic two-stage nature of the effective capture rate. Finally, we discuss the mobility behaviour observed in the samples studied by us.

2. Theoretical descriptions

2.1. Previous approaches

For a quantitative description of spatial correlation of donors charges in HgSe:Fe two types of approach have been employed (Wilamowski *et al* 1988, 1990a, Swiatek *et al* 1988; see also reviews by Kossut *et al* (1990a, b) and Jantsch and Wilamowski (1990)): the first one is based on numerical simulations of the partially occupied system of donors interacting via the screened Coulomb potential. The second one is analytical and it relies on an *ad hoc* introduced step-like form of the pair correlation function. Both approaches can easily be extended to include both the positive and negative charge states of the donor as appropriate for the negative-U system of DX centres (Kossut *et al* 1990a).

In the case of HgSe, a correlated occupancy is clearly evident from the numerical simulations. The resulting pair correlation function $g(r)$ is close to 1 for large distances which means that there is no long-range order. For r smaller than a certain critical radius r_c, however, $g(r)$ drops rapidly to zero implying a short-range order. Thus, instead of a perfect charge superlattice assumed by Mycielski, the numerical simulations indicate that only a short-range correlation exists comparable to that in a liquid or a glass (Wilamowski 1990). This short-ranged nature of the correlation is fully compatible with experimental findings (see, for example, Pool *et al* 1987).

The numerical procedure allows also a direct evaluation of the distribution of donor states in energy, that is, the density of impurity states under the influence of the many-body Coulomb interaction. The results for HgSe indicate the existence of a Coulomb gap, that is, a region of low density of defect states close to the Fermi level, which is expected since it takes some minimum energy to put an electron onto one of the 'lattice' sites. This Coulomb gap also represents the gain due to correlation.

Two major difficulties arise if the analysis is based on the numerical simulation method. In order to calculate the effect of correlation on the mobility, simulation of a large system is necessary in order to obtain meaningful results. For the same reason, it is not easy to extract information about the Fermi level position because of large numerical errors in the derivative of the total Coulomb energy E_C with respect to the concentration of free carriers n, $\partial E_C/\partial n$. In those cases a simple analytical model proved to be of considerable advantage (Wilamowski *et al* 1988, 1990a, Kossut *et al* 1990). In this model, the pair correlation function is approximated by a step function $g(r) = \Theta(r - r_c)$ which equals 0 for $r < r_c$ and 1 for larger distances. The correlation radius r_c can be obtained from simple considerations involving the probabilities of finding positively charged and neutral donors in the sample. Knowing $g(r)$ analytically one can calculate, for example, the structure factor required to obtain the mobility and the energy gain (per charged donor) due to the ordering. The latter is given by

$$E_c = \frac{1}{2} \int n^2 (1 - g(r)) eV(r) \, d^3 r \qquad (1)$$

where $eV(r)$ is the electrostatic potential due to the charged impurities. The Fermi level can be also calculated from the ground state condition $dE_{tot}/dn = 0$, where E_{tot} contains, apart from E_C given by equation (1), also the kinetic energy of the conduction electrons and the energy of the electrons localized on the impurities.

Since the analytical approach neglects the correlation in positions of further distant neighbours it has several drawbacks:

(i) it does not allow us to calculate the potential fluctuations;

(ii) its applicability is limited to weakly correlated systems;

(iii) the analytical solution for r_c is valid only if the fraction of occupied impurities is small;

(iv) so far this method has been formulated only in the $T = 0$ limit;

(v) it becomes very complicated for systems where a few types of charge states are involved.

Therefore, we put forward here an alternative approach to the problem of Coulomb interaction of charged donors in $Al_xGa_{1-x}As$. Instead of considering the pair correlation function as in the numerical studies or in the analytical model we will assume a completely random distribution of charged centres and we treat the problem of correlation in terms of screening as described below.

2.2. The model of impurity self-screening

As a first step, we look for an effective, self-consistent, screening radius, where the screening by the impurities themselves is also taken into account (discussed, for example, by Larsen (1975) and by Morgan (1965)). In the Thomas–Fermi approximation the total screening radius λ is given by

$$\lambda^2 = \varepsilon_s / \kappa^2 \qquad (2)$$

where ε_s is the static dielectric constant and κ is the total Thomas–Fermi momentum given by the sum of contributions from different sources of screening

$$\kappa^2 = \sum \kappa_\alpha^2. \qquad (3)$$

Each κ_α^2 is determined by an appropriate density of states (DOS), $\rho_\alpha(E)$

$$\kappa_\alpha^2 = 4\pi e^2 \int \left(-\frac{df}{dE} \right) \rho_\alpha(E) \, dE. \qquad (4)$$

Here, $f(E)$ represents the distribution function.

In order to evaluate the screening radius, the real DOS distribution functions $\rho_\alpha(E)$ and the actual position of the

Fermi level E_F have to be known. Here we assume that the sum in equation (3) contains the contributions from the conduction band and that from the impurity states. The conduction band density of states is calculated in the effective-mass approximation. For the impurity DOS, $\rho_\alpha(E)$, we assume a Gaussian distribution with the second moment σ given by the variance of the potential fluctuations. For randomly distributed centres the value of σ can be expressed by

$$\sigma^2 = 2\pi e^4 N_i \lambda / \varepsilon_s^2. \qquad (5)$$

The set of equations (2)–(5) together with the condition of charge neutrality represents a self-consistent problem: the screening length λ depends on E_F and the defect density of states, $\rho_\alpha(E)$, which is a function of σ. Now σ, according to equation (5), depends on λ again, so self-consistency of the solution must be achieved, which in practice requires numerical methods. Finally, we can use the self-consistent value of the Fermi energy to calculate the concentration of the electrons in the conduction band and the screening radius, λ, which is necessary to evaluate the electron mobility.

The present approach does not include effects of the Coulomb gap since the distributions of positive and negative charges in space are not considered explicitly. On the other hand, a rough estimate of the value of the Coulomb gap is given by the gain in the total Coulomb energy due to charge correlation per charged impurity E_c/N_i. The effects due to the Coulomb gap are of importance only if its value is large compared with the potential fluctuations σ and the temperature, $E_c/N_i \sim \sigma$ and $E_C/N_i > kT$. Having these conditions in mind we conclude that the self-screening approach should not be applied to HgSe:Fe at low temperatures. It can be easily applied, however, to $Al_xGa_{1-x}As$ where, because of the negative U and a high compensation, the total number of charged centres N_i is large. The Coulomb gap E_C/N_i is thus small whereas σ is big (cf equation (5)). In addition, because of the existence of high barriers between the impurity and the band states, the spatial redistribution of the impurity changes (i.e. the process leading to the self-screening) takes place at high temperature.

This model has considerable advantages. It takes finite temperature explicitly into account. It allows us to include different types of charged centres, provided that their level energies are known. The model is not difficult to solve and it can be readily used to describe various physical quantities. In the following sections, we make use of this model in order to analyse equilibrium transport properties (section 3) and their kinetics (section 4). In particular, the peculiar temperature dependence of the mobility of $Al_xGa_{1-x}As$:Si can be explained in terms of the impurity self-screening model (section 5).

3. Electron concentration, quasi-Fermi level and the energy diagram

The analysis of the electron concentration in $Al_xGa_{1-x}As$ with DX centres (obtained from the Hall

Table 1. Energy of the DX state due to Si and Sn in GaAs relative to the conduction band minimum at the Γ point, $E_{DX}^{i=0} - E_\Gamma$, and its pressure coefficient $\gamma = \mathrm{d}(E_{DX}^{i=0} - E_\Gamma)/\mathrm{d}p$ as determined from the pressure dependence of the Hall concentration. The values derived by Maude *et al* (1987) without considering the spatial correlation of DX charges are also given.

N_{DX} (cm^{-3})	$E_{DX}^{i=0} - E_\Gamma$(meV) no correlation	$E_{DX}^{i=0} - E_\Gamma$(meV) with correlation	γ(meV kbar^{-1}) with correlation
Si: 1.1×10^{18}	234		
Si: 6.8×10^{18}†		300	-12
Si: 4.6×10^{18}	174		
Sn: 2.2×10^{19}	287		
Sn: 7.0×10^{18}	204	320	-14
Sn: 3.2×10^{18}	159		

† Data of Suski *et al* (1990)

voltage measurements) as well as the analysis of the data dealing with kinetics of the capture and emission processes requires knowledge of the values of the DX ground state energies for each possible Al environment ($i = 0, 1, 2, 3$) and of the barriers effective in capture and emission. We derive a consistent set of these parameters by making use of the results for various alloy compositions and hydrostatic pressures.

First, we consider the case of GaAs ($x = 0$) in order to avoid the complications brought about by alloy splitting. Maude *et al* (1989) derived values of the energy of the resonant DX state from their electron concentration versus hydrostatic pressure data without taking the inter-impurity interaction into account. They have also determined the pressure coefficient of this level. The energies derived by them show a strong dependence on the doping level. Taking account of the Coulomb potential fluctuations and the resulting level broadening, the experimental data can be fitted with a single value of the DX level energy, $E_{DX}^{i=0}$, and its pressure coefficient, independent of the dopant concentration. The results of the evaluation, based on the step-like pair correlation function, are given in table 1 for Si- and Sn-related DX centres. The energies of the DX state given in table 1 are higher than their extrapolated values determined by other authors. This is understandable in view of the fact that the present

determination accounts for the finite width of the DX level as well as for its two-electron nature. As a consequence, the Fermi energy is pinned at the energies that are notably lower than the peak position of the impurity DOS.

The value of E_{DX}^i for $i = 0$ is given also in table 2 where a correction is made for the temperature dependence of the DX energy (the Fermi level position in experiments of Maude *et al* (1987) was frozen at $\simeq 100$ K, while table 2 gives the values extrapolated to $T = 0$).

Let us turn now to DX centres in Al$_x$Ga$_{1-x}$As alloys. As mentioned above, we have to deal in this case with the alloy splitting. The splitting is directly observable in experiments studying the thermal emission of electrons from the DX state to the conduction band (Mooney *et al* 1988, Piotrzkowski *et al* 1991). The values of the emission barriers resulting from these experiments (after being reanalysed by Jantsch *et al* (1991)) are displayed in table 2. Knowing the emission barrier and E_{DX} for $i = 0$ (i.e. for the DX configuration with no Al atoms in the neighbourhood) we can calculate the height of the barrier $E_B^{i=0} = E_{DX}^{i=0} + E_E^{i=0}/2$. The resulting value is given in the last column of table 2.

Now the question arises whether the differences in the emission barrier E_E^i in the first column of table 2 originate from differences of the ground state energy of the DX centre, E_{DX}^i, or of their barrier height, E_B^i. To discriminate between these two extreme cases we have performed two types of measurements of the Hall concentration. In the first one, the sample was illuminated by an LED during the measurements and n versus T dependences were studied under conditions of steady-state photoconductivity. The result for a AlGaAs sample with $x = 0.31$ and $N_{Si} = 3 \times 10^{18}$ cm^{-3}, is shown in figure 1(a) by asterisks. Secondly, the sample, after being cooled down to liquid-helium temperature in darkness was illuminated by an LED until saturation of the persistent photoconductivity (PPC) was reached. Then, the light was switched off and measurements were taken while the temperature was raised at a constant rate. The results, shown in figure 1(a) by crosses, display two distinct

Table 2. Activation energy for two-electron emission E_E, ground state energy of the DX centre E_{DX} at $T = 0$ (in contrast to the values given in table 1 which correspond to $T \approx 100$ K at which the correlation freezes), and the top of the barrier E_B (all energies in meV).

i	E_E	E_{DX}	$E_B = E_{DX} + E_E/2$
0	314 ± 3	293 ± 10	450 ± 10
1	380 ± 5	265 ± 10	455 ± 10
2	428 ± 5	246 ± 10	460 ± 10
3	460 ± 10	235 ± 20	465 ± 10

$\mathrm{d}(E_{DX} - E_\Gamma)/\mathrm{d}T = (0.15 \pm 0.05)\mathrm{d}E_g/\mathrm{d}T$
$\mathrm{d}(E_{DX} - E_\Gamma)/\mathrm{d}p = (12 \pm 1)$ meV kbar$^{-1} = 1.1\,\mathrm{d}E_g/\mathrm{d}p$
$\mathrm{d}(E_{DX} - E_\Gamma)/\mathrm{d}x = 940$ meV $= (0.70 \pm 0.03)\mathrm{d}E_g/\mathrm{d}x$

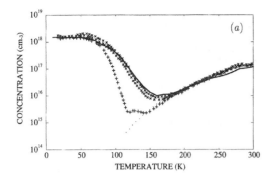

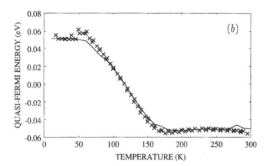

Figure 1. Conduction electron concentration (*a*) and resulting position of the quasi-Fermi level (*b*) relative to the Γ minimum of the conduction band of an $Al_xGa_{1-x}As$ sample ($x = 0.31$) doped with Si (3×10^{18} cm^{-3}). Asterisks show the behaviour under illumination with an LED, the crosses were obtained during warm-up in darkness after illumination at low temperature. The full curves correspond to the result of the calculation using the parameters from table 2. For comparison we present also experimental points obtained during cooling in darkness (dots). The region where all three types of experimental points converge corresponds to thermal equilibrium conditions.

humps. The same results are presented in figure 1(*b*) where, instead of the electron concentration, we have plotted the quasi-Fermi level as a function of temperature. The curve taken under the illumination, extrapolated to $T = 0$, gives the minimum value of the top of the barrier (Jantsch *et al* 1991) while the value of its relatively flat portion at high temperatures is determined by the filling of energetically lowest DX sublevels. The difference between the quasi-Fermi energy at $T = 0$ and that at high temperatures provides information about a possible splitting of the top of the barrier. A fitting procedure applied to the experimental temperature dependence of n (or E_F^*), with the splitting of the top of the barriers treated as an adjustable parameter, results in a value of the barrier splitting equal to 5 ± 5 meV. The temperature coefficient of the E_{DX} energy was the second fitting parameter and its value is found equal to $d(E_{DX} - E_\Gamma)/dT = (0.15 \pm 0.05)\,dE_g/dT$. The full curves in figure 1 show the quality of the fit. In this way we obtained the values given in the last column of table 2, and—using the relation between E_E^i, E_B^i and E_{DX}^i—the entries for E_{DX}^i for $i = 1$ and $i = 2$. We see that the alloy splitting of the top

of the barrier is significantly smaller than the broadening due to potential fluctuations ($\simeq 50$ meV). In other words, the capture probability is practically the same for each DX configuration in the alloy.

A confirmation of the derived value of the alloy splitting is provided by our second type of experiment. Results are shown by crosses in figure 1. The difference of the heights of the steps of E_F^* appearing between $T = 120$ K and 150 K is a direct measure of the splitting between the E_{DX}^i for $i = 1$ and $i = 2$. The value obtained in this way of about 20 meV is in good agreement with that derived above and given in table 2. The pinning of E_F^* to the $i = 1$ and $i = 2$ sublevels, responsible for the occurrence of the steps, is related to a specific two-stage character of the capture kinetics to be discussed in the next section.

To confirm further the above determination of values of $E_{DX}^i(i = 1, 2)$ and to estimate the value for $i = 3$ we consider the carrier concentration or, rather, its apparent activation energy, $E_{act}^*(x)$ at thermal equilibrium at high temperatures. Experimental data of Chand *et al* (1984), Mizuta and Mori (1988) and our own data are shown in figure 2. The results of fitting the fluctuation model described in the previous section are shown also in figure 2. The values of $E_{DX}^i(i = 0, 1, 2)$ are taken from table 2 and $E_{DX}^{i=3}$ and the Al concentration dependence of E_{DX} (common for all i values), $d(E_{DX} - E_\Gamma)/dx$, are treated as adjustable parameters. The best-fit values of the parameters are given in table 2. These values obviously are different from those obtained previously, when Coulomb interaction was omitted (Chand *et al* 1984, Morgan 1991).

Figure 3 summarizes the results for the level energies and the barriers of the Si-related DX centres in $Al_xGa_{1-x}As$ as functions of alloy composition and of hydrostatic pressure.

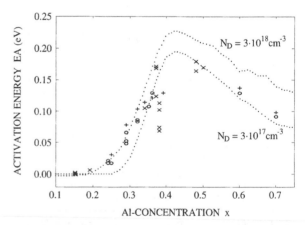

Figure 2. Activation energy of the electron concentration in the region 200–300 K as a function of Al content in $Al_xGa_{1-x}As$. Experimental points were taken from Chand *et al* (1984) (+), Mizuta and Mori (1988) (○) and our data (×). The scatter of points for $x = 0.25$ reflects the sensitivity of the activation energy to the doping and compensation levels. The curves are calculated taking the potential fluctuations into account as described in section 2 for two concentrations of the dopants N_D.

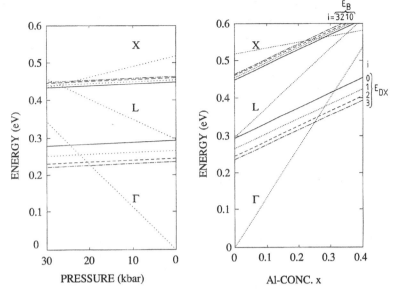

Figure 3. DX centre ground state energy and the barrier energy in $Al_xGa_{1-x}As$ (with the alloy splitting shown) as functions of alloy composition x (right-hand side) and in GaAs as functions of the hydrostatic pressure (left-hand side). The positions of various conduction band minima are also shown (dotted lines). The crosses indicate the position of the neutral D^0 state.

4. Two-stage nature of the electron capture kinetics

Investigating the capture kinetics we employed two methods: (a) we measured the Hall effect warming up a sample which was previously illuminated at a low temperature (figure 1) or (b) we studied isothermal decays after photoexcitation. Nearly always in the latter method highly non-exponential decays of the free carrier concentration were found (Jantsch et al 1990). In this section we discuss three possible sources of the non-exponential nature of the decay:

(i) a sizable change of the quasi-Fermi level E_F^* during the process of capture;

(ii) a spread of the capture barrier energies due to the potential fluctuations which cause a substantial spread of the capture probabilities by various donors;

(iii) the alloy splitting of characteristic donor energies E_{DX} and E_B, which leads to a complicated, two-stage mechanism of the relaxation of the system to equilibrium.

Our discussion makes use of the energy diagram derived in the preceding section, shown in figure 3. Moreover, we assume that the DX state is a two-electron state with a negative U. Knowing also that the emission barrier of the Si DX centre in $Al_xGa_{1-x}As$ is independent of Fermi energy (Piotrzkowski et al 1990) the capture rate per empty donor can be shown to be proportional to $\exp[-2(E_B - E_F^*)/kT]$ (see Jantsch et al 1991). The activation energy for capture is $2(E_B - E_F^*)$ and depends thus directly on the Fermi level position.

Let us discuss the first of the mentioned sources of the non-exponentiality of the photoconductivity decay. Illumination of the sample at low temperature induces an increase of the concentration of the conduction electrons by photoionization of the DX centres. The quasi-Fermi level increases until photoionization and the recapture rates compensate each other. When the light is switched off, E_F^* decreases by a certain amount ΔE_F^* as the carriers are recaptured onto the DX states (note that ΔE_F^* can be as large as 100 meV—see figure 1(b)). Simultaneously, however, the effective capture barrier increases and therefore the capture rate decreases. As a consequence, nonlinear terms appear in the rate equation resulting in nonexponential solutions.

The second mechanism leading to non-exponential behaviour is related to the fluctuations $\delta V(r)$ of the resulting Coulomb potential (with a width of the distribution $\sigma \approx 50$ meV) due to charged donors and acceptors. The fluctuations of the potential are reflected by the spatial fluctuations of the conduction band edge E_Γ and the energies of the DX centres (E_{DX} and E_B). The Fermi level, in contrast, remains constant in real space. The activation energy governing the effectiveness of the capture process is given by $2(E_B - E_F^*)$. This energy fluctuates in the same manner as $2\delta V(r)$ in a complete contrast to the emission barrier, which is not affected by δV.

The role of both effects discussed above is important here. This becomes evident from the following considerations: the ratio of decay rates at the beginning and at the end of the transient is given by $\exp(-2\Delta E_F^*/kT)$. Similarly, the ratio of the decay rates related to the capture by two donors located at points where the potential differs by 2σ is $\exp(-4\sigma/kT)$. Both expressions reach values of 10^9 at 100 K. Therefore, the degree of the non-exponentiality is considerable even at this elevated temperature. At lower temperatures the degree of the non-exponentiality of the decay is still bigger which leads to a truly persistent behaviour of the photoconductivity after its small initial decay.

As indicated in the table 2 there are only small differences between E_B^i for various i although the ground state energies, E_{DX}^i, and the emission barriers are quite different. As a consequence, the capture probability is practically equal for the four types of DX centre. Let us consider again the photoconductivity decay shown by crosses in figure 1. In this experiment, the electrons are first photoexcited at a low temperature to the conduction band and then the sample is warmed up. When the temperature is slightly raised the capture probability increases. The ensuing capture of the electrons leads to a lowering of the Fermi level, and hence to an increase of the capture barrier. In the first stage of the transient at temperatures lower than 110 K the emission barrier is still higher than the capture barrier and the only process that takes place is electron capture. The capture rate is self-regulating and decreases strongly in time. The resulting fractional occupation of various types of DX levels per donor is equal and does not depend on the energy E_{DX}^i. Their occupation is, therefore, far from thermal equilibrium.

The first stage discussed above prevails until the quasi-Fermi level moves with increasing temperature into the energetic vicinity of the large DOS associated with the DX levels making the capture and emission barriers comparable in magnitude. Then a second stage of the relaxation of the system begins: DX states start to re-emit the electrons trapped on them. A temporary balance develops between the total capture process and the emission from a given DX producing a pseudo-equilibrium state with the quasi-Fermi level pinned to some particular DX sublevel. The pinning of the quasi-Fermi level is responsible for the step-like features in the $E_F^*(T)$ dependence seen in figure 1. The first one, at $T = 120$ K, corresponds to the pinning of the quasi-Fermi level to $i = 1$ sublevel (for $x = 0.31$ the step due to the $i = 0$ state, whose emission barrier is the lowest, is hardly observable since the abundance of this state is only about 1%). With the Fermi level pinned to the $i = 1$ state the re-emission and the subsequent capture causes a net decrease of the population of this state while the populations of $i = 2$ and $i = 3$ states increase. This situation persists until there are electrons available in $i = 1$ state. Then, the quasi-Fermi level jumps and becomes pinned to $i = 2$ state resulting in the appearance of the second step at $T \approx 105$ K in figure 1. The mechanism just described provides effective means of the thermalization of the populations of various DX sublevels. In conclusion, the steps on the $E_F^*(T)$ curves are in reality related to the alloy splitting of the emission barrier and mark the onset of the re-emission from consecutive DX sublevels during the capture transient.

The two-stage character of the relaxation of the photoexcited system described above is an additional source of the non-exponential capture kinetics observed also in the isothermal decays. It is also responsible for the complex electron mobility dependences in the next section.

5. Electron mobility

As discussed earlier (for example, Wilamowski *et al* 1990a, Kossut *et al* 1990a) proper understanding of the electron mobility requires consideration of the effect of charge correlation. Within the present model of impurity self-screening, the mobility (limited by charged impurity scattering) can be calculated using the standard formulae using, however, the expression for the screening length given by equations (2)–(5). Since the energies of the DX states are already determined (figure 3) the only quantities entering this expression are the concentrations of DX centres N_D and the concentration of compensating donors N_A. Our initial calculations indicated that assuming $N_A/N_D \approx 0.2$–0.4 one obtains good agreement of the mobility with the observed values.

Calculation of the mobility in the alloys, however, aimed at a quantitative interpretation of the data is very complicated. First of all, it has to take into account the self-screening by four different DX states. Furthermore, considering the mobility in steady-state photoconductivity there appears also the problem of the relative populations of the four DX states which can be far from thermal equilibrium, as discussed above. During the transients, the situation becomes even more difficult since a complicated differential–integral set of kinetic equations has to be solved. Therefore, we shall not attempt here to interpret the observed mobility behaviour in detail. Instead, we point out its extreme sensitivity to minute changes of the parameters and/or initial conditions.

Figure 4 illustrates this sensitivity by showing mobility transients in two samples grown in the same run. They differ only by the substrate material used for their preparation (Cr doped and undoped GaAs). The mobility in these two samples is quite different: at temperatures where one sample shows a small decrease the other increases by the factor of 2. This particular temperature range coincides exactly with the onset of the second stage of the capture process. We attribute this quite different mobility behaviour in samples that are seemingly similar to the influence of the four DX sublevels. Each of these sublevels contributes to the screening. If only one DX level is present (as in GaAs) and it is fully occupied then emptying this level leads initially to an increase of the screening efficiency since the DOS at the quasi-Fermi level grows. Later on, E_F^* passes the maximum of the DOS and screening becomes less efficient again. Likewise, filling the empty single state causes the screening efficiency also initially to grow and then to weaken. In $Al_xGa_{1-x}As$ with four DX sublevels, electrons are transferred from one sublevel to the others via re-emission and recapture

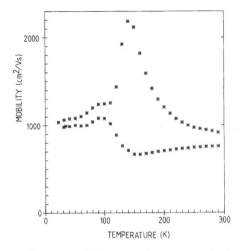

Figure 4. Electron mobility versus temperature for two similar AlGaAs ($x = 0.25$) samples. The samples were grown in the same run and differ only in the substrate material. The measurements were taken during warm-up after illumination at low-temperature.

and the total screening efficiency is subject to an interference of four non-monotonic dependences. This 'interference' is extremely sensitive to the donor concentration and may produce a mobility behaviour as shown in figure 4.

6. Conclusions

Because of the negative U of DX centres all donors in n-type III–V compounds are charged. Therefore fluctuations of the Coulomb potential are particularly large—typically the amplitude is 50 meV, larger than kT and comparable to the alloy splitting of DX centres in $Al_xGa_{1-x}As$. This fact has to be taken into account in the evaluation of many experiments. Within the framework of the impurity self-screening approach proposed here, a number of experimental results in steady-state photoconductivity and its kinetics and also the apparent activation energy at high temperatures close to thermal equilibrium can be understood quantitatively. Self-consistent treatment of the latter problem allows the evaluation of both the ground state energies and the barrier height of the Si DX centres in $Al_xGa_{1-x}As$, and we have determined these energies as a function of alloy composition and pressure. We have shown that the capture barriers for the four Si DX centres in $Al_xGa_{1-x}As$ are practically equal, in contrast to the ground state energies which differ by as much as 140 meV (between the $i = 0$ and the $i = 3$ situation). This gives rise to a rather complicated two-stage capture kinetics which is responsible for the strong non-exponentiality of capture transients. The electron mobility exhibits rather complicated behaviour which can be understood qualitatively in terms of impurity self-screening.

Acknowledgments

We are indebted to K Köhler and K Lübke for generously providing samples and to G Brunthaler and A Falk for many helpful discussions and to U Hannesschlager for expert technical assistance. Work supported by 'Fonds zur Förderung der Wissenschaftlichen Forschung' Austria.

References

Brunthaler G and Köhler K 1990 *Appl. Phys. Lett.* **57** 2225
Chadi D J and Chang K J 1988 *Phys. Rev. Lett.* **61** 837
——1989 *Phys. Rev.* B **39** 10366
Chand N, Henderson T, Klem J, Masselink W T, Fisher R, Chang Y C and Morkoc H 1984 *Phys. Rev.* **30** 4481
Dobrowolski W, Dybko K, Mycielski A, Mycielski J, Wrobel J, Piechota S, Palozewska M, Szymczak H and Wilamowski Z 1986 *Proc. 18th Int. Conf. on the Physics of Semiconductors* ed O Engström (Singapore: World Scientific) p 1743
Efros A L and Shklovskii B I 1985 *Electron–Electron Interactions in Disordered Systems* ed A L Efros and M Pollak (Amsterdam: North-Holland) Ch 5
Jantsch W, Ostermayer G, Brunthaler G, Stöger G, Wöckinger and Wilamowski Z 1990 *Proc. 20th Int. Conf. on the Physics of Semiconductors* ed E M Anastassakis and J D Joannopoulos (Singapore: World Scientific) p 485
Jantsch W and Wilamowski Z 1990 *Localization and Confinement of Electrons in Semiconductors (Springer Series in Solid-State Sciences* **97**) ed F Kuchlar *et al* (Berlin: Springer) p 137
Jantsch W, Wilamowski Z and Ostermayer G 1991 *Semicond. Sci. Technol.* **6** B47
Kossut J, Dobrowolski W, Wilamowski Z, Dietl T and Swiatek K 1990b *Semicond. Sci. Technol.* **5** S260
Kossut J, Wilamowski Z, Dietl T and Swiatek K 1990a *Proc. 20th Int. Conf. on the Physics of Semiconductors* ed E M Anastassakis and J D Joannopoulos (Singapore: World Scientific) p 613
Larsen D M 1975 *Phys. Rev.* B **11** 3904
Maude D K, Eaves L, Foster T and Portal J C 1989 *Phys. Rev. Lett.* **62** 1922
Maude D K, Portal J C, Dmowski L, Foster T, Eaves L, Heiblum N M, Harris J J and Beal R B 1987 *Phys. Rev. Lett.* **59** 815
Mizuta M and Mori M 1988 *Phys. Rev.* B **37** 1043
Mooney P M 1990 *Proc. 20th Int. Conf. on the Physics of Semiconductors* ed E M Anastassakis and J D Joannopoulos (Singapore: World Scientific) p 2600
Mooney P M, Theis T N and Wright S L 1988 *Appl. Phys. Lett.* **53** 2546
Morgan T 1965 *Phys. Rev.* **139** A343
——1991 *J. Electron. Mater.* **20** 63
Mosser V, Contreras S, Piotrzkowski R, Lorenzini, Robert J L and Rochette J F 1991 *Semicond. Sci. Technol.* **6** 505, at press
Mycielski J 1986 *Solid State Commun.* **60** 165
O'Reilly E P 1989 *Appl. Phys. Lett.* **55** 1409
Piotrzkowski R L, Litwin-Staszewska E, Robert J L, Mosser V and Lorenzini P 1991 *Semicond. Sci. Technol.* **6** 500, at press
Piotrzkowski R, Suski T, Wisniewski P, Ploog K and Knecht J 1990 *J. Appl. Phys.* **68** 3377
Pool F, Kossut J, Debska U and Reifenberger R 1987 *Phys. Rev.* B **38** 3900

Suski T, Wisniewski P, Litwin-Staszewska, Kossut J,
 Wilamowski Z, Dietl T, Swiatek K, Ploog K and Knecht
 J 1990 *Semicond. Sci. Technol.* **5** 261
Swiatek K, Wilamowski Z, Dietl T and Kossut J 1988 *Proc.
 19th Int. Conf. on the Physics of Semiconductors* ed W
 Zawadzki (Warsaw: Institute of Physics, Polish
 Academy of Sciences) p 1571
Wilamowski Z 1990 *Acta Phys. Polon.* **A77** 133
Wilamowski Z, Jantsch W and Hendorfer G 1990b *Semicond.
 Sci. Technol.* **5** 266

Wilamowski Z, Kossut J, Suski T, Wisniewski P and
 Dmowski L 1991 *Semicond. Sci. Technol.* **6**
Wilamowski Z, Mycielski A, Jantsch W and Hendorfer G
 1988 *Proc. 19th Int. Conf. on the Physics of
 Semiconductors* ed W Zawadzki (Warsaw: Institute of
 Physics, Polish Academy of Sciences) p 1225
Wilamowski Z, Swiatek K, Dietl T and Kossut J 1990a *Solid
 State Commun.* **74** 833

Semicond. Sci. Technol. **6** (1991) B47–B50. Printed in the UK

Photoconductivity saturation of AlGaAs : Si —a new criterion for negative *U*

W Jantsch†, Z Wilamowski‡ and G Ostermayer†

† Institut für Experimentalphysik, Johannes-Kepler-Universität, A–4040 Linz-Auhof, Austria
‡ Institute of Physics of the Polish Academy of Sciences. Al. Lotnikow 32/46, 02–668 Warsaw, Poland

Abstract. For heavily doped samples of $Al_xGa_{1-x}As$: Si ($x \cong 0.3$) the persistent photoconductivity is limited by the finite height of the capture barrier. This effect allows the determination of the capture barrier. The result is compatible with literature data for the emission barrier only for a negative intra-atomic Hubbard correlation energy. Modelling shows that the thermal emission from the Si DX centre is a two-electron process.

1. Introduction

The most prominent effect of DX centres is persistent photoconductivity (PPC) which is explained in terms of a large barrier both for capture and emission (Lang 1986, Mooney 1990). The complexity of PPC becomes evident in investigations of kinetics, as observed in DLTS (Mooney *et al* 1991), the temperature-dependent Hall effect (Brunthaler and Köhler 1990), isothermal transient photoconductivity (Ostermayer 1991) and pressure-dependent transport (Piotrzkowski *et al* 1991). Interpretation of these investigations reveals the existence of up to four different DX centres for group IV donors in $Al_xGa_{1-x}As$ with different level energies. This multiplicity is expected for a displacement of the substitutional donor along a $\langle 111 \rangle$ direction which was postulated for the deep ground state on the basis of *ab initio* calculations by Chadi and Chang (1988, 1989). In this distorted state the donor approaches three group III neighbours, out of which $i = 0, 1, 2$ or 3 can be aluminium. In addition, we expect considerable broadening of these four levels due to Coulomb potential fluctuations caused by other nearby charged defects (Wilamowski *et al* 1991).

The most debated question concerns the nature of the deep ground state of DX centres and, especially, the sign of the intra-atomic Hubbard correlation energy *U*. Recent theoretical work predicts a negative value of *U* (Chadi and Chang 1988, 1989, Dabrowski *et al* 1990). So far, attempts to describe the kinetics in terms of negative *U* do not allow an unequivocal interpretation. Recently, some authors inferred a negative charge state of the DX centres from various experiments and took this as evidence for a negative *U* value. This conclusion however, does not appear justified in view of a model with small positive *U*, which allows also negatively charged DX centres, depending on the Fermi level position (Jantsch *et al* 1990).

In this paper, we report investigations of the steady state photoconductivity of heavily doped AlGaAs : Si, which provide a new type of criterion for negative *U*. This criterion is based on the comparison of experimental values for the capture and the emission barriers and their dependence on the free-electron quasi-Fermi level E_F^*.

In the case of a degenerate electron gas and $U > 0$, the effective capture barrier is given by $E_B - E_F^*$ as indicated schematically in figure 1(*a*), where E_B is the

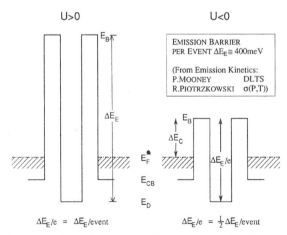

Figure 1. Schematic representation of capture (ΔE_C) and emission (ΔE_E) barrier heights in a one-electron picture in the case of positive *U* and negative *U*.

energy at the top of the barrier. The barrier for emission is given by $E_B - E_D$ where E_D is the donor level energy of the ground state. At low temperatures, illumination with photon energies above the photoionization threshold causes PPC: the quasi-Fermi level E_F^* increases until it reaches E_B, if there are a sufficient number of uncompensated donors and if there are no additional excited donor states below E_B which could also pin E_F^*. So we expect that E_F^* extrapolates towards E_B at low temperatures T.

With increasing T, the distribution function of the free carriers becomes broader in energy and an increasing number of electrons are captured by the deep DX centres. Thus E_F^* decreases, and it decreases in an almost linear fashion as can be seen by considering steady state conditions. Finally, at high temperatures, re-emission becomes efficient, and the system approaches thermal equilibrium. Then E_F^* is pinned to the impurity density of states close to E_D for highly doped samples. In summary, we expect a variation of E_F^* between E_B at low T and E_D at high T. This span corresponds to the emission energy ΔE_E for $U > 0$.

2. Experiment

Experimental results are given in figure 2 for a sample of $Al_x Ga_{1-x} As$:Si grown by MBE on semi-insulating GaAs. The sample thickness is 2 μm and there is an undoped buffer layer with the same $x \cong 0.31$ and a thickness of about 0.5 μm in order to avoid effects due to modulation doping of a two-dimensional channel at the interface. The Si concentration is nominally 2×10^{18} cm^{-3}. Figure 2 shows the position of E_F^* as calculated from the free-carrier concentration. The latter is obtained from Hall-effect measurements under illumination by an infrared-emitting diode with a peak wavelength of 0.92 μm. Three sets of data are plotted for different illumination intensities. For comparison results obtained during cool-

Table 1. Activation energy for two-electron emission ΔE_E, the ground state of energy of the DX centre E_{DX} at $T = 0$ and the top of the barrier E_B per electron (all energies in meV).

i	ΔE_E	E_{DX}	$E_B = E_{DX} + E_E/2$
0	314 ± 3^a	293 ± 10^b	450 ± 10
1	380 ± 5^a	265 ± 10^b	455 ± 10
2	428 ± 5^a	246 ± 10^b	460 ± 10
3	460 ± 10^b	235 ± 20^b	465 ± 10

[a] From figure 4.
[b] From Wilamowski *et al* (1991).

ing in darkness are given (low values between 110 and 160 K).

We observe a pinning of E_F^* for $T > 160$ K and a linear increase for lower temperatures as expected. Only for $T < 50$ K, E_F^* saturates again, most likely due to an additional excited donor state as discussed below. Extrapolating the linear part (160 K $> T >$ 50 K) of $E_F^*(T)$ for $T \rightarrow 0$, we obtain E_B and the total variation of E_F^*, which should be compared with the emission energy ΔE_E. This value amounts to 200 meV. Literature values for the emission energy, in contrast, are typically twice as large (see table 1) which rules out a positive value of U. As we shall show in the following, the missing factor of two can be understood in terms of negative U and a two-electron emission barrier.

3. The two-electron DX model

In the following we consider a negative U, D$^-$ ground state of the donor. We also assume that emission is a two-electron process as can be inferred from the schematic total energy diagram given in figure 3, which agrees in its essential features with the predictions of *ab initio* local density approximation calculations by Dabrowski *et al* (1990). In figure 3, the total energy is given as a function of the configuration parameter Q which describes the

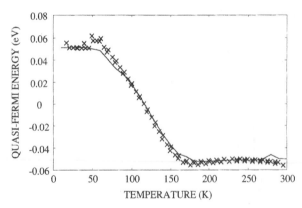

Figure 2. Quasi-Fermi level as obtained from the free-electron concentration of an $Al_x Ga_{1-x} As$ sample ($x = 0.31$, $N_D^{tot} \cong 2 \times 10^{18}$ cm^{-3}) under illumination by an infrared-light-emitting diode (three different illumination intensities) as a function of temperature. Experimental data, crosses; theory; full curve.

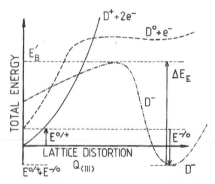

Figure 3. Schematic total energy diagram as a function of the local displacement of a substitutional donor along a $\langle 111 \rangle$ direction for the ionized D$^+$ state, the neutral donor D^0 state and a negative two-electron donor state D$^-$ (after Dabrowski *et al* 1990). The conduction band edge E_{CB} is chosen as zero.

displacement of the substitutional donor. The three curves given correspond to (i) an ionized donor and two electrons at the bottom of the conduction band ($D^+ + 2e^-$), (ii) the neutral donor plus one conduction electron ($D^0 + e^-$) and (iii) the negatively charged D^- state, respectively. The lowest energy is realized for the distorted two-electron D^- state. According to this diagram, thermal emission is a two-electron process: the donor has to be excited to the two-electron barrier energy E'_B in its D^- state before emission is possible. Once the donor has passed the barrier E_B, it crosses the ($D^+ + 2e^-$) curve and it will then emit both electrons spontaneously or, depending on the transition probabilities, continue on the D^- curve until it crosses the D^0 curve and emits then either both electrons or first one and then the other, again depending on the relative magnitude of the matrix elements.

The two-electron nature of the barrier is reflected in figure 3 by the position of the intersections on the left-hand side of the top of the barrier. This position is strongly supported by the following observations: the emission barrier of the Si DX centre practically does not depend on pressure or alloy composition (Calleja *et al* 1990, Theis and Mooney 1990, Piotrzkowski *et al* 1991). In figure 3 pressure or an increase in x causes an upward shift of the D^+ curve which moves the intersection only further towards the left—the barrier for emission is not affected.

Emission here is a thermally activated process with an activation energy of

$$\Delta E_E = E'_B - (E^{0/+} + E^{-/0}) \qquad (1)$$

where $E^{0/+}$ and $E^{-/0}$ stand for the 'ionization' energies of the neutral and the negatively charged donor, respectively. Here ΔE_E is the thermodynamic energy required for the two-electron emission process: the top of the barrier may be viewed as an excited state of the deep D^- ground state whose occupation is described by a Boltzmann factor. Therefore we define the thermal emission rate as $e_{D^-}^{th}(T) = e_D^\infty \exp(-\Delta E_E/kT)$, in contrast to the standard expression used in DLTS which results from the definition of the capture rate and thus contains an additional factor T^2. Within the present definition, the activation energy obtained from an Arrhenius plot of the emission rate is just ΔE_E, which is the energy required for two-electron emission. Experimental data taken from the literature (Mooney *et al* 1991, Piotrzkowski *et al* 1991) are given in figure 4, where we have omitted the T^2 factor. As can be seen in figure 4, the present definition describes the experimental data very well for many orders of magnitude. In addition, this definition also removes the problem of different pre-exponential factors (Mooney *et al* 1991). Results for the activation energies for emission are close to 400 meV as given in table 1 for the Si DX centres with different i.

In the present model, the thermal capture rate $c_D^{th}(T)$ can be calculated from detailed balance considerations and the definition of $e_{D^-}^{th}(T)$ instead of the traditional way, where $c_D^{th}(T)$ is defined and $e_{D^-}^{th}(T)$ is calculated. This approach is more feasible here since the capture

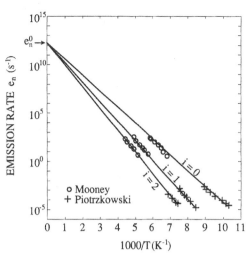

Figure 4. Arrhenius plot of the emission rate of Si DX centres in Al$_x$Ga$_{1-x}$As obtained from DLTS (Mooney *et al* 1991) and high-pressure transport kinetics (Piotrzkowski *et al* 1991).

process is more complicated than the emission. The latter process practically is limited by the emission barrier only. Steady state is described by the equality of capture and emission rates

$$e_{D^-}^{th} N_{D^-} + e_{D^-}^{opt} \Phi N_{D^-} = c_D^{th}(N_D^{tot} - N_{D^-}) \qquad (2)$$

where Φ is the photon flux, $e^{opt}(h\nu)$ the photoionization cross section and N_D^{tot} the total donor concentration. The capture coefficient is obtained from the thermal equilibrium conditions: $\Phi = 0$ and N_{D^-} equal to the thermal equilibrium occupancy of the D^- state

$$N_{D^-} = N_D^{tot}\left[\exp\left(-\frac{E^{0/+} - E_F}{kT}\right)\exp\left(-\frac{E^{-/0} - E_F}{kT}\right)\right]$$
$$\times \left[1 + 2\exp\left(-\frac{E^{0/+} - E_F}{kT}\right)\right.$$
$$\left. + \exp\left(-\frac{E^{0/+} - E_F}{kT}\right)\exp\left(-\frac{E^{-/0} - E_F}{kT}\right)\right]^{-1}. \qquad (3)$$

Here we have considered only the one- and two-electron ground states; a possible occupation of excited donor states is not included. For c_D^{th} we obtain

$$c_D^{th}(E_F^*) = \frac{e_D^\infty}{2} f(E^{0/+})\exp\left(-\frac{E'_B - E^{0/+} - E_F^*}{kT}\right) \qquad (4)$$

where, according to the level ordering of figure 3, $f(E^{0/+})$ is the thermal occupancy of an undistorted (A_1) neutral donor state. If $E^{0/+} > E_F^*$, then equation (4) can be approximated by (where $E'_B = 2E_B$)

$$c_D^{th}(E_F^*) = e_D^\infty \exp\left(-\frac{2(E_B - E_F^*)}{kT}\right) \qquad (5)$$

which defines the activation energy for capture. This approximation is used in figure 1(*b*). The capture coefficient in equations (4) and (5) depends explicitly on the Fermi level. Inserting expressions (3) and (4) into equation (2) we can evaluate the quasi-Fermi level in the high- and the low-temperature limits

$$E_F^* \to \frac{E_B'}{2} \qquad \text{for } T \to 0$$

$$E_F^* \to \tfrac{1}{2}(E^{0/+} + E^{-/0}) \qquad \text{for high } T \qquad (6)$$

and the difference of these two limiting values is $\tfrac{1}{2}[E_B' - (E^{0/+} + E^{-/0})]$, which, according to equation (1), is equal to $\tfrac{1}{2}\Delta E_E$ (see also figure 1(b)). In other words: for $U < 0$, the possible variation of E_F^* in the photoconductivity experiment is only half of the emission energy, in contrast to the case of a normal donor with $U > 0$. Intuitively, the factor $\tfrac{1}{2}$ in the case of $U < 0$ can be attributed to the fact that emission here is a two-electron process, whereas capture occurs stepwise—first one electron is captured and then the second. Emission experiments thus measure the activation energy per event which should be divided by two for a comparison with the capture energy as indicated in figure 1(b).

A comparison of the experimental variation of E_F^* as given in figure 2 with values for the emission barrier clearly supports the negative-U model and the hypothesis of two-electron emission for the Si DX centre in $Al_xGa_{1-x}As$. The saturation of E_F^* for $T < 50$ K deserves further attention: as noted above, a pinning of E_F^* may be taken as evidence for extra density of states. The localized, undistorted A_1 neutral donor state, which was obtained by Dabrowski *et al* (1990) in the microscopic theory (see also figure 3) and which has been inferred by Dmochowski *et al* (1991) from photoluminescence and far-infrared spectroscopy in GaAs under high hydrostatic pressure could have this effect: if the energy $E^{0/+}$ is smaller than the barrier energy E_B as we have assumed in figure 3 in agreement with the results by Dabrowski *et al* (1990), then E_F^* will be pinned at this energy rather than at E_B as $T \to 0$.

Further complications arise from the alloy splitting and the broadening of DX centres due to Coulomb potential fluctuations (see Wilamowski *et al* 1991): as can be seen in figure 4 and in table 1, the emission barrier depends on the number of Al neighbours and the splitting amounts to about 30–40 meV for neighbouring levels and the broadening is of comparable magnitude. In order to account also for this situation we consider the steady-state conditions for the complete system including the broadened $i = 1, 2$ and 3 DX ground states. The $i = 0$ state can be omitted here because of its negligible abundance (0.005 for $x = 0.31$). The ground state energies given also in table 1 are obtained from other experiments as described by Wilamowski *et al* (1991). In order to explain also the low-temperature saturation of E_F^* we have included an extra donor state at $E_{CB} + 120$ meV which may be just $E^{0/+}$. Results of this calculation are also given in figure 2 by the full curve which reproduces the experiments practically in a quantitative way.

4. Summary

In summary, we have shown that the steady-state photoconductivity of heavily doped $Al_xGa_{1-x}As$:Si exhibits a temperature-dependent saturation which can be explained quantitatively in terms of the negative-U model only, considering literature data for the emission rate. The model also requires two-electron emission: within the Chadi and Chang model, it indicates that the Si$^-$ donor atom has to pass through the triangle of As atoms in its negative charge state on its way to the substitutional site before it can emit its electrons in agreement with the results of Dabrowski *et al* (1990). The observed low-temperature saturations of E_F^* can be understood in terms of an additional resonant donor state.

Acknowledgments

We thank J Dabrowski and M Scheffler for making unpublished material available to us, K Köhler for generously providing samples, G Brunthaler and J Kossut for helpful discussions and U Hannesschläger for expert technical assistance. Work supported by Fonds zur Förderung der Wissenschaftlichen Forschung, Austria. One of us (ZW) acknowledges support during a two month stay in Linz by the Bundesministerium für Wissenschaft und Forschung, Vienna.

References

Brunthaler G and Köhler K 1991 *Appl. Phys. Lett.* **57** 2225
Calleja E, Garcia F, Gomez A, Muñoz E, Mooney P M, Morgan T N and Wright S L 1990 *Appl. Phys. Lett.* **56** 934
Chadi D J and Chang K J 1988 *Phys. Rev. Lett.* **61** 873
——1989 *Phys. Rev. Lett.* B **39** 10063
Dabrowski J, Scheffler M and Strehlow R 1990 *Proc. 20th Int. Conf. on the Physics of Semiconductors* ed E M Anastassakis and J D Joannopoulos (Singapore: World Scientific) p 489 (and private communication)
Dmochowski J E, Wang P D and Stradling R A 1991 *Semicond. Sci. Technol.* **6** 118
Jantsch W, Ostermayer G, Brunthaler G, Stöger G, Wöckinger J and Wilamowski Z 1990 *Proc. 20th Int. Conf. on the Physics of Semiconductors* ed E M Anastassakis and J D Joannopoulos (Singapore: World Scientific) p 485
Lang D 1986 *Deep Centers in Semiconductors* ed S Pantelides (New York: Gordon and Breach) p 489
Mooney P M 1990 *J. Appl. Phys.* **67** R1
Mooney P M, Theis T N and Calleja E 1991 *J. Electron. Mater.* **20** 23
Ostermayer G 1991 *Diploma Thesis* Johannes Kepler Universität, Linz
Piotrzkowski R L, Litwin-Staszewska E, Robert J L, Mosser V and Lorenzini P 1991 *Semicond. Sci. Technol.* **6** at press
Theis T N and Mooney P M 1990 *Mater. Res. Soc. Symp. Proc.* **163** 729
Wilamowski Z, Kossut J, Jantsch W and Ostermayer G 1991 *Semicond. Sci. Technol.* **6** B38

Semicond. Sci. Technol. **6** (1991) B51–B57. Printed in the UK

Ionization and capture kinetics of DX centres in AlGaAs and GaSb: approach for a negative-U defect

L Dobaczewski†‡ and P Kaczor‡

† Department of Electrical Engineering and Electronics and Centre for Electronic Materials, University of Manchester Institute Science and Technology, PO Box 88, Manchester M60 1QD, UK
‡ Institute of Physics, Polish Academy of Sciences, al. Lotnikow 32/46, 02–668 Warsaw, Poland

Abstract. In this paper the consequences of an idea in which DX centres in semiconductors form a negative-U system for the approach to commonly used DLTS and photoionization measurements are discussed. For the negative-U model of DX centres to be valid the carrier exchange between the DX states and the conduction band must occur via an intermediate one-electron D^0 state. In such a system the neutral D^0 state must be thermodynamically unstable, but obviously should play a role in all carrier capture and emission processes ($D^- \leftrightarrows D^0 + e^- \leftrightarrows D^+ + 2e^-$). The experimental evidence for the existence of such an intermediate state is presented. It is based upon the detailed observations of the temperature evolution of the photoionization transients of the DX centres in $Al_xGa_{1-x}As:Te$ as well as the capacitance transients of the thermal emission process from DX centres in GaSb:S. Rate equations for photoionization and isothermal DLTS experiments on a defect forming a negative-U system are presented.

1. Introduction

The physical mechanisms leading to metastability phenomena observed in DX centres in III–V semiconducting compounds are still a subject of intensive discussion. Recently, Chadi and Chang [1], and independently Morgan [2], postulated that an explanation of these effects is substitutional–interstitial defect motion. In the Chadi and Chang model the defect in such a configuration is stabilized by a capture of two electrons, and thus must have a negative electron correlation energy, U ($U < 0$) [3]. The ground state of DX forms a negatively charged DX^- acceptor state.

A key problem in the DX puzzle is experimental verification of the theoretical models of the defect and the microscopic mechanism leading to the metastability effects. The standard techniques, sensitive to the local environment of the defect, appear to fail in the case of the DX centre ground state. EXAFS and Mössbauer spectroscopy produce somewhat ambiguous results, since both techniques suffer from the same drawbacks; both require a high concentration of DX centres (some clustering and compensation are then inevitable). During the measurements, large numbers of electron–hole pairs are created and thus hole capture at low temperatures [4] must lead

to the formation of the ionized D^+ undistorted states. This problem is also a real reason for the inapplicability of photoluminescence techniques to the DX centre ground state.

The use of electron paramagnetic resonance could provide the most direct information on the centre symmetry. However, all known attempts produce no ESR spectrum related to the ground state of the DX centre [5]. Broadening does not seem to be the obstacle as, even for infrared absorption measurements at 20 T, magnetic fields [6] fail to produce the expected spectrum. In contrast, all these attempts clearly reveal the metastable spectrum of the photoinduced X shallow state [5–7].

Somewhat indirect information about the nearest-neighbour environment of the DX centre can be obtained from the analysis of the structure seen in the DLTS spectra of the AlGaAs alloy with low Al content [8]. This, as well as the DLTS structure seen in the ordered AlGaAs alloys [9], confirms a localized picture of the DX centre in reasonable agreement with the Chadi and Chang or Morgan models.

Until now, the main source of information about the ground state of DX centres came from Hall and DLTS measurements as well as studies of the photoionization process. The Hall data were not able to conclusively confirm or reject the positive-U or negative-U models of DX [10–12]. Disproportion of the one-electron D^0 states in semiconductors should lead to the self-compensation of the DX defects. Therefore, whatever the doping, crystals must always behave as though well compensated. Here D^- states should act as very deep acceptors in addition to all the other standard acceptors in the sample. This must result in a single activation energy in the temperature dependence of the carrier concentration, a very characteristic phenomenon of the DX centres [11, 12].

In this presentation, it is shown that studies of non-stationary processes, i.e. photoionization, thermal ionization and electron capture, can easily reveal the differences between these two models. For the negative-U model of DX centres to be valid, the carrier exchange between the DX states and the conduction band must occur via an intermediate one-electron D^0 state. In such a system the neutral D^0 state must be thermodynamically unstable, but obviously should play a role in all carrier capture and emission processes ($D^- \rightleftharpoons D^0 + e^- \rightleftharpoons D^+ + 2e^-$). This state need not necessarily be the well known hydrogenic excited state of the DX centres which is associated with either Γ or X conduction band minima.

In this review the experimental evidence for the existence of such an intermediate state is presented. It is based upon the detailed analysis of rate equations for ionization and capture processes of a two-electron defect forming a negative-U system. The model presented is illustrated by the results of observations of photoionization transients of the DX centres in $Al_xGa_{1-x}As$:Te as well as capacitance transients of the thermal emission process from DX centres in GaSb:S.

In both cases, the thermal and optical ionization transients were found to be strongly non-exponential.

The photoionization of $Al_xGa_{1-x}As$:Te [13] at low temperatures and high light intensities and the thermal ionization of GaSb:S [14, 15] were successfully modelled by two mono-exponential functions (figure 1) with the ratio of the two exponential components strongly dependent on temperature or light intensity. The most extreme case of non-exponential behaviour can be observed at higher temperatures and at relatively high photon fluxes, when the photoionization transients for both DX defects exhibit 'overshoots' (figure 2).

A variety of phenomenological models (e.g. two different donors with the strong compensation due to native or unintentional acceptors, one double donor, a single charged centre with a long-lived excited state) have been considered in order to explain the data presented in figures 1 and 2 [17]. For some of these models an acceptable fit can be obtained for each individual kinetic. However, only the assumption that the two energy states of DX correspond to the two charge states of a negative-U-type defect can provide a consistent explanation for *all* the experimental data.

2. Rate equations for two-step ionization and capture processes

The photoionization, thermal ionization and electron capture kinetics for the defect can be derived from a set of rate equations describing carrier exchange between a two-electron defect and the conduction band. For such a defect the electron concentration n is given by

$$n = N_D - N_A - N_1 - 2N_2 \qquad (1)$$

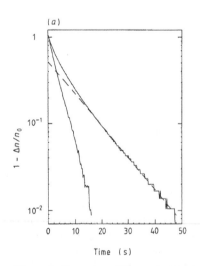

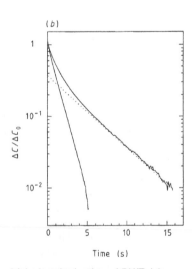

Figure 1. Examples of the non-exponential transients of (a) photoionization of DX(Te) in AlGaAs ($\lambda = 2.1~\mu m$, $T = 84$ K, $x = 0.55$) and (b) thermal ionization of DX(S) in GaSb ($T = 128$ K). A fit of a monoexponential function (broken line) to the tail of the transient gives the lower of the emission constants. Subtracting this fit from the transient leads to a new monoexponential function, with a higher emission constant. Such a decomposition of the transient is possible only when equations (2) are linear, i.e. the C_1 and C_2 parameters are negligible in comparison with e_1 and e_2.

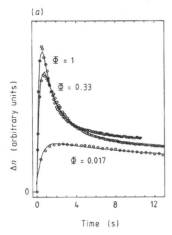 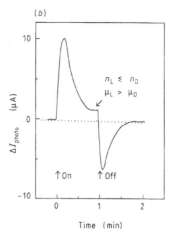

Figure 2. At higher temperatures, i.e. above the PPC regime, the extreme examples of the non-exponential behaviour of the photoionization transients for (a) DX(Te) in $Al_{0.35}Ga_{0.65}As$ (for different photon fluxes Φ) ($T = 129$ K, $\lambda = 1.3$ μm) and (b) DX(S) in GaSb ($\lambda = 1.8$ μm) [16] are observed.

and the rate equations are

$$dN/dt = -e_1N_1 + e_2N_2 + c_1(N_D - N_1 - N_2) - c_2N_1 \tag{2a}$$

$$dN_2/dt = -e_2N_2 + c_2N_1 \tag{2b}$$

where N_i denotes concentration of defects possessing i electrons. The emission rates e_1 and e_2 are the sums of the thermal (e_t) and optical emission rates ($e_0 = \sigma_0\Phi$, where σ_0 is the photoionization cross section and Φ is the photon flux). The proportionality of the capture rates $c_{1,2}$ to the electron concentration may result in non-linearity of these equations. When the number of electrons participating in the process is only a fraction of all electrons in the crystal, these equations remain linear and can be solved analytically. The solution of this set of equations depends critically on the initial conditions and, therefore, on the sign of U, as well as the type of experiment being performed, i.e. photoconductivity transients, isothermal DLTS (thermal emission process) or electron capture process.

2.1. Photoconductivity transients ($Al_xGa_{1-x}As:Te$)

The photoionization process should be studied at relatively low temperatures, where it can be assumed that $e_{t1,2} = 0$. If this condition is not fulfilled, for very low photon fluxes thermal emission may participate in the total ionization process as well as photoionization. This makes quantitative analysis of the results less straightforward. Moreover, at higher temperatures it is not possible to achieve a complete freeze-out of all electrons on the centre, and the initial conditions for the process are temperature dependent.

If the sample is slowly cooled in darkness, electrons freeze-out on the defects. According to equilibrium statistics for the positive- and negative-U defects, the initial

conditions depend on the sign of U in the following way

$$N_1 = [D^0] = 0 \quad \text{and}$$

$$N_2 = [D^-] = \tfrac{1}{2}(N_D - N_A) \qquad \text{for } U < 0 \tag{3a}$$

$$N_1 = N_D - N_A \text{ and } N_2 = 0 \qquad \text{for } U > 0. \tag{3b}$$

In photoconductivity experiments, the photoexcited electrons modify the bulk conductivity, and thus the capture rates $c_{1,2}$ are not constant. During the fitting procedure these parameters can be redefined to make them independent of the electron concentration: $c_i = C_in/N_D$ ($i = 1, 2$) and furthermore the C_i parameter can be used as a constant.

The photoionization kinetics derived from the rate equations (1) and (2) with the initial conditions for a negative-U defect (equation (3a)), describe perfectly the photoionization transients at temperatures where DX is metastable and at higher temperature (129 K) where it is not and the 'overshoots' behaviour is observed (figure 2(a)). For each temperature the transients for all photon fluxes were fitted simultaneously and only for the negative-U model of DX were all fitting parameters, i.e. the capture rates $c_{1,2}$ and photoionization cross sections $\sigma_{o1,2}$, independent of the light intensity. The curves in figure 2(a) represent the fits to the experimental data.

It was also found that the capture rate c_1 depends strongly on temperature and vanishes for $T < 72$ K. This represents the inability of the empty DX centre (D^+ charge state) to recapture the first electron at low temperatures. An increase of c_1 with temperature exactly reflects an activation character of the electron capture cross section already found for Te-related DX centres in AlGaAs [18]. Interestingly, the capture rate c_2, i.e. the capture rate of the second electron by the DX centre in the neutral charge state D^0, depends very weakly on temperature and even for $T = 40$ K was found to be close to 1 s^{-1}. This means that even in the persistent photoconductivity (PPC) regime the photogenerated D^0 state efficiently captures the second electron. The large and weakly temperature-dependent electron capture rate c_2 indicates also that the capture of the second electron occurs while the DX centre is already in the relaxed state.

2.1.1. Photoionization process at very low temperatures $(e_{t1,2} = c_{1,2} = 0)$.

At low temperatures or for very high photon fluxes, Φ, capture rates are very small compared with emission rates. Thus, equations (2) become linear and can be solved analytically

$$N_1(t) = \tfrac{1}{2}(N_D - N_A)\frac{e_2}{(e_2 - e_1)}\left[\exp(-e_1 t) - \exp(-e_2 t)\right]$$
(4a)

$$N_2(t) = \tfrac{1}{2}(N_D - N_A)\exp(-e_2 t).$$
(4b)

In photoconductivity experiments one observes a change in the electron concentration

$$n(t) = (N_D - N_A)\left(1 - \frac{\tfrac{1}{2}e_2 - e_1}{e_2 - e_1}\exp(-e_2 t)\right.$$

$$\left. - \frac{\tfrac{1}{2}e_2}{e_2 - e_1}\exp(-e_1 t)\right)$$
(4c)

Figure 3 shows the occupation of the defect D^- and D^0 charge states, as well as the electron concentration given by equations (4). As one can see, the occupation of the intermediate state D^0 for the photoionization process, depending on the ratio between e_1 and e_2, may reach even higher values for a certain period of time. An observation of the defect in this charge state is not

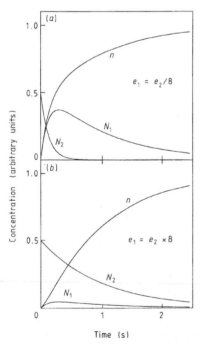

Figure 3. Occupations of two charge states of a defect forming a negative-U system (N_1, N_2) and concentration of electrons in the conduction band (n) derived from rate equations (1), (2) and initial conditions (3a) for the case $c_1 = c_2 = 0$.

possible for equilibrium $(t = 0)$ and metastable $(t \to \infty)$ conditions. However, for some experimental techniques the equilibrium conditions are not necessary, and thus this charge state may be detected by, for example, non-stationary ESR, or absorption.

A commonly used approach to study the photoionization process is to observe the initial slope of the photoconductivity transients. According to equation (4c) this slope is given by

$$dn/dt(t = 0) = \tfrac{1}{2}e_2(N_D - N_A).$$
(5)

Equation (5) shows that the initial-slope method is adequate only for the investigation of the photoionization process of the D^- defect charge state. If, in the experiment before sample illumination, any pre-illumination is applied, then the initial conditions (3a) are not valid, and the initial-slope method gives values of the photoionization rate as a complicated relation between e_1 and e_2.

2.2. Isothermal DLTS: thermal emission and electron capture processes $(e_{o1,2} = 0, \text{GaSb:S})$

To describe thermal emission and electron capture processes, the same set of rate equations (1) and (2) can be used with the initial conditions reflecting the specific character of the isothermal DLTS experiment. In this technique the capacitance transients due to recharging of the defects in the diode depletion region are observed. At the beginning of the filling pulse, all of the defects in this region are in the D^+ charge state. Any decrease in diode reverse bias populates the defects in a thin layer of the space charge region with electrons. This capture process is observed during a time equal to t_f (a filling pulse). The initial conditions for the capture process are the following

$$N_1 = 0 \quad \text{and} \quad N_2 = 0.$$
(6)

If the volume of the space charge layer, where this capture process occurs, is much smaller than the volume of the crystal outside the depletion region, then the number of captured electrons is much smaller than the number of electrons in the crystal. In this case, the capture process does not affect the electron concentration in the conduction band, and thus the capture parameters $c_{1,2}$ can be regarded as constant. The rate equations (1), (2) with initial conditions (6) can be solved analytically

$$N_1(t) = (N_D - N_A)\frac{c_1}{PM}\left[\lambda_2(e_2 - \lambda_1)\exp(-\lambda_1 t)\right.$$

$$\left. - \lambda_1(e_2 - \lambda_2)\exp(-\lambda_2 t) + e_2 P\right]$$
(7a)

$$N_2(t) = (N_D - N_A)\frac{c_1 c_2}{PM}\left[\lambda_2\exp(-\lambda_1 t)\right.$$

$$\left. - \lambda_1\exp(-\lambda_2 t) + P\right]$$
(7b)

where

$$P = \left[(e_2 - e_1 - c_1 - c_2)^2 + 4c_2(e_2 - c_1)\right]^{1/2}$$

$$M = c_1 c_2 + (e_1 + c_1)e_2$$

and

$$\lambda_{1,2} = \tfrac{1}{2}(e_1 + e_2 + c_1 + c_2 \pm P).$$

the capture process cannot be observed directly in the DLTS experiment because it occurs outside the actual depletion region of the diode. However, it does result in a certain occupation of both charge states:

$$N_1(t = t_f) = N_{01} \quad \text{and} \quad N_2(t = t_f) = N_{02}. \quad (8)$$

After the filling pulse, the diode reverse bias is increased and the thermal emission process, represented by the decay of the diode capacitance, is observed. This process can again be modelled by the rate equations (2). However, in this case, the initial conditions are given by (8).

In the space charge region, the electric field makes the carrier recapture process ($c_{1,2} = 0$) impossible. Thus, the emission process is described by the following formulae

$$N_1(t) = [N_{02}e_2/(e_2 - e_1) - N_{01}]$$
$$\times \exp(-e_1 t) - N_{02}e_2/(e_2 - e_1)\exp(-e_2 t) \quad (9a)$$
$$N_2(t) = N_{02}\exp(-e_2 t). \quad (9b)$$

The capacitance transients are caused by the total change of charge within the space charge region. Thus they can be described by the sum of $N_1(t)$ and $N_2(t)$. The fitting of the transients, observed for the filling pulse width, t_f, to the two-exponential function $N_1(t) + N_2(t)$ gives the values of the thermal emission rates $e_1 = e_{t1}$ and $e_2 = e_{t2}$ and pre-exponential amplitudes N_{01} and N_{02}. Repeating this fitting for different filling pulses reproduces the capture kinetics (equations (7), (8)) and gives the values of capture constants c_1 and c_2.

The above procedure was applied for the case of DX centres in GaSb:S. This system possesses certain advantages in comparison with the DX centres in AlGaAs. In contrast to the alloy, the parameters describing the thermal emission process for the DX centres in binary semiconductors are not broadened by the alloy fluctuations. The observed thermal emission process at constant temperature, according to equations (9), can be decomposed with great accuracy into a sum of two individual exponential transients. The amplitudes N_1 and N_2, measured as a function of the filling pulse width (figure 4), represent the capture process by the two energy levels, and have been fitted using kinetics inferred from the rate equations (7), (8). It can be clearly seen that at low temperatures one of the charge states becomes thermodynamically unstable (curve $N_1(t_f)$), although it can bind an electron for a certain period of time.

The Arrhenius plot of the emission rates e_{t1} and e_{t2} measured at different temperatures gave the activation energies of the thermal emission process and the capture cross section for both energy states: $E_1 = 0.25$ eV, $\sigma_{1\infty} = 4 \times 10^{-14}$ cm^2, $E_2 = 0.22$ eV, and $\sigma_{2\infty} = 8 \times 10^{-16}$ cm [14, 15]. The large difference between the values of $\sigma_{1\infty}$ and $\sigma_{2\infty}$ indicates again that both energy states differ in charge state. This is because the defect in the D^0 state when empty (the D^+ charge state) has the long-range Coulomb potential whilst the initial state for the electron capture process for the D^- state is the D^0 state which has no long-range potential. The value of E_1

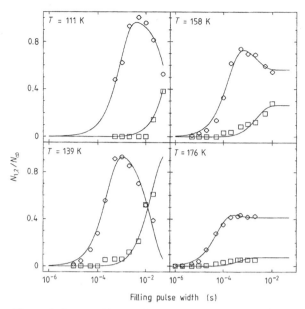

Figure 4. Capture kinetics for two energy levels associated with DX(S) in GaSb (N_1, circles; N_2, squares). The full curves are the best fit to the experimental points, where the trapping kinetics was inferred from the rate equations (7), (8).

confirms the supposition that the intermediate state for the thermal ionization process cannot be a hydrogen-like energy state of the defect.

The asymptotic solutions of the rate equations (2) for steady state conditions ($dN_{1,2}/dt = 0$) must give the occupations of the one- and two-electron energy states equivalent to that obtained from more general considerations [11]. It is easy to show that the ratio e_{t1}/e_{t2} must be related to the value of the correlation energy U according to the formula $e_{t1}/e_{t2} = (\sigma_1/\sigma_2)\exp(-U/kT)$. One can expect that the capture cross section for the one-electron state, D^0, will always be larger than that for the two-electron state D^- ($\sigma_1 > \sigma_2$). Thus, for the case $U < 0$, one always gets $e_{t1} > e_{t2}$. Despite the fact that this relation is observed for DX in GaSb:S, it does not itself mean that $U < 0$.

In practice, for DX centres, the capture kinetics (equations (8)) are mainly governed by capture constants, i.e. $c_{1,2} \gg e_{t1,2}$. If the capture rate is too fast to be observed, even for the shortest filling pulses, the amplitude N_{01} will be almost negligible. Also, the case $e_{t1} > e_{t2}$ does not favour the observation of the thermal emission from the D^0 charge state, and consequently it may happen that the capacitance transients are found to be almost perfectly exponential.

3. Discussion

The analysis presented shows that for the case of a defect forming a negative-U system the capture and emission processes have to be described in terms of the two-exponential kinetics. Relative amplitudes and time constants of these kinetics are an interplay between four parameters involving $e_{t1,2}$ and $c_{1,2}$. For the case of the DX centre at least three of them ($e_{t1,2}$ and c_1) are strongly temperature dependent. Performing the DLTS experiment with the temperature scan gives a peak which is a convolution of the two-step capture and two-step emission processes, which both vary strongly with temperature. Practically, it may result in DLTS peak broadening effects or even the occurrence of two separate peaks with relative amplitudes strongly dependent on the filling pulse and the temperature. Any quantitative DLTS lineshape analysis for the case of DX must take into account all of these effects.

For the case of the DX centres in AlGaAs and other ternary compounds a direct deconvolution of the thermal emission process (equation (9)) into two separate processes may not be possible due to unknowns in defect parameter broadening effects caused by random alloy fluctuations. Thus, the emission as well as capture constants can only be regarded here in the sense of their average values.

Numerous authors have analysed the thermal emission process of electrons from the DX centre and found it to be strongly non-exponential [18–21]. Usually, random alloy fluctuations were suspected of being the reason for this phenomenon. On the other hand, this explanation is in obvious contradiction to the widely observed independence of the thermal [22] and optical [23, 24] ionization processes on the crystal alloy composition. The recent finding that the sulphur atoms in AlGaAs crystals are mainly decorated by aluminium atoms [25] may indicate that for the case of the DX centres, the defect's local surroundings may differ substantially from that which is expected from the virtual crystal approximation.

For the defect forming a negative-U system the capture and emission processes are two-step processes and so they are, in principle, non-exponential. This indirect way of changing the defect charge state is the centre's intrinsic feature and may be the fundamental reason for the observation of the non-exponential processes. When the negative-U defect is analysed in the way permitted only for a one-electron centre, then the capture and ionization parameters must be seen as 'broadened', i.e. kinetics are not perfectly exponential. The generally used approach to the broadening effects must then be revised.

In the analysis of the photoionization and thermal ionization transients based upon appropriate rate equations it was assumed that no other short-lived energy states of DX participate in these processes. A limiting factor was the sampling rate (approximately 30 samples/s for photoionization and 10^5 samples/s for isothermal DLTS) of the experimental apparatus making it impossible to record very fast processes. Consequently, it was assumed that no short-lived energy states of DX participate in these processes. However, these states, if present, may participate very effectively in the processes of electron exchange between DX and the conduction band, and definitely influence the spectral and temperature dependences of the observed ionization and capture

processes. Thus, the ionization and capture parameters obtained from the analysis presented are valid for no longer than the assumed scenario of the processes is correct. In this approach, only two main processes are considered: $D^- \leftrightharpoons D^0$ and $D^0 \leftrightharpoons D^+$. If these processes are more complicated than simple ionization and electron capture processes, then the parameters we obtained must contain some information about this fact. For instance, any intra-centre transition or defect transformation (reconfiguration) which does not result in the appearance of an electron in the conduction band will not be seen in the type of experiment discussed here. On the other hand, this process will definitely have an influence on the values of the parameters obtained.

4. Summary

In this study the consequences of the negative-U concept of the DX centres in the approach to the intensively investigated defect ionization and electron capture processes are presented. A detailed analysis of the photoionization process of the DX(Te) centres in $Al_xGa_{1-x}As$ ($0.25 < x < 0.55$) as well as the thermal emission and capture for the DX(S) in GaSb revealed that the ionization and capture processes do indeed go through two steps: $D^- \leftrightharpoons D^0 + e^- \leftrightharpoons D^+ + 2e^-$, since it is a consequence of the negative-U character of the defect. The intermediate state of the process is not the effective-mass X- or Γ-like excited state of DX, but the neutral D^0, localized state of the defect. The presented rate equations with the initial conditions characteristic for a defect forming a negative-U system quantitatively describe the observed ionization and capture kinetics. Commonly observed non-exponentiality of the thermal ionization and electron capture processes for the DX centres in ternary semiconductors may in reality be caused not only by alloy fluctuations in the defect's surroundings but also by a negative-U character of the defect. An obvious consequence of the negative-U concept of DX is a necessity to reconsider DLTS and photoionization data for the centres as well as a revision of defect parameters obtained using these methods.

Acknowledgments

The authors would like to thank R Pritchard for valuable comments on the manuscript. This work has been financially supported by SERC grants in the UK, and by the programme CPBP 01.05 in Poland.

References

[1] Chadi D J and Chang K J 1988 *Phys. Rev. Lett.* **61** 873; 1989 *Phys. Rev.* B **39** 10063
[2] Morgan T N 1989 *Mater. Sci. Forum* **38–41** 1079
[3] Anderson P W 1975 *Phys. Rev. Lett.* **34** 953; Watkins G D 1984 *Festkörperprobleme (Advances in Solid State Physics 24)* ed P Grosse (Braunschweig: Vieweg-Verlag) p 163

[4] Brunthaler G, Ploog K and Jantsch W 1989 *Phys. Rev. Lett.* **63** 2276

[5] Mooney P M, Wilkening W, Kaufmann U and Kuech T F 1989 *Phys. Rev. B* **39** 5554

[6] Knap W, Brunel J-C and Martinez M (private communication)

[7] Glaser E, Kennedy T A and Molnar B 1989 *Shallow Impurities in Semiconductors 1988 (Inst. Phys. Conf. Ser. 95)* ed B Monemar (Bristol: Institute of Physics) p 233

[8] Mooney P M, Theis T N and Calleja E J 1991 *J. Electron. Mater.* **20** 23

[9] Baba T, Mizuta M, Fujisawa T, Yoshino Y and Kukimoto H 1989 *Japan. J. Appl. Phys.* **28** L891

[10] Theis T N 1989 *Shallow Impurities in Semiconductors, 1988 (Inst. Phys. Conf. Ser. 95)* ed B Monemar (Bristol: Institute of Physics) p 307

[11] Dmochowski J E and Dobaczewski L 1989 *Semicond. Sci. Technol.* **4** 579

[12] Khachaturyan K, Weber E R and Kaminska M 1989 *Mater. Sci. Forum* **38–41** 1067

[13] Dobaczewski L and Kaczor P 1991 *Phys. Rev. Lett.* **66** 68

[14] Dobaczewski L, Kaczor P, Langer J M, Peaker A R and Poole I 1990 *Proc. 20th Int. Conf. on the Physics of Semiconductors* ed E M Anastassakis and J D Joannopoulos (Singapore: World Scientific) p 497

[15] Dobaczewski L, Kaczor P, Karczewski G and Poole I 1991 *Acta Phys. Pol.* at press

[16] Poole I unpublished

[17] Dobaczewski L and Kaczor P unpublished

[18] Dobaczewski L and Langer J M 1987 *Mater. Sci. Forum* **10–12** 399

[19] Calleja E, Gomez A I and Munoz E 1986 *Solid State Electron.* **29** 83

[20] Munoz E, Gomez A, Calleja E, Crieddo J J, Herrero J M and Sandoval F 1987 *Mater. Sci. Forum* **10–12** 411

[21] Calleja E, Mooney P N, Wright S L and Heiblum M 1986 *Appl. Phys. Lett.* **49** 657

[22] Mooney P M, Caswell N S and Wright S L 1987 *J. Appl. Phys.* **62** 4786

[23] Lang D V, Logan R A and Jaros M 1979 *Phys. Rev. B* **19** 1015

[24] Legros R, Mooney P M and Wright S L 1987 *Phys. Rev. B* **35** 7505;
Mooney P M, Northrop G A, Morgan T N and Grimmeiss H G 1988 *Phys. Rev. B* **37** 8298

[25] Rowe J E, Sette F, Pearton S J and Poate J M 1990 *Physics of DX Centers in GaAs Alloys (Solid State Phenomena 10)* ed J C Bourgoin (Vaduz: Sci. Tech. Publications) p 283

Semicond. Sci. Technol. **6** (1991) B58–B61. Printed in the UK

Local-environment dependence of the DX centre in GaAlAs: alloy and superlattice studies

S Contreras†‡, V Mosser‡, R Piotrzkowski§, P Lorenzini†, J Sicart†, P Jeanjean†, J L Robert† and W Zawadzki ‖

† Groupe d'Etudes des Semiconducteurs, UA 357, USTL, 34095 Montpellier Cedex, France
‡ Schlumberger Montrouge Recherche, 50 av. J. Jaurès, BP 620–05 F 95542 Montrouge, France
§ High Pressure Research Center, Polish Academy of Sciences, 01.142 Warsaw, Poland
‖ Institute of Physics, Polish Academy of Sciences, 02.668 Warsaw, Poland

Abstract. We present new experimental data on the capture of electrons onto Si-induced impurity states, which give evidence of the existence of several configurations for the DX centre with a large lattice relaxation. The investigated samples were Si-doped GaAlAs alloys and GaAs/AlAs short-period superlattices, grown by molecular beam epitaxy, with different aluminium contents. Persistent photoconductivity in the samples has been investigated by means of Hall measurements under pressures up to 15 kbar, between 4.2 and 400 K.

In GaAlAs alloys several steps have been observed on the thermostimulated capture curves between 90 and 140 K. The mechanism responsible for the presence of these steps is not only related to capture processes, but also results from the equilibrium between capture and emission processes, involving the different states of the DX centre. Our data confirm that the DX centre presents several configurations, each of them with discrete emission and capture barriers, related to the local environment of the Si atom.

For the short-period superlattices selectively doped in GaAs layers or in AlAs layers, the thermal annealing curves decrease monotonically, as expected for a single barrier between 90 and 140 K. In uniformly doped GaAs/AlAs superlattices the $n(T)$ curve shows a two-step behaviour, indicating that two barriers are involved in the capture process. These results confirm the existence of several configurations for the DX centres in short-period superlattices. The selective doping of short-period superlattices in AlAs layers made it possible to study the all-Al environment of the Si atom, unobserved until now.

1. Introduction

Si donors in GaAs and in $Ga_{1-x}Al_xAs$ alloys possess interesting properties related to metastability and large lattice relaxation effects. In spite of numerous experiments performed on these III–V compounds to explain the unusual properties exhibited by the so-called DX centre, there is no complete representation of its microscopic structure. A large lattice relaxation has been predicted by the pseudo-potential calculation of Chadi and Chang [1]. In this model, the substitutional Si atom breaks a bond with the nearest neighbour As atom and moves towards an interstitial site. In his vacancy–interstitial model, Morgan [2] proposes that the Si atom moves through the basis of the surrounding As tetrahedron to an interstitial position. In both approaches, the relaxation is trigonal and leads to four possible configurations of the DX centre.

The following experiments are consistent with the above picture. It has been recently shown in DLTS experiments that the thermal emission energies of electrons from DX states in dilute GaAlAs alloys have discrete values, related to the local Ga versus Al environment of the Si atom [3, 4]. Discrete DLTS peaks have also been observed in AlAs/GaAs ordered microstructures [5]. In these experiments, the configurations have been revealed by their emission activation energy, and a configuration in which the Si atom has all Al atoms as nearest neighbours has not yet been observed. In this paper, we report experimental data on the thermostimulated cap-

0268-1242/91/100B58+04 $03.50 © 1991 IOP Publishing Ltd

ture of electrons onto Si induced impurity states in GaAlAs alloy and superlattices. They offer evidence for the existence of several configurations for the DX centre induced by the local environment of the Si atom.

The samples were GaAlAs alloys and GaAs/AlAs short-period superlattices grown by molecular beam epitaxy and doped with Si. The $Al_xGa_{1-x}As$ layers were separated from the semi-insulating GaAs substrate by a large undoped spacer to avoid 2D effects. The Al content ($x = 0.33$) was checked by double x-ray diffraction. The effective doping density N_D-N_A was measured at 300 K by the C-V method, using a polaron profiler (N_D-N_A = 1.2×10^{18} cm^{-3}). The short-period GaAs/AlAs superlattices had periods P between 30 and 50 Å. The Al content ($0.34 < x < 0.42$) was determined by simple and double x-ray diffraction. The growth temperature T_g was between 580 and 650 °C. The silicon doping was uniform (sample 3) or selective in GaAs (sample 1) or AlAs (sample 2) layers. All the samples were patterned in a double bridge or clover-leaf van der Pauw geometry. The Hall measurements were performed between 4.2 and 300 K reversing both the current and the magnetic field. Hydrostatic pressure up to 15 kbar, generated in a gaseous medium, was applied to change the carrier concentration [6].

2. GaAlAs alloys

Figure 1 shows the electron concentration as a function of the temperature, under various pressures. At zero pressure, the dark-carrier density first decreases with temperature and then stabilizes at a constant value as soon as the metastability temperature is reached. The same behaviour is observed under pressure. The carrier density is lower in the metastability temperature range, indicating that the remaining carriers do not originate in another hypothetical shallow impurity. Under LED illumination at 4.2 K, the carrier density increases to the

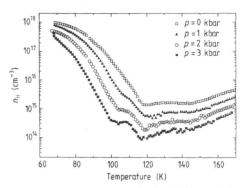

Figure 2. Experimental thermostimulated decay of the Hall carrier density n_H in $Al_{0.33}Ga_{0.67}As$ after photoionization at $p = 0$ (top curve), 1, 2 and 3 kbar. The temperature sweep rate is 1 K min^{-1}.

value N_D-N_A. After switching the LED off, the carrier density does not change for hours as long as the temperature is lower than 65 K. This is the well known persistent photoconductivity effect (PPC).

When the sample is warmed up at a constant rate (typically 1 K min^{-1}) the Hall carrier concentration decreases (figure 2). The $n(T)$ curves show a multistep decrease which appears to be more pronounced at high pressure. Between 90 and 140 K, three shoulders are clearly observed on the curves. A tentative explanation of the existence of the first plateau has been proposed by Brunthaler and Köhler [7]. They believe that the first strong decrease of the carrier density corresponds to trapping on only one configuration, other configurations still being ionized because of their higher capture barriers.

Our experiments under pressure qualify this assumption. If the first plateau corresponds to the saturation of the first DX configuration only, the corresponding carrier concentration should be independent of the pressure. This is not the case, and such an interpretation must be discarded. Consequently, the competition between capture and emission processes for the various configurations is responsible for the several shoulders observed on the $n_H(T)$ curves.

3. Short-period superlattices

The use of short period superlattices (SPS) with selective doping allows us to separate the local environments of the Si atom (all Ga nearest neighbours from an all Al nearest neighbours situation). We have investigated the thermal annealing of PPC for the three types of superlattices described above. The same kind of experiment as in the alloy has been performed on these microstructures. As seen in figure 3, the dark-electron concentration behaves similarly to that in the alloy, even in the superlattices doped in the GaAs layers. As observed in the alloy, the electron concentration increases when the sample is illuminated at low temperature. Due to the barrier for thermal capture, the PPC effect is observed for the three types of superlattice.

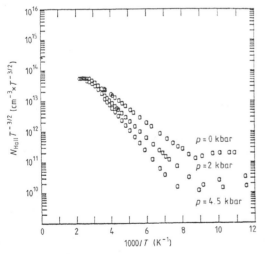

Figure 1. Arrhenius plot of the Hall carrier concentration in $Al_{0.33}Ga_{0.67}As$ for three different applied pressures: $p = 0$ (top curve), 2 and 4.5 kbar.

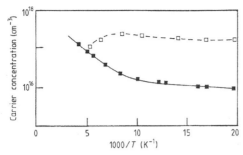

Figure 3. Typical curve of the Hall carrier concentration in SPSS against 1000/*T* in darkness (full squares) and under light illumination (open squares).

Figure 4 shows the thermal annealing curves of PPC for superlattices selectively doped with silicon in GaAs (1) or AlAs (2) layers. In the two cases, the carrier concentration decreases monotonically when the temperature is increased above 80–90 K. The annealing curve for the sample (2) is shifted towards higher temperature indicating that the activation energy E_c for the electron capture is lower in sample 1 (Si atom having all Ga nearest neighbours) than in the sample 2 (Si atom having all Al nearest neighbours). Figure 5 shows the thermal annealing for the uniformly doped superlattice. Two steps are observed, indicating clearly the presence of two discrete configurations of the DX centre.

4. Discussion

To estimate the difference between the activation energies E_c for the electron capture, corresponding to the thermal shift shown in figure 4, we assume that thermal emission is negligible at the onset of the capture process. Thus, after an incremental time Δt, the carrier concentration decreases exponentially

$$\Delta n = A \exp\left(-\frac{E_c}{kT}\right). \tag{1}$$

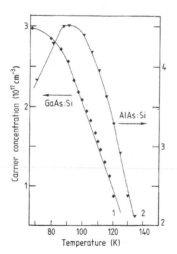

Figure 4. PPC thermal annealing curves in selectively doped SPSS: (1) doping in GaAs layers; (2) doping in AlAs layers.

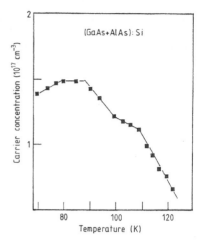

Figure 5. PPC thermal annealing curve in an uniformly doped SPS.

Treating T, E_c and A as independent variables, calculating the full differential $\delta(\Delta n)$ and equating it to zero, we obtain

$$\frac{\Delta E_c}{E_c} = \frac{\Delta T}{T} + \frac{kT}{E_c}\frac{\Delta A}{A}. \tag{2}$$

The activation energy E_c is related to the absolute energy E_B of the barrier and to the Fermi energy E_F through the following equations.
In the case of negative-U model for the DX centre [8]

$$E_c = E_B - 2E_F. \tag{3}$$

In the case of single-donor model

$$E_c = E_B - E_F. \tag{4}$$

Equation (3) can also be written

$$E_c = E_e + (E_{DX^-} - 2E_{cond}) - 2(E_F - E_{cond}) \tag{5}$$

in which E_e represents the barrier height for the emission $E_B - E_{DX}$ and E_{cond} denotes the energy of the conduction miniband.
In the same way, equation (4) can be written as

$$E_c = E_e + (E_{DX} - E_{cond}) - (E_F - E_{cond}). \tag{6}$$

$E_F - E_{cond}$ is calculated knowing the electron density and the effective mass of carriers. $E_{DX^-} - 2E_{cond}$ is twice the activation energy E_a of the carrier concentration in the negative-U model, while $E_{DX} - E_{cond}$ is the activation energy E_a of the carrier concentration in the single-donor model. The calculation of the conduction miniband for the two superlattices shows that we are dealing with an E_X pseudo-direct gap [9]. This result is confirmed by the fact that the carrier concentrations have the same low sensitivity to the applied pressure. Thus, an estimation of $E_F - E_{cond}$ leads at the onset of the capture process to a value nearly equal to 16 meV for the two samples. The emission energy E_e has been found in superlattices in the range 320–430 meV [5]. In equation (1), the term $(kT/E_c)(\Delta A/A) \approx 0.01$ is negligible in comparison with $\Delta T/T \approx 0.2$. The temperature shift measurement ($\Delta T = 20$ K) leads to ΔE_c lying in the range 60–80 meV [10].

Figure 6. Representation of the DX centre in a negative-U model: emission energies become larger with the increasing number of Al nearest neighbours.

On the basis of the results presented above, we propose a model for the DX centre due to the Si donor surrounded by various combinations of Ga and Al atoms as nearest neighbours. In order to determine the absolute difference ΔE_B in energy for the barriers, we can write in the negative-U model

$$\Delta E_c = \Delta E_B - 2\Delta E_{cond} + 2\Delta(E_{cond} - E_F) \qquad (7)$$

in which $\Delta E_B = E_B(2) - E_B(1)$ and $\Delta E_{cond} = E_{cond}(2) - E_{cond}(1)$.

$E_{cond} - E_F$ has been found to be nearly the same for the two samples. ΔE_{cond} is calculated from the energy diagram [9] to be around 35–40 meV. Thus, ΔE_B can be considered as equal to zero.

In the case of the single-donor model, we found in the same way ΔE_B in the range 30–40 meV.

In the following representation of the DX centre for a negative-U model, figure 6 accounts for the experimental fact that emission energies become larger with increasing number of Al nearest neighbours. In a coordinate configuration representation, involving an intermediate state for capture and emission processes, the main consequence of the proposed picture is that the lattice relaxation becomes larger as the number of Al nearest neighbours increases [10].

5. Conclusion

In summary, we have observed several DX states related to the Si donors in bulk GaAlAs and GaAs/AlAs superlattices. In the alloy, up to three discrete structures have been observed in the thermostimulated capture curves under pressure. The experimental results demonstrate clearly that capture and emission processes for the various configurations are competing. In GaAs/AlAs superlattices, the selective doping allows us to separate the all Ga nearest neighbours from the all Al nearest neighbours environment of the Si atom and to determine the absolute difference in the barrier energy, corresponding to these two extreme configurations of the DX centre.

We find that the assumption of an intrinsic barrier E_B for the studied DX configurations is verified only in the framework of the double-donor $U < 0$ model.

Acknowledgments

The authors thank Dr J F Rochette (Picogiga F-91940 Les Ulis, France), Dr F Mollot and Dr R Planel (Laboratoire de Microstructures et Microélectronique F-92220 Bagneux, France) for providing the samples.

References

[1] Chadi D J and Chang K J 1988 *Phys. Rev. Lett.* **61** 873; 1989 *Phys. Rev.* B **39** 10063
[2] Morgan T N 1989 *Mater. Sci. Forum* **38–41** 1079
[3] Mooney P M, Theis T N and Wright S L 1989 *Mater. Sci. Forum* **38–41** 1067
[4] Calleja E, Gomez A, Criado J and Muñoz E 1989 *Mater Sci. Forum* **38–41** 1115
[5] Baba T, Mizuta M, Fujisawa T, Yoshino J and Kukimoto H 1989 *Japan. J. Appl. Phys.* **28** L891
[6] Contreras S 1989 *PhD Thesis* USTL, Montpellier, France.
[7] Brunthaler A and Köhler K 1990 *Appl. Phys.* A **50** 515
[8] Look D C 1981 *Phys. Rev.* B **24** 5852
[9] Bastard G 1988 *Wave Mechanics Applied to Semiconductor Heterostructures* (Les Ulis, France: Editions de Physique)
[10] Sicart J, Jeanjean P, Robert J L, Zawadzki W, Mollot F and Planel R 1991 *Phys. Rev.* B **43** 7351

Semicond. Sci. Technol. **6** (1991) B62–B65. Printed in the UK

Direct study of kinetics of free-electron capture on DX centres

I E Itskevich and V D Kulakovskii

Institute of Solid State Physics of the USSR Academy of Sciences, 142432, Chernogolovka, Moscow District, USSR

Abstract. Direct investigations of the relaxation of the persistent photoconductivity (PPC) in $Al_{0.29}Ga_{0.71}As$: Te have been performed with the use of hydrostatic pressure P. The rate of free-electron relaxation, dn_e/dt, exhibits a dramatic decrease while the Fermi energy of the electrons, E_F, lowers at constant temperature T. In a wide range of E_F the dependence $\ln(dn_e/dt)$ appears to be linear in E_F with the slope about $3\,kT^{-1}$. The height of the capture barrier E_B has been found to increase linearly with reduction of E_F, the $-\Delta E_B/\Delta E_F$ ratio being 3.2 at $P = 1$ bar and 2.5 at $P = 4$ kbar, i.e. larger than expected for both the neutral and negatively charged DX centre models.

1. Introduction

The capture of electrons on DX centres, or the relaxation of persistent photoconductivity (PPC) has been studied for many years [1, 2]. Nevertheless the only well established feature of the relaxation is its non-exponentiality, the other details being the subject of controversies. Many different reasons have been discussed for the non-exponentiality, such as a shift of the Fermi level position during the relaxation [3–6], a broadening of the capture barriers [3, 4, 7], a spatial separation of electrons and ionized donors (or holes) [8, 9], etc.

We have undertaken a thorough direct investigation of the relaxation of the PPC. Free-electron capture on DX centres is to be described by some differential equation, the relaxation rate, dn_e/dt, being a function of a number of parameters but not of time. The solution of this equation gives the relaxation curve that depends on the parameters in some complicated way. Note that as the relaxation is non-exponential it is not correct to use the concept of relaxation time. Usually the experimental curves (plotted on suitable axes) are compared with numerically obtained data, and fitting procedures give the values of parameters [4, 6, 7, 10–12]. In such a case it is difficult to assess the possible mistakes and the limits where the concrete rate equation is valid. We have preferred to examine the derivative of the experimental curves, dn_e/dt, as it is much more sensitive to possible distortions than the integral curve $n_e(t)$ and hence can give more information.

2. Experimental details

We performed our measurements on an $Al_{0.29}Ga_{0.71}As$: Te sample grown by liquid phase epi-taxy on a semi-insulating GaAs substrate. The sample was thick (12 μm) to reduce interface effects. The Te dopant concentration was $2 \times 10^{18}\,cm^{-3}$. The scheme for the Hall measurements was prepared on the sample with use of photolithography and chemical etching. The sample was installed in a low-temperature fixed-pressure cell [13]. For illumination the GaAs light-emitting diode (LED) was placed inside the cell near the sample. The cell in the vacuum tube was placed into the solenoid inside the He cryostat, so transport and Hall measurements could be made at any temperature. The frequency of transport AC measurements was about 20 Hz.

As the relaxation rate appeared to be very sensitive to temperature, reliable temperature control was necessary. The temperature, T, was measured by the calibrated thermometer resistor that was situated inside the high-pressure cell just near the sample. The heating manganin wire was wound around the pressure cell, so the temperatures of the thermometer and the sample were expected to be the same.

The experimental procedure was the following one. After the LED was switched off the resistance R_{XX} versus time was measured. Each curve was recorded over 20–50 min while the temperature was kept constant with an accuracy of about 0.05–0.1 K. The sample was graduated. The free-electron mobility was found to depend on temperature and electron concentration rather weakly, so the concentration could be determined at every point of the $R_{XX}(t)$ curves. The relaxation was studied in the temperature range 60–120 K. Practically at every temperature a large interval of n_e could not be covered as the relaxation became too slow. Sometimes after 30–40 minutes' recording at given T the sample was heated a bit to accelerate the relaxation and then cooled down to the same temperature. This allowed us to cover a wider range of free-electron concentrations at a given temperature.

0268-1242/91/100B62+04 $03.50 © 1991 IOP Publishing Ltd

To obtain good Arrhenius plots a small step in T (about 1–1.5 K) was used.

Our experiments were held at hydrostatic pressure $P = 4$ kbar and at $P = 1$ bar. At a given pressure there is a certain range of electron concentrations and therefore of Fermi energies available for relaxation studies. Limits of this range are related to the mutual positions of donor levels and the Γ valley bottom. The lower limit corresponds to the position of the DX level, and the upper one to that of the donor level related to the X valley. Pressure modifies the positions of the levels, so it effectively changes the available Fermi energy range.

3. Results and discussion

The measured curves $R_{XX}(t)$ were transformed into $n_e(t)$ curves, then smoothed and differentiated. At a given temperature the Fermi energy was calculated for every value of n_e. This enabled us to plot the dependences found as $\ln(dn_e/dt)$ versus E_F. Examples of such dependences are shown in figure 1 for $P = 1$ bar and figure 2 for $P = 4$ kbar. For both pressures there is a wide range of Fermi energies and temperatures where the plotted dependences are linear. This implies $dn_e/dt \propto \exp(\alpha E_F/kT)$. We relate this fact to the domination of a single relaxation process.

Figure 1 shows other important features of the dependences. Firstly, 10–15% of free electrons relax at a constant rate at the beginning of the relaxation. The origin of this phenomenon remains unclear. It could be related partly to the X donor level that pins the Fermi energy of free electrons, but this explanation seems to be invalid at lower Fermi energies.

Secondly, at lower temperatures (when we deal with high E_F values) another relaxation process is observed. It is more pronounced at $P = 1$ bar but can be clearly seen at $P = 4$ kbar too. The coexistence of two relaxation processes results in the superlinear $\ln(dn_e/dt)$ versus E_F dependence. The rate of the second process is related to the Fermi energy more strongly than that for the first one. For the highest values of E_F at a given pressure this process affects decisively the relaxation picture, but at lower Fermi energies it becomes negligible. Practically it is observed in the case when the relaxation is slow from the very beginning. This second relaxation process can correspond, for example, to the recombination with the spatially separated charges.

Finally, the relaxation rate decreases quickly at the lowest values of E_F at a given P. This is obviously related to the activation of electron emission from the DX centres.

Let us consider the linear parts of the dependences above. The well known model proposed by Mooney et al [4] implies that electron capture on the DX centre occurs through the intermediate state. The occupancy of this state results from the thermal quasi-equilibrium of the electron system excluding DX centres. If we neglect the broadening of the capture barrier, the rate equation has the simplest form

$$dn_e/dt = \kappa n_{DX}^+ \exp[-(E_B - \alpha E_F)/kT] \qquad (1)$$

where κ is the rate constant, n_{DX}^+ is the number of ionized DX centres and E_B is the capture barrier height relative to the bottom of the conduction band. As the effective barrier $E_B^* = E_B - \alpha E_F$ is much greater than kT, Boltzmann statistics is valid. The coefficient α appears when we take the charge state of the DX centre into account. It was pointed out recently [5] that $\alpha = 2$ for the negatively charged DX centre model with two electrons to be trapped by one donor; for the neutral DX centre model $\alpha = 1$. Equation (1) implies that $\ln(dn_e/dt)$ (or $\ln(n_e^{-1} dn_e/dt)$ more precisely) is linear in E_F. The relaxation is strongly non-exponential but E_B^* has a quite certain value for a given Fermi energy. Therefore the relaxation rate must have a thermoactivated character.

The broadening of the relaxation barrier height (for example, due to fluctuations of the Al content) makes the rate equation more complicated. For equation (1) the broadening can be introduced by the distribution of the density of the ionized DX centres states in energy [4]. This distribution evolves inhomogeneously in time, as the thermoactivated relaxation strongly selects the lowest barriers. Practically it can be presented by the effective dependence of barrier height E_B on Fermi energy. For each Fermi energy there is a certain value of the effective barrier height; it depends on temperature (at given E_F) but rather weakly for the temperatures of our

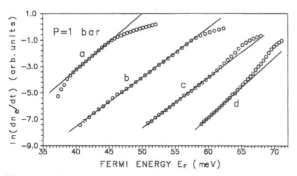

Figure 1. Examples of the dependence of the logarithm of the electron relaxation rate, $\ln(dn_e/dt)$, on the Fermi energy, E_F, at pressure $P = 1$ bar. The temperatures are: a, 99.8 K; b, 88.1 K; c, 79.6 K d, 73.3 K.

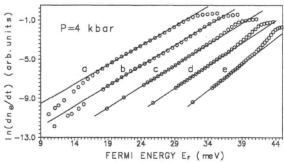

Figure 2. The same as figure 1 at the pressure $P = 4$ kbar. The temperatures are: a, 84.2 K; b, 79.6 K; c, 75.4 K; d, 71.1 K; e, 66.6 K.

experiment. So the thermoactivated character of the relaxation rate must persist.

In our limits of temperature and other parameters the change in the effective barrier height is close to linear in E_F. So the equation (1) remains valid but with a slightly larger α. By the order of magnitude the effective increase in α is the barrier distribution width over the electron Fermi energy when all the donors are ionized.

Within the linear parts of the experimental curves in figure 1, their slopes are α/kT in terms of equation (1). The values of α found are shown in figure 3. In a certain temperature range the values of α vary from 2.8 to 3.1 both at $P = 1$ bar and at $P = 4$ kbar. The increased α at lower temperatures means that two relaxation processes described above are not properly resolved (especially at $P = 1$ bar). Similarly at higher T the emission from the DX centres affects the slope of the experimental curve. So we must suppose that the experimental curves do really correspond to the single relaxation process in the narrower range of temperatures with α being approximately constant.

The values of α found are rather large. The increase in α due to the barrier broadening is taken as not greater than 0.6 for the neutral DX centre model; if the DX centres are negatively charged, the increase in α is halved. The question is whether the slope of $\ln(dn_e/dt)$ versus E_F or of $\ln(n_e^{-1} dn_e/dt)$ versus E_F is to be taken as α/kT (corresponding to the negatively charged or neutral DX centre model respectively). In the latter case the curves in figure 1 do not change qualitatively, although α becomes smaller: from 2.5 to 2.8. Again these values of α appear to be much greater than would be expected for the latter model.

For determining the barrier energy E_B^* we have constructed the Arrhenius plots for the relaxation rate dn_e/dt at fixed Fermi energies. Figure 4 shows that the plots of $\ln(dn_e/dt)$ at constant E_F are excellently linear in reciprocal temperature within the wide range of dn_e/dt. This confirms the activated character of the relaxation process and the validity of the energy barrier concept.

The values of E_B^* obtained from such plots are $E_B^* = E_B - \alpha E_F$ in terms of equation (1). Figure 5 shows the determined dependence E_B^* versus E_F for both pressures.

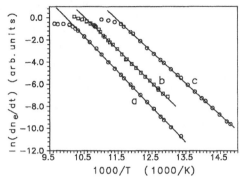

Figure 4. Examples of the Arrhenius plots for the relaxation rate at fixed Fermi energy: curve b ($\square$), $P = 1$ bar, $E_F = 49$ meV; curves a and c ($\bigcirc$), $P = 4$ kbar, $E_F = 18$ meV and $E_F = 30$ meV respectively.

The dependence is linear and rather strong, the slope $-dE_B/dE_F$ equals 3.2 ± 0.1 for $P = 1$ bar and 2.5 ± 0.1 for $P = 4$ kbar. It corresponds qualitatively to the values of α discussed above. The value $-dE_B/dE_F = 2.5$ might be explained within the model of negatively charged DX centres, but the value $-dE_B/dE_F = 3.2$ is too large even for this case. Our data support this model better than the neutral DX centres one, but really the relaxation processes seem not to be a reliable foundation for definition of the DX centre charge state, as many other things can affect the relaxation.

Figure 5 shows that at the same E_F (in our case at $E_F \simeq 40$ meV) the barrier energy depends rather strongly

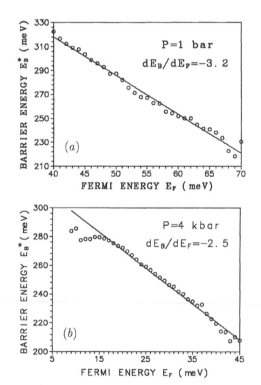

Figure 5. The dependence of determined barrier energy, E_B^*, on electron Fermi energy, E_F, at pressures $P = 1$ bar (a) and $P = 4$ kbar (b).

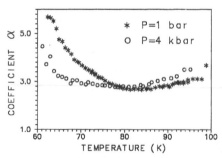

Figure 3. The dependence of the coefficient α on electron Fermi energy. α is determined as $kT \ln(dn_e/dt)$ being the slope of the experimental curves in figures 1 and 2 over $1/kT$. $*$, $P = 1$ bar; $\bigcirc$, $P = 4$ kbar.

on pressure too. $\Delta E_B/\Delta P$ is more than 20 meV kbar^{-1}, i.e. two times greater than the shift of the Γ conduction band with P. It agrees with the recent transport measurements of [5], but contradicts the DLTS data of [14, 15], where the effect of Fermi energy on the effective barrier height seems not to be taken into account. Our result correlates also with the reported dependence of E_B on Al concentration, x_{Al}, being much greater than dE_Γ/dx_{Al} [4, 16].

4. Conclusion

In conclusion, we have carried out a direct investigation of PPC relaxation at pressures 1 bar and 4 kbar. In some ranges of temperature and concentration the logarithm of the relaxation rate is linear in Fermi energy with the slope of about $3/kT$. At fixed Fermi energy the relaxation rate exhibits a thermoactivated character, being linear in the reciprocal temperature. The effective barrier energy decreases linearly with the Fermi energy. The reduction is found to be stronger than would be expected in terms of both the neutral and negatively charged DX centre models. Our data give more support to the latter model.

References

[1] Lang D V and Logan R A 1977 *Phys. Rev. Lett.* **39** 635
[2] Lang D V, Logan R A and Jaros M 1979 *Phys. Rev. B* **19** 1015
[3] Caswell N S, Mooney P M, Wright S L and Solomon P M 1986 *Appl. Phys. Lett.* **48** 1093
[4] Mooney P M, Caswell N S and Wright S L 1987 *J. Appl. Phys.* **62** 4786
[5] Mosser V, Contreras S, Piotrzkowski R, Lorenzini Ph, Robert J L, Rochette J F and Marty A 1991 *Semicond. Sci. Technol.* **6** 505
[6] Izpura I and Muñoz E 1989 *Appl. Phys. Lett.* **55** 1732
[7] Zukotynsky S, Ng P C N and Pindor A J 1987 *Phys. Rev. Lett.* **59** 2810
[8] Queisser H J and Theodorou D E 1986 *Phys. Rev. B* **33** 4027
[9] He L X, Martin K P and Higgins R J 1989 *Phys. Rev. B* **39** 13276
[10] Lin J Y, Dissanayake A, Brown G and Jiang H X 1990 *Phys. Rev. B* **42** 5855
[11] Bourgoin J C, Feng S L and von Bardeleben H J 1988 *Appl. Phys. Lett.* **53** 1841
[12] Dobson T W, Scalvi L V A and Wager J F 1990 *J. Appl. Phys.* **68** 601
[13] Itskevich E S 1963 *Sov. Phys.–Prib. Tekh. Eksp.* no 4 148
[14] Li M F, Shan W, Yu P Y, Hansen W L, Weber E R and Bauser E 1988 *Appl. Phys. Lett.* **53** 1195
[15] Shan W, Yu P Y, Li M F, Hansen W L and Bauser E 1989 *Phys. Rev. B* **40** 7831
[16] Mooney P M, Calleja E, Wright S L and Heiblum M 1986 *Mater. Sci. Forum* **10-12** 417

Semicond. Sci. Technol. **6** (1991) B66–B69. Printed in the UK

X-ray diffractometer as a tool for examining lattice relaxation phenomena

M Leszczynski†, G Kowalski‡, M Kaminska‡, T Suski† and
E R Weber§

† High Pressure Research Center, UNIPRESS, Polish Academy of Sciences, ul.
Sokolowska 29/37, 01–142 Warszawa, Poland
‡ Institute of Experimental Physics, Warsaw University ul. Hoza 69, 00–681
Waszawa, Poland
§ Department of Materials Science and Materials Engineering, Center for Advanced
Materials, Lawrence Berkeley Laboratory, University of California, Berkeley, CA
94720, USA

Abstract. The use of an x-ray diffractometer is proposed for examining the lattice
relaxation effects accompanying the transfer of DX centres and EL2 defects
between their stable and metastable states. The comparison of experimental
results for GaAlAs:Te (with DX centres), semi-insulating GaAs and
low-temperature GaAs (with EL2 defects) is given.

1. Introduction

Despite the great amount of experimental and theoretical
work done on properties of DX and EL2 centres their
microscopic models are still controversial. One of the
most intriguing features of these two technologically
important defects is the existance of their stable and
metastable configurations. In the case of DX centres,
deep levels associated with donors in III–V compounds,
the major problem is the 'amount' of lattice relaxation
occurring when an electron (two electrons [1]) is localized on the centre. The experimental evidence and
theoretical work either indicate large lattice relaxation
(LLR) [1–4] or small lattice relaxation (SLR) [5, 6].

For EL2 defects the most controversial problem is
whether they are formed by an isolated arsenic antisite
As_{Ga} [7] or by As_{Ga} coupled with another defect [8]. The
transition between the stable and metastable state of the
EL2 defect (EL2 $\leftrightarrow$ EL2*) occurs without charge transfer
to the conduction band and is believed to be accompanied by LLR. However, there is no direct experimental
result characterizing this lattice perturbation.

The tool that is commonly used for examining lattice
strains is an x-ray diffractometer. An instrument specially
designed for low-temperature measurements enabled the
first observations to be obtained of strain creation and
annihilation by changing temperature and illumination
for GaAlAs:Te [9] and semi-insulating (SI) GaAs [10].

The aim of this paper is to compare effects caused by
DX centres, EL2 defects and EL2-like defects in low-
temperature (LT) GaAs. The latter material was obtained
by MBE at a very low (180–200 °C) growth temperature,

which enabled a 1 % excess of arsenic to be obtained. The
optical absorption measurements gave a two to three
order higher concentration of EL2-like defects in comparison with SI GaAs. The majority (but not all) of the
properties of these defects are similar to those of
EL2 [11].

2. Experiment

The rocking curves were obtained with a double-crystal
diffractometer set in $(+, -)$ non-dispersive mode. For
high-quality GaAs crystals the apparatus enables one to
obtain (004) Cu K$_{\alpha 1}$ reflection peaks having the full
width at half maximum (FWHM) about 10″ (close to the
theoretical value).

The accuracy of FWHM establishment was 1″ and any
broadening of peaks by more than 2″ could be related to
a presence of internal strains, either as-grown or induced
by temperature–illumination processes applied in this
work. The measurements were performed with a continu-
ous-flow cryostat at 295 K and 77 K. The temperature
stability was about 0.1 K.

All measurements were repeated several times and for
many samples of every kind of material examined.
Figures show typical changes of x-ray rocking curves
during the experiments. The values of FWHM inserted in
the figures are the average values over several measurements.

For GaAlAs:Te samples the Hall concentration of
electrons was measured in order to determine the
number of electrons trapped by DX centres during the

0268-1242/91/100B66 + 04 $03.50 © 1991 IOP Publishing Ltd

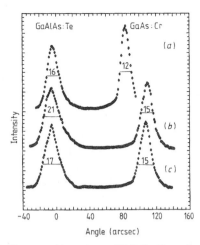

Figure 1. X-ray rocking curves (004) Cu K_{α_1} reflection of $Ga_{0.75}Al_{0.25}As$:Te epilayer: (a) 295 K, (b) 77 K after cooling in the dark and (c) 77 K after cooling in the dark and 4 min illumination with white light. The FWHM are the average values over several measurements.

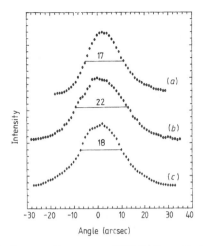

Figure 2. X-ray rocking curves (004) Cu K_{α_1} reflection of SI GaAs: (a) 295 K, (b) 77 K after cooling in the dark and (c) 77 K after cooling in the dark and 4 min illumination with white light. The FWHM are the average values over several measurements.

illumination–cooling processes. Moreover, it was checked that the x-ray beam used for measurements did not influence the electron concentration.

2.1. GaAlAs:Te

Two kinds of LPE grown GaAlAs:Te ($n \simeq 10^{18}$ cm^{-3}) layers grown on semi-insulating GaAs:Cr were examined.

For $x = 0.15$ samples the DX level [12] is resonant with the conduction band and no changes either of free-electron concentration or of FWHM were observed in experiments in which the samples were cooled down to 77 K in the dark or under illumination.

For $x = 0.25$ samples the DX level is below the Fermi level and cooling in the dark induced the trapping of about 1×10^{17} cm^{-3} electrons by DX centres (which was observed by a reduction in the Hall concentration of electrons). This was accompanied by an increase of FWHM by about 5″ (figure 1). Illumination with white light, causing a persistent photoconductivity (PPC) effect, did recover the value of FWHM to its small, room-temperature value. When the samples were cooled in light (DX centre empty) no changes of FWHM were observed.

2.2. Semi-insulating GaAs

Several samples of semi-insulating, undoped GaAs containing about 2×10^{16} cm^{-3} EL2 defects were investigated. The pattern of FWHM changes was similar to that of $Ga_{0.75}Al_{0.25}As$:Te.

Cooling samples down to 77 K in the dark caused an increment of FWHM by about 5″. After 2–4 min illumination with white light (EL2 supposed to be transferred to the metastable configuration) the FWHM decreased to the room-temperature value (figure 2).

It is reasonable to deduce that the metastable configuration induces less perturbation of the surrounding lattice, or at least that this perturbation becomes more localized.

Cooling of the SI GaAs samples in the light or for n-GaAs samples of $n \simeq 10^{18}$ cm^{-3} brought about no distinct changes of FWHM.

In figure 3 the temperature 'recovery' from the metastable state is shown. After transferring the defects into the metastable configuration (cooling in dark plus illumination at 77 K) the samples were heated up to a certain temperature, then cooled down to 77 K and examined (being kept in the dark all the time). The recovery to having a broad peak occurred at about 110 K, which agrees well with other experiments showing the transfer from the metastable to the stable state (e.g. absorption measurements [13]).

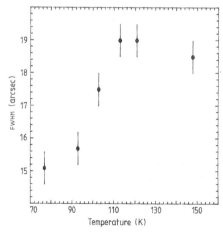

Figure 3. Thermal recovery of the diffraction peak of SI GaAs sample (one other than in figure 2).

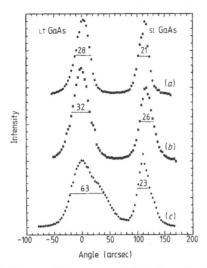

Figure 4. X-ray rocking curves (004) Cu K$_{\alpha 1}$ reflection of LT GaAs epilayers on a SI GaAs substrate: (*a*) 295 K, (*b*) 77 K after cooling in the dark and (*c*) 77 K after cooling in the dark and 4 min illumination with white light. The FWHM are the average values over several measurements.

2.3. Low-temperature GaAs

The epilayers 1–3 μm thick, grown at 180–200 °C by MBE, possessing 1 % arsenic excess, were investigated. The concentration of EL2-like defects was 10^{19}–10^{20} cm, as established by absorption measurements.

In figure 4 the results for one of the samples are shown. Only slight (taking into account very high EL2 concentration) broadening of peaks was observed after cooling in the dark. Illumination with white light caused the peak to broaden towards the lower values of lattice constants. It can be seen that a part of defects causing epilayer lattice expansion is not sensitive to the white light radiation.

Thermal recovery from the metastable state occurred at about 120 K, which is in agreement with absorption measurements [14].

When cooling in light no significant changes of FWHM were observed, whereas the absorption measurements showed that only half of the EL2 defects may be transferred to the metastable configuration.

3. Conclusions

The presented results prove that an x-ray diffractometer can be applied to examination of the lattice relaxation effects accompanying the transfer between the stable and metastable configurations of DX and EL2 centres.

Supposing the simplest model of an elastic matrix with uniformly spaced defects (for GaAlAs:Te one DX centre per about 40 unit cells in line, for SI GaAs one EL2 defect per about 50 and for LT GaAs one EL2 defect per about 10 unit cells in line) observed changes of FWHM can be caused by the increment or diminution of unit cells

containing defects by 1–5 %. This means that the changes of FWHM can be related qualitatively to LLR. More accurate estimations need more information about the 'real structure' of the materials examined.

The results for GaAlAs:Te seem to be understood. The changes of FWHM follow the transitions between stable and metastable configurations of the DX centre. This would mean the confirmation of LLR models proposed to explain other observations.

The results of investigations of EL2 metastability are not quite so clear. The following experimental facts indicate that the observed changes of lattice strains can be connected with EL2 defects:

(i) there are no changes of FWHM in *n*-GaAs samples containing no EL2 defects ($n \geq 10^{18}$ cm^{-3});

(ii) the illumination after cooling in the dark changes the FWHM;

(iii) the changes are persistent at 77 K;

(iv) the temperature of thermal recovery is 110–120 K, the same as observed in absorption measurements.

But the other observations are not explicable yet. These are as follows:

(i) cooling SI GaAs samples in the dark (EL2 in its stable configuration) is accompanied by FWHM increase. No former experiment can be connected with this observation;

(ii) why, if (i) is related to EL2, cooling LT GaAs in the dark causes only a small change of FWHM.

The differences between properties of EL2 defects in SI GaAs and LT GaAs can arise from as-grown strains in LT GaAs (but it is still not clear [11] what kinds of defect cause such a large lattice expansion in as-grown LT GaAs layers) or from an interaction between closely spaced El2 and other defects. The occurrence of such differences has been confirmed recently by absorption measurements [14].

Acknowledgment

The authors are indebted to Drs F W Smith and A R Calawa of MIT for kindly supplying LT GaAs samples.

References

[1] Chadi D J and Chang K J 1989 *Phys. Rev.* B **39** 10063
[2] Dabrowski J, Scheffler M and Strehler R 1990 *Proc. 20th Int. Conf. on the Physics of Semiconductors* ed E M Anastassakis and J D Joannopoulos (Singapore: World Scientific) p 489
[3] Mooney P M, Northrop G A, Morgan T N and Grimmeiss H G 1988 *Phys. Rev.* B **37** 8298
[4] Morgan T N 1989 *Defects in Semiconductors* ed G Ferenczi (Switzerland: Trans. Tech.) p 1079
[5] Bourgoin J C, Feng S L and von Bardeleben H J 1989 *Phys. Rev.* B **40** 7663;

Yamaguchi E, Shiraishi K and Ohno T *Proc. 20th Int. Conf. on the Physics of Semiconductors* ed E M Anastassakis and J D Joannopoulos (Singapore: World Scientifc) p 501

[6] Kitano T and Mizuta M 1987 *Japan. J. Appl. Phys.* **26** L1806

[7] Kaminska M, Skowronski M and Kuszko W 1985 *Phys. Rev. Lett.* **55** 2204

[8] Bourgoin J C, von Bardeleben H J and Stievenard D J 1988 *J. Appl. Phys.* **64** R65

[9] Leszczynski M, Suski T and Kowalski G 1991 *Semicond. Sci. Technol.* **6** 59

[10] Kowalski G and Leszczynski M unpublished

[11] Smith F W 1990 *PhD Thesis* MIT

[12] As it has been demonstrated (Wisniewski P *et al* 1991 *Semicond. Sci. Technol.* **6** B146) the DX centre exhibits a multicomponent structure in GaAlAs alloys. In most experiments an effective DX level is observed.

[13] Vincent G, Bois D and Chantre A 1982 *J. Appl. Phys.* **53** 3643

[14] Kaminska M *et al* unpublished

Semicond. Sci. Technol. **6** (1991) B70–B77. Printed in the UK

Bistability, local symmetries and charge states of Sn-related donors in Al$_x$Ga$_{1-x}$As and GaAs under pressure studied by Mössbauer spectroscopy

D L Williamson† and Pierre Gibart‡

† Department of Physics, Colorado School of Mines, Golden, CO 80401, USA
‡ Laboratoire de Physique du Solide et Energie Solaire, CNRS, Sophia Antipolis, F–06560 Valbonne, France

Abstract. The bistable character of Sn donors Al$_x$Ga$_{1-x}$As for $x > 0.2$ or in GaAs under pressure > 2.4 GPa has been studied by ^{119}Sn Mössbauer spectroscopy (MS). The shallow Sn donor state and the deep Sn DX state are observed to exist simultaneously and are readily distinguishable due to significantly different electronic configurations as manifested in their isomer shifts. An upper limit on the amount of non-cubic local lattice distortion at the Sn DX centre has been obtained through establishment of an upper limit on the local electric field gradient as determined from the quadrupole interaction. The MS data from GaAs under pressure, coupled with magnetotransport data, provide strong evidence that the Sn DX centre localizes at least two electrons in its ground state.

1. Introduction

Substitutional donors in Al$_x$Ga$_{1-x}$As with $x > 0.2$ induce deep levels with very unusual properties. These donor-related deep levels exhibit a persistent photoconductivity (PPC) effect at low temperatures and they show a large Stokes shift between optical and thermal ionization energies. There are basically two classes of models to account for the occurrence of deep states in n-type Al$_x$Ga$_{1-x}$As. The first, which seems appropriate for substitutional donors, would argue that DX levels are effective-mass states (EMS) associated with a higher conduction band (L or X) and which are deepened due to intervalley mixing and central cell corrections [1]. This description involves no or small lattice relaxation (SLR). Supporting this EMS model, four donor levels, D_1 to D_4, have been predicted by theory and have been observed in photoluminescence spectra of Si-doped Al$_x$Ga$_{1-x}$As [2–4]. However, features like PPC cannot be explained by the EMS model. The DX centre exhibits a bistable nature: deep for $x > 0.2$ or in GaAs for pressures > 2.4 GPa, and shallow otherwise; for $x < 0.2$ or in GaAs for pressures < 2.4 GPa the DX state becomes resonant in the Γ conduction band, while the ground state of the substitutional donor is shallow.

In the second class of models, the deep-state behaviour of donors in Al$_x$Ga$_{1-x}$As is understood in the limit of strongly localized states. These models imply most often

a large lattice relaxation (LLR) [5] but possibly an SLR [6]. In addition to well known features (PPC, large Stokes shift) which support a localized state description of donors, recent results unambiguously show that the behaviour of DX centres depends on the local symmetry [7]. Furthermore, a chemical shift due to the nature of the donor was shown in Te-doped GaAs [8, 9] and GaAs$_{1-x}$P$_x$ [10].

In the localized state description, a qualitative understanding is provided by a bonding–antibonding scheme [11]. A substitutional donor D in the Ga site builds four bonding and four antibonding states corresponding to the D–As bonds, and the interaction between these states leads to a splitting into a A_1 component and a triply degenerate T_2 component. Several theoretical calculations of this type have been made and discussed in detail [11]. One of the most recent models was proposed by Chadi and Chang [12] who claim that the DX centre is, in its ground state, a negatively charged donor. In an extension of calculations done for the As$_{Ga}$ antisite in GaAs, the authors calculate the total energy of Si-doped GaAs in an 18-atom supercell of GaAs. Their conclusion is that the most stable state of the Si is a negatively charged donor, D$^-$, with a broken bond configuration. The distortion consists of an off-centre displacement of the group IV donor along a $\langle 111 \rangle$ direction. The group IV atom then has three covalent bonds with three As neighbours and two non-bonding electrons. In addition there are two electrons on the dangling bond of the As

0268-1242/91/100B70 + 08 $03.50 © 1991 IOP Publishing Ltd

atoms corresponding to the broken Si–As bond. This model is clearly of the LLR type. The ground state being D^-, two donor dopant atoms are required to form a single DX centre and a single shallow ionized donor state, d^+ according to

$$2D^0 \rightarrow d^+ + D^-. \qquad (1)$$

The single-electron neutral state, D^0, is unstable and thus may be observed only under non-equilibrium conditions such as photoexcitation.

Recently, Yamaguchi et al [13, 14] determined the most stable lattice configuration of the DX centre from a first-principles method. Systematic calculations of total energy and Hellman–Feynman forces were done in 64-atom supercells as a function of atomic displacements. Distortions of T_d, D_{2d} (Oshiyama and Ohnishi's proposal [15]), and C_{3v} (Chadi and Chang's model [12]) symmetries were considered and it was concluded that a slight T_d distortion yields the most stable configuration. D_{2d} and C_{3v} were found to be either unstable or metastable. Moreover, the energy of the DX centre, the A_1 state, strongly depends upon the local environment and the nature of the substitutional donor. When calculating the probability of an optical transition from the A_1 state to the conduction bands, it was found that the transition from A_1 to T_2 is dominant compared with transitions from A_1 to any extended state because the A_1 and T_2 states are strongly localized on the same donor atom. The large optical threshold was therefore proposed to be the $A_1 \rightarrow T_2$ transition.

Although the original work of Lang et al [5] proposed donors that formed DX centres by complexing with some unknown native defect, most experiments have since been interpreted to show that the substitutional donor itself is actually the DX centre. However, the assumption of an 'X' complexing with the donors continues to receive support. Van Vechten [16] argued that the LLR which is associated with the PPC is nearest-neighbour hopping according to

$$D^+V_{As} + 3e^- \rightarrow D^+Ga(\text{or } Al)^{2-}_{As} V^-_{Ga}. \qquad (2)$$

Hence, according to this model, the DX defect complex is the metal-atom-on-As-site antisite plus a Ga vacancy. The high-pressure appearance of DX centres in GaAs is attributed to adequate concentrations of vacancies present in doped GaAs [17].

Currently, the experimental status regarding the large and small lattice relaxation models for DX centres remains controversial but there is considerable very recent evidence that the charge state is of the two-electron, negative-U type. We summarize these recent experiments in the next section before detailing our Mössbauer experiments.

2. Experimental determination of the atomic-scale nature and charge state of DX centres

Among the few experimental techniques able to explore the local structure of defects at the atomic scale, extended x-ray absorption fine structure (EXAFS) measurements can provide a direct determination of the bond lengths around a donor in $Al_xGa_{1-x}As$. In highly doped (Se, Sn) $Al_xGa_{1-x}As$, two EXAFS studies did not detect the existence of LLR [18, 19]. However, Rowe et al [20] showed that in S-implanted $Al_xGa_{1-x}As$, Al preferentially clusters around the donor first-neighbour shell for $x = 0.20$–0.45 and is depleted in the second shell. Note that in the Chadi and Chang D^- model, only one fourth of the D–As bonds are expected to be greater than the mean D–As value and a maximum of 50% of the donors will exist as DX centres. Thus, only one eighth of the D–As distances are expected to be modified in the D^- state. In Van Vechten's model a maximum of 33% of the donors could exist as DX centres. Thus, detection of DX states by EXAFS, already a difficult task at 10^{18} to 10^{19} cm^{-3} doping levels, may require more precise experiments.

Very recent characterization of the substitutionality of Te and Sn donors in $Al_xGa_{1-x}As$ by particle-induced x-ray emission and ion-beam channelling methods found no off-centre displacement larger than 0.014 nm [21].

Assuming a neutral, single-electron DX ground state, a paramagnetic susceptibilty should exist. On the other hand, a two-electron ground state is expected to be diamagnetic. Magnetic susceptibility measurements have yielded contradictory results [22, 23]. Other paramagnetic impurities or defects from the substrate or even the epitaxial layer could give parasitic paramagnetic signal.

In principle, electron paramagnetic resonance (EPR) is a powerful method for investigation of the electronic and atomic configurations of paramagnetic defects. The measured parameters include the g tensor, the hyperfine interaction tensor and the fine-structure terms, which lead to an extremely detailed picture of the electronic and atomic scale structure of the defect as well as the spin concentration. EPR has been applied recently to DX centres in $Al_xGa_{1-x}As$, and in all reported investigations [24–26] the only detectable EPR signal came from photoexcited DX centres. Very recently, two experiments provided new insight as to the nature of these photoexcited states [27, 28]. With thick Sn-doped $Al_xGa_{1-x}As$ an EPR spectrum with a resolved hyperfine (HF) doublet related to the ^{117}Sn and ^{119}Sn isotopes was observed [27]. This proves that the photoionized DX centre is paramagnetic, and it was found that the Sn HF interaction of this paramagnetic state corresponds to a spin density of about 13% relative to an Sn 5s orbital. Such a localization is typical of a deep defect. Consequently the photoionization involves an intracentre transition which was interpreted as the $D^- \rightarrow D^0 + e^-$ transition. The HF interaction was further examined with a ^{119}Sn isotopically enriched Sn-doped $Al_{0.3}Ga_{0.7}As$ sample [28]. From the HF constant a 23% localization of the 5s Sn orbital was deduced, in reasonable agreement with the above experiment. The intracenter transition was proposed to be the $A_1 \rightarrow T_2$ transition. Both crucial experiments [27, 28] showed unambiguously that the photoexcited state of the DX centre is paramagnetic. Thus, on the basis of EPR experiments three charge states of the donors are suggested: (i) D^-, a two-electron, diamagnetic ground state;

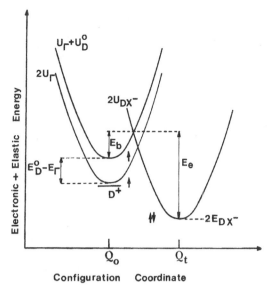

Figure 1. Configuration coordinate diagram for the negatively charged two-electron state DX centre in $Al_xGa_{1-x}As$ or GaAs under pressure. The configuration coordinate Q represents the change in the configuration around a donor in the ground state and after photoionization. E_e is the activation energy for emission and E_b the activation energy for capture from the one-electron excited state. E_{DX^-} is the average energy per electron of the DX^- state. The dominant capture process is assumed to occur via the excited one-electron state D^0 (here assumed to have no LLR) according to $2U_\Gamma \rightarrow U_\Gamma + U_{DX^0} \rightarrow 2U_{DX^-}$. Hence the total capture energy E_c is the sum of E_b and the activation energy needed to populate the excited state, $E_{D^0} - E_\Gamma$.

(ii) photoexcited D^0, a paramagnetic, metastable state;
(iii) d^+, the shallow, ionized, diamagnetic, donor state.
Figure 1 suggests these three states in the form of a configuration coordinate diagram.

In the model proposed by Yamaguchi et al [13, 14], the $A_1(ab)$ ground state is expected to be paramagnetic. The photoexcited paramagnetic state is assumed to be the $T_2(ab)$ state which subsequently yields the shallow d^+ donor state together with an electron in the Γ band.

Finally we mention very recent results from local vibrational mode spectroscopy measurements of Si-doped GaAs at high pressure [29]. The DX centres were observed to have a slightly lower Si vibrational frequency and the relative intensity of the shallow donor and DX centre peaks was consistent with two electrons bound to the Si DX centre.

Mössbauer spectroscopy is another technique that offers a means of determining electronic and atomic-scale structure of defects in semiconductors. However, because of the relatively low concentrations of donors, such experiments are difficult and time-consuming. Spectra that are only partially resolved must be interpreted with care and supported by systematic and complementary studies. Below we summarize results obtained during the last few years by ^{119}Sn Mössbauer spectroscopy applied to the determination of local symmetries, electronic con-

figurations and charge states of Sn-related donors in $Al_xGa_{1-x}As$.

3. Mössbauer spectroscopy of defects in semiconductors

Over 40 of the elements in the periodic table have been demonstrated to exhibit the nuclear gamma-ray resonance phenomenon known as the Mössbauer effect. Of these, approximately ten or so are practical as spectroscopic tools for the study of solids. A defect in a semiconductor may be investigated by Mössbauer spectroscopy provided one of these Mössbauer-active atoms is an intimate part of that defect. Changes in the environment of the Mössbauer atom beyond the second-neighbour coordination sphere are not usually detectable. This is due to the nature of the hyperfine interactions that produce the energy shifts and splittings in the Mössbauer resonance: the total electronic wavefunction together with the electromagnetic fields *at the nucleus of the Mössbauer-active atom* determines these interactions. There are three hyperfine interactions accessible through the Mössbauer effect:

(i) the electric monopole interaction, better known as the isomer shift, δ, which is proportional to the total electron density at the nucleus, $\rho(0)$;
(ii) the quadrupole interaction which yields the electric field gradient tensor at the nucleus
(iii) the magnetic dipole interaction, which is proportional to the total magnetic field at the nucleus produced by a net electron spin density and any applied field.

Of the atoms known to exhibit DX centre behaviour in $Al_xGa_{1-x}As$, the elements Ge, Sn and Te have isotopes that are Mössbauer-active, ^{73}Ge, ^{119}Sn and ^{125}Te. Our work has focussed on ^{119}Sn since the ^{73}Ge Mössbauer effect poses extreme experimental difficulties [30] and the ^{125}Te resonance does not offer as a good a resolution as the ^{119}Sn resonance. Further details of Mössbauer spectroscopy applied to defects in semiconductors, including a discussion of the source versus absorber experimental methods, can be found in a recent review article [31].

All of our experiments with ^{119}Sn to date [32–36] have been absorber studies done with GaAs and $Al_xGa_{1-x}As$ epitaxial layers grown by liquid-phase epitaxy (LPE) [32] or metal–organic vapour-phase epitaxy (MOVPE) [33–36] and doped during growth with 72–93% ^{119}Sn-enriched Sn material. The absorber thickness needs to be 50–150 μm thick and the growth of such samples is not a straightforward task. As already discussed elsewhere [37], LPE and molecular beam epitaxy (MBE) cannot provide such thick layers of $Al_xGa_{1-x}As$, and MOVPE is the appropriate choice. In steady-state growth conditions, i.e. all flows kept constant for 10 to 30 hours, the value of x is constant throughout the layer. The Sn metal–organic source, $^{119}Sn(CH_3)_4$, is prepared and purified from a few grams of enriched tin. Steady bubbling of H_2 through a small stainless steel cylinder containing the liquid tetramethyl tin at −49 °C provides the Sn-containing vapour. The concentration of Sn in a 60 μm thick, $x = 0.3$ sample was measured across the thickness of the layer on a carefully cleaved surface by electron microprobe analysis and found to be uniform. Such very long runs require reinforced safety procedures. MOVPE has provided 30–60 μm thick layers for the Möss-bauer studies and up to 100 μm thick samples have been grown for EPR, EXAFS and magneto-optical measurements [19, 25-28]. The critical thickness for perfect epitaxy by elastic accommodation (pseudomorphism) is far beyond these values, and the thick layers appear to be curved with a radius that decreases with increasing layer thickness.

4. Results

4.1. Shallow donor sites in ^{119}Sn-doped $Al_xGa_{1-x}As$

Figure 2 shows Mössbauer absorption spectra obtained at liquid-nitrogen temperature from ^{119}Sn-doped $Al_xGa_{1-x}As$ samples ranging from $x = 0$ to $x = 1$. The Mössbauer resonance of the shallow, substitutional donor site, Sn_{Ga}, has been identified from previous source [38-40] and absorber [32, 33] experiments with GaAs. The isomer shift of Sn_{Ga} is $\delta = 1.8$ mm s^{-1} (relative to $CaSnO_3$) and the quadrupole splitting is $\Delta = 0$ mm s^{-1}, consistent with the T_d symmetry. The absorber experiments on GaAs demonstrated agreement between the free-carrier concentration deduced from Hall measurements and the density of Sn_{Ga} sites up to 5×10^{18} cm^{-3}. For higher impurity concentrations or for annealing under As-rich atmospheres, a compensating species appears, most likely $(Sn_{Ga}-V_{Ga})^-$, which has Mössbauer parameters nearly identical to those of Sn_{Ga}. More recently, evidence for a compensation ratio of $N_A/N_D \simeq 1/6$ has been obtained by comparison of Hall and Shubnikov–de Haas results from Sn-doped GaAs [36]. Although all measurements with Sn-doped GaAs to date have been done at temperatures where the shallow donor is ionized, i.e. Sn_{Ga}^+, a detectable isomer shift is not likely to result for the neutral ground state due to the extended nature of the weakly bound electron wavefunction.

Analysis of the data in figure 2 yields an isomer shift for line 1 of 1.75–1.8 mm s^{-1} and essentially no line broadening consistent with $\Delta \simeq 0$ over the entire range of x [34]. The presence of Al atoms in the next-neighbour shell clearly has little effect on the substitutional Sn site resonance. Note that the shallow donor site, Sn_{III}, exists in significant concentrations at all x, even at values of x where maximum DX formation occurs. This will be discussed in more detail below.

4.2. DX centres in ^{119}Sn-doped $Al_xGa_{1-x}As$

The absorption spectra shown in figure 2 for $x = 0.30$, 0.40 and 0.43 have an enhanced resonance signal on the positive velocity side of the Sn_{III} resonance. This new resonance could be fitted with two additional lines, 2 and 3, as discussed in detail elsewhere [34, 35]. Among the

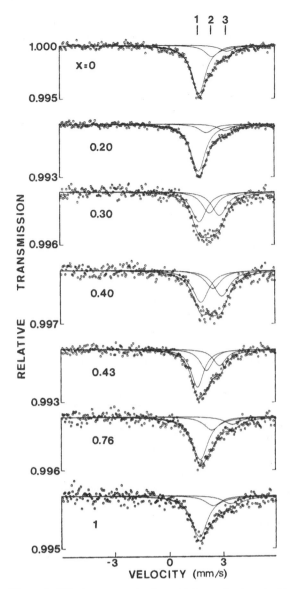

Figure 2. Mössbauer spectra from $Al_xGa_{1-x}As$ doped with ^{119}Sn. The full curves passing through the data are least-squares fits of the three Lorentzian lines indicated (from [34]).

possible interpretations proposed, the one assigning the average velocity position of lines 2 and 3 as the average isomer shift of the Sn DX centre, $\delta = 2.67 \pm 0.05$ mm s^{-1}, has been confirmed in our recent high-pressure experiments [36], which yield $\delta = 2.60 \pm 0.05$ mm s^{-1} for Sn DX centres in GaAs. Although it is tempting to attribute the splitting of lines 2 and 3 to a quadrupole splitting of $\Delta = 0.6$ mm s^{-1}, this cannot be concluded since the lines are not resolved and an alternative intepretation is that a broadening is produced by a distribution of isomer shifts from a family of Sn DX centres due to different (Al, Ga) near-neighbour environments.

An important feature of the data in figure 2 that was not appreciated in our original discussions [34, 35] is the relative population of DX centres compared with shallow donor Sn sites. These populations are directly obtainable from the relative resonance areas of lines 1, 2 and 3 (subtracting the non-active Sn species resonance which, unfortunately, interferes with the DX resonance). These experimental populations can be compared with those expected on the basis of one-electron and two-electron DX ground states as shown in figure 3. The experimental fractions are clearly in better agreement with the model where each Sn DX centre localizes two electrons according to

$$2Sn^0 \rightarrow Sn_{Ga}^+ + Sn_{DX}^-. \tag{3}$$

That is, two doped Sn atoms are needed to form one DX centre (unless electrons are supplied by some other dopants or defects). Relation (3) is identical to the Chadi and Chang model [11], relation (1).

These same data [34] were recently interpreted by Van Vechten to support his antisite DX model that requires localization of three electrons [15]. Two specific problems with the interpretation are the assignment of only line 3 as the DX centre and the absence of a modified Mössbauer resonance for the proposed D^+V_{As} shallow state (relation (2)) as opposed to the above identification of Sn_{Ga}^+ as the shallow state. A nearest-neighbour As vacancy should produce a significant quadrupole splitting and an isomer shift of the Sn_{Ga} resonance. In fact, ion implantation studies have earlier proposed the parameters of the $Sn_{Ga}V_{As}$ site as $\delta = 2.8$ mm s^{-1} and $\Delta = 0.4$ mm s^{-1} [41], quite different from those of Sn_{Ga} cited above.

4.3. DX centres in ^{119}Sn-doped GaAs under pressure

High-pressure Mössbauer experiments have recently been reported [36] that confirm the previous identifica-

tion of Sn DX centres in $Al_xGa_{1-x}As$ [34, 35], as noted above, and provide even stronger support for localization of more than one electron in the Sn DX ground state. A ^{119}Sn-doped layer was grown by MOVPE, characterized by Hall measurements to have a free-carrier concentration of 6×10^{18} cm^{-3} (corrected to 4×10^{18} cm^{-3} on the basis of Shubnikov–de Haas measurement on similarly prepared samples), and characterized by Mössbauer measurements to have a total Sn content of 8×10^{18} cm^{-3}. As shown in figure 4, application of high pressure to this sample produces a new resonance on the positive velocity side of the shallow donor resonance similar to the effect observed in figure 2 upon increasing x. Also, consistent with the behaviour versus x, the shallow donor resonance fraction remains at nearly 50%

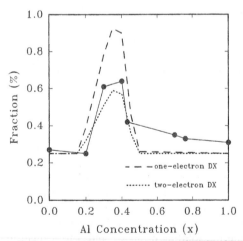

Figure 3. Summed Mössbauer resonance fraction from lines 2 and 3 (full circles and line) which is attributed to Sn DX centres and an approximately constant level of electrically inactive Sn. The two broken curves are calculations based on a multivalley system with a DX activation energy of 100 meV, a concentration of 5×10^{18} electrically active Sn cm^{-3}, a constant level of 25% (1.7×10^{18} Sn cm^{-3}) non-active Sn, and the assumptions of either one- or two- electron localization at the DX centre (experimental data from [34]).

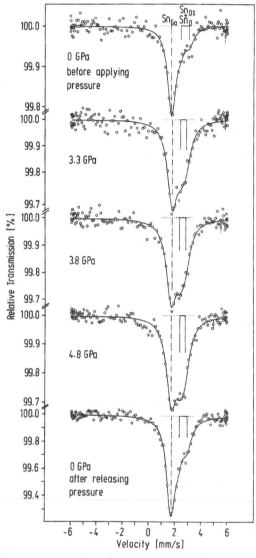

Figure 4. Mössbauer spectra from ^{119}Sn-doped GaAs under pressure. The full curves passing through the data are least-squares fits of a superposition of a single line (Sn_{Ga} site) and a doublet (Sn DX centres + non-active Sn) (from [36]).

even at the highest pressure of 4.8 GPa, where all electrons are expected to be localized on DX centres. Straightforward analysis and careful consideration of experimental uncertainties led to the result that 2.5 ± 0.9 electrons are localized on each Sn DX centre [36].

The fits shown in figure 4 utilized a quadrupole doublet to fit the pressure-induced resonance appearing at 2.6 mm s^{-1}, and yielded a value of $\Delta \simeq 0.4$ mm s^{-1}. However, equally satisfactory fits were obtained with a single line at this same position superimposed upon a doublet due to the electrically inactive Sn [36].

5. Discussion

For ^{119}Sn Mössbauer spectroscopy, changes in $\rho(0)$ are produced mainly by changes in the 5s-like valence electron occupation number n_s, although substantial changes in 5p-like and 5d-like occupation numbers n_p and n_d, can also change $\rho(0)$ because of a shielding effect on the 5 s electrons. In solids the occupation numbers of valence electrons are non-integers and realistic values should be deduced from band structure calculations. The widely accepted structure of Sn (and other group IV elements) in GaAs as sp^3 is far from being correct. Recently Svane and Antoncik [42] calculated $\rho(0)$ in several compounds containing Sn in an effort to establish the proportionality constant between δ and $\rho(0)$. Their calibration will be used as a basis for discussing the electronic structure of Sn in $Al_xGa_{1-x}As$. Figure 5 shows the δ versus $\rho(0)$ correlation for different tin compounds together with the calculated electronic configurations. Since a detailed theoretical investigation of the electronic structure of Sn in $Al_xGa_{1-x}As$ is beyond the scope of the present paper, we use an approximation discussed by Antoncik [43]

$$\rho(0) = 61.0\, n_s - 8.25\, n_s^2 - 3.3\, n_s n_p \qquad (4)$$

which is found to give reasonable agreement between δ and the theoretical fit of Svane and Antoncik [42] (square symbols in figure 5). Equation (4) will be used in the following discussion as an approximate means of estimating the Sn electronic structure. We emphasize that such estimates are only semi-quantitative.

5.1. Electronic structure of shallow Sn donors in $Al_xGa_{1-x}As$

The electronic configuration of Ga and As in GaAs deduced from a tight binding approximation are $5s^{1.48}5p^{1.38}$ and $5s^{1.54}5p^{3.60}$, respectively [44]. When an atom acts as an electrically active impurity in a semiconductor the correct evaluation of its electronic configuration should include local volume compression or expansion effects. Furthermore, the electronic structure of the host semiconductor and its influence on the impurity should be taken into account. These features have been discussed by Antoncik and Gu [45] and the following conclusions were drawn:

(i) In the case of Sn_{Ga} in GaAs it is of very little importance whether this level is occupied or not, the wavefunction associated with a shallow donor is highly delocalized and does not contribute significantly to $\rho(0)$.

(ii) The best agreement with δ for Sn_{Ga} corresponds to Sn donors having three valence electrons and transferring no charge to the neighbouring As atoms.

(iii) The occupancy numbers of Sn_{Ga} in GaAs are $5s^{1.28}5p^{1.72}$ and the corresponding value of δ based on the calibration in figure 5 and equation (4) is plotted in figure 5.

Antoncik and Gu [45] discussed the limits of validity of their evaluations; they argue that the lack of relevant information on the compression effect does not allow one

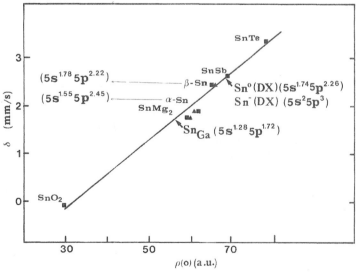

Figure 5. Measured isomer shifts of tin compounds versus calculated electron contact densities, $\rho(0)$. Occupation numbers are given according to Svane and Antoncik [42] or as calculated from a tight-binding approximation [44].

to give a unique value for the extra potential energy term in the Hamiltonian caused by the Sn impurity. A redistribution of the electrons of the Sn atom could also be made under the constraint that the total number of s and p electrons is held constant; in this case the occupation numbers for Sn_{Ga} were found to be $5s^{1.19}5p^{1.67}$ [44] which gives the same value for $\rho(0)$ as the above configuration (according to equation (4)).

5.2. Electronic structure and charge state of Sn DX centre in $Al_xGa_{1-x}As$

As suggested by various atomic-scale DX models, the hybridization of s and p orbitals on the DX centre is quite different from the shallow donor and also different from donors tied to the X valley (like Sn in AlAs). Thus photoionization of a DX centre involves not only the transfer of electrons to the Γ valley, but also a bond reconstruction. Based on the average isomer shift of the Sn DX centre of 2.67 mm s^{-1}, figure 5 and equation (4) yield an electronic configuration of $5s^{1.74}5p^{2.26}$ assuming localization of one electron. However, as discussed in sections 4.2 and 4.3, the Mössbauer data provide strong evidence for the localization of two or three electrons on the DX centre in its ground state. Assuming two-electron localization (relation (3)), the electronic configuration is estimated from figure 5 and equation (4) to be $5s^25p^3$.

5.3. Local symmetry of Sn DX centre

The existence of a non-zero quadrupole splitting at the Sn DX centre site would constitute proof of a local distortion of the substitutional Sn site upon DX centre formation. The size of the quadrupole splitting would help distinguish between an LLR and an SLR model.

Evidence for quadrupole splittings were obtained from both types of Mössbauer experiments: $\Delta = 0.6$ mm s^{-1} from $Al_xGa_{1-x}As$ and 0.4 mm s^{-1} from GaAs at high pressure. However, these values are not decisive since alternative interpretations or fits were possible. The issue of a large symmetry-breaking lattice relaxation therefore remains uncertain. The value of $\Delta = 0.6$ mm s^{-1} represents an upper limit for the amount of local distortion that may be associated with the LLR. Theoretical predictions of Δ for various models of LLR are not yet available.

The issue of a quadrupole splitting versus a distribution of isomer shifts can, in principle, be resolved by application of a large magnetic field [46. 47]. Preliminary results from the application of a 6 T field to the $x = 0.3$ sample suggest that the quadrupole splitting interpretation is more likely [48].

6. Conclusions and comments

Mössbauer spectroscopy has provided new and unique information on the atomic-scale nature of DX centres. All of the Mössbauer data from ^{119}Sn-doped $Al_xGa_{1-x}As$ can best be interpreted in terms of the negative-U DX model of Chadi and Chang [12], which has the primary features of a two-electron DX ground state with a broken bond configuration. The ionized shallow Sn donor and the Sn DX centre are observed to exist simultaneously at liquid-nitrogen temperatures with approximately equal concentrations for $x = 0.3$-0.4 and for $x = 0$ at pressures greater than 2 GPa, consistent with a two-electron DX ground state. The difference in isomer shifts of the two Sn states is consistent with a substantial change in electronic structure associated with the broken bond model, and an estimate of the Sn DX configuration yields $5s^25p^3$. Based on the evidence of a quadrupole splitting at the Sn DX site, the local symmetry is very likely less than cubic, again consistent with a broken bond model that yields a non-cubic distortion of the LLR type.

An intermediate excited paramagnetic state has been detected by EPR [25-28] and may correspond to the unstable, one-electron DX^0 state. The local lattice relaxation of this state is controversial. It would be of great interest to obtain new information on this metastable paramagnetic state and the completely ionized DX state (PPC state) produced via photoexcitation of DX centres. Mössbauer experiments designed to explore these states of the Sn DX centre are currently in progress.

Finally, we note that our absorption experiments require relatively high doping levels. Samples with lower concentrations of Sn are needed to avoid the non-active Sn sites (the resonance of which overlaps the DX resonance) and minimize compensation by shallow acceptors (the resonance of which overlaps the shallow donor site resonance) and therefore to maximize the fraction of DX centres. Such samples might be prepared as radioactive sources if higher specific activity ^{119m}Sn were to become available. Source experiments could lead to spectra with significantly better statistical quality than obtained with our absorbers (figure 2 and figure 4).

Acknowledgments

The authors are grateful to J C Portal for magnetotransport measurements under high pressure and to J Moser for the preliminary high-magnetic-field results. Support from NSF (grant no DMR-8902512) and from cooperative programs CNRS-NSF, DFG-CNRS, and ESPRIT (contract no 3168) is gratefully acknowledged.

References

[1] Bourgoin J C and Mauger A 1988 *Appl. Phys. Lett.* **53** 749
[2] Henning J C M and Ansems J P M 1987 *Semicond. Sci. Technol.* **2** 1
[3] Henning J C M, Ansems J P M and Roksnoer 1988 *Semicond. Sci. Technol.* **3** 361
[4] Gil B, Leroux M, Contour J P and Chaix C 1991 *Phys. Rev. B* **43** 12335
[5] Lang D V, Logan R A and Jaros M 1979 *Phys. Rev. B* **19** 1015

[6] Hjalmarson H P and Drumond T J 1986 *Appl. Phys. Lett.* **48** 656

[7] Baba T, Mizuta M, Fujisawa T, Yoshino J and Kukimoto H 1989 *Japan. J. Appl. Phys.* **28** L891

[8] Suski T, Piotrzkowski R, Wisniewski P, Litwin-Staszewska E and Dmowski L 1989 *Phys. Rev. B* **40** 4012

[9] Sallese J M, Lavielle D, Singleton J, Leycuras A, Grenet J C, Gibart P and Portal J C 1990 *Phys. Status Solidi* a **119** K41

[10] Sallese J M private communication

[11] Lannoo M 1990 *Physics of DX Centres in GaAs Alloys* ed J C Bourgoin (Vaduz: Sci. Tech. Publications) p 209

[12] Chadi D J and Chang K J 1988 *Phys. Rev. Lett.* **61** 873; 1989 *Phys. Rev. B* **39** 10063

[13] Yamaguchi E 1986 *Japan. J. Appl. Phys.* **25** L643

[14] Yamaguchi E, Shiraishi K and Ohno T 1991 unpublished

[15] Oshiyama A and Ohnishi S 1986 *Phys. Rev. B* **33** 4320

[16] Van Vechten J A 1985 *Mater. Res. Soc. Symp. Proc.* **46** 83

[17] Van Vechten J A 1989 *J. Phys.: Condensed Matter* **1** 5171

[18] Kitano T and Mizuta M 1987 *Japan. J. Appl. Phys.* **26** L1806; 1988 *Appl. Phys. Lett.* **52** 126

[19] Hayes T M, Williamson D L, Outzourhit A, Small P, Gibart P and Rudra A 1989 *J. Electron. Mater.* **18** 207

[20] Rowe J E, Sette F, Pearton S J and Poate J M 1990 *Physics of DX Centers in GaAs Alloys* ed J C Bourgoin (Vaduz: Sci. Tech. Publications) p 283

[21] Yu K M, Khachaturyan K, Weber E R, Lee H P and Colas E G 1991 *Phys. Rev. B* **43** 2462

[22] Kachaturyan K A, Awschalom D D, Rozen J R and Weber E R 1989 *Phys. Rev. Lett.* **63** 1311

[23] Katsumoto S, Matsunaga N, Yochida Y, Sugiyama K and Kobayashi S *Japan. J. Appl. Phys.* **29** L1572

[24] Mooney P M, Wilkening W, Kaufmann K and Kuech T F 1989 *Phys. Rev. B* **39** 5554

[25] von Bardeleben H J, Bourgoin J C, Basmaji P and Gibart P 1989 *Phys. Rev. B* **40** 5892

[26] von Bardeleben H J, Zazoui M, Alaya S and Gibart P 1990 *Phys. Rev. B* **42** 1500

[27] Fockele M, Spaeth J M and Gibart P 1990 *Mater. Sci. Forum* **65–66** 443

[28] von Bardeleben H J, Bourgoin J C, Delerue C and Lannoo M 1991 unpublished

[29] Wolk J A, Kruger M B, Heyman J N, Walukiewicz W, Jeanloz R and Haller E E 1991 *Phys. Rev. Lett.* **66** 774

[30] Pfieffer L, Raghavan R S, Lichtenwalner C P and Cullis A G 1975 *Phys. Rev. B* **12** 4793; Pfieffer L 1977 *Nucl. Instrum. Meth.* **140** 57

[31] Williamson D L and Gibart P 1990 *Physics of DX Centers in GaAs Alloys* ed J C Bourgoin (Vaduz: Sci. Tech. Publications) p 163

[32] Williamson D L 1986 *J. Appl. Phys.* **60** 3466

[33] Williamson D L, Gibart P, El Jani B and N'Guessan K 1987 *J. Appl. Phys.* **62** 1739

[34] Gibart P, Williamson D L, El Jani B and Basmaji P 1988 *Phys. Rev. B* **38** 1885

[35] Gibart P, Williamson D L, El Jani B and Basmaji P 1988 *Gallium Arsenide and Related Compounds 1987* (*IOP Conf. Ser. 91*) ed A Christou and H S Ruprecht (Bristol: Institute of Physics) p 377

[36] Gibart P, Williamson D L, Moser J and Basmaji P 1990 *Phys. Rev. Lett.* **65** 1144

[37] Gibart P and Basmaji P 1990 *Physics of DX Centers in GaAs Alloys* ed J C Bourgoin (Vaduz: Sci. Tech. Publication) p 1

[38] Weyer G, Petersen J W, Damgaard S, Nielsen H L and Heinemeier J 1980 *Phys. Rev. Lett.* **44** 155

[39] Nielsen O H, Larsen F K, Damgaard S, Petersen J W and Weyer G 1983 *Z. Phys. B* **52** 99

[40] Weyer G, Damgaard S, Petersen J W and Heinemeier J 1980 *J. Phys. C: Solid State Phys.* **13** L181

[41] Bonde-Nielsen K, Grunn H, Haas H, Pedersen F T and Weyer G 1985 *13th Int. Conf. on Defects in Semiconductors* vol 14a ed L C Kimerling and J M Parsey Jr (New York: AIME) p 1065

[42] Svane A and Antoncik E 1986 *Phys. Rev. B* **34** 1944; 1987 *Phys. Rev. B* **35** 4611

[43] Antoncik E 1977 *Phys. Status Solidi* b **179** 605

[44] Gu B L 1982 *PhD Thesis* Aarhus University

[45] Antoncik E and Gu B L 1982 *Phys. Scr.* **25** 836

[46] Goldanskii V I and Herber R 1968 *Chemical Applications of Mössbauer Spectroscopy* (New York: Academic)

[47] Gorlich E A, Latka K and Moser J 1989 *Hyperfine Inter.* **50** 723

[48] Moser J private communication

Semicond. Sci. Technol. **6** (1991) B78–B83. Printed in the UK

Observation of a local vibrational mode of DX centres in Si doped GaAs

J A Wolk†‡, M B Kruger†, J N Heyman†, W Walukiewicz‡, R Jeanloz§ and E E Haller‡∥

† Department of Physics, University of California, Berkeley, CA 94720, USA
‡ Center for Advanced Materials, Material Science Division, Lawrence Berkeley Laboratory, Berkeley, CA 94720, USA
§ Department of Geology and Geophysics, University of California, Berkeley, CA 94720, USA
∥ Department of Material Science and Mineral Engineering, University of California, Berkeley, CA 94720, USA

Abstract. We report the observation of a new local vibrational mode (LVM) in hydrostatically stressed, Si-doped GaAs. The corresponding infrared absorption peak is distinct from the Si_{Ga} shallow-donor LVM peak, which is the only other LVM peak observed in our samples, and is assigned to the Si DX centre. Analysis of the relative intensities of the Si DX LVM and the Si shallow-donor LVM peaks has been combined with Hall effect and resistivity analysis to infer that the Si DX centre is negatively charged.

1. Introduction

The DX centre is a deep-level defect found in several n-type III–V semiconductors and their alloys. For example, it has been observed in $Al_xGa_{1-x}As$ for $x \geq 0.22$ [1–3] and GaAs under hydrostatic pressure [4] greater than approximately 20 kbar. It is characterized by several unusual physical properties, including a large Stokes shift (~ 1 eV) [3] and persistent photoconductivity [3]. Experiments have demonstrated that this defect is related to an isolated substitutional donor [5] whose shallow electronic level becomes deep under the above-mentioned conditions of pressure or alloying. The mechanism for this transformation is not yet fully understood, but recent experimental and theoretical investigations have suggested that it occurs via an intermediate excited state [6–8]. While many recent experiments have been performed in order to determine the microscopic structure and charge state of this defect, both remain the subject of controversy.

A recent theoretical model [9, 10], proposed to explain the behaviour of DX centres, suggests that this defect is a negative-U centre [11]. This implies that, although the impurity is only a single donor, its energy is lower when it binds two electrons rather than one. A lattice relaxation accompanies the localization of the second electron, and the decrease in energy caused by this relaxation is sufficient to overcome Coulomb repulsion. Similar models have been used to describe intersti-

tial boron [12] and the Si–A [13] centre in Si and an oxygen–arsenic vacancy complex in GaAs [14], all of which have been found to exhibit negative-U behaviour.

Most, but not all, experiments testing the validity of the negative-U model for the DX centre have either supported the model or been inconclusive. Many experimenters have tried to find evidence of a lattice relaxation in the neighbourhood of the DX centres. The result of an experiment in $Al_xGa_{1-x}As$:Si found several closely spaced DLTS peaks whose energies did not depend on x, but whose relative intensities did [15]. This was interpreted as demonstrating that the Si atoms had moved to interstitial positions where the average number of Al neighbours they had was a function of alloy content. However, EXAFS measurements gave no evidence for a change in the nearest-neighbour distance for Sn and Se substitutional donor atoms [16, 17], thus implying that there is no large lattice relaxation.

Other experimenters have attempted to determine the charge state of the DX centre. The results of a recent experiment using deep-level transient spectroscopy on samples co-doped with germanium and silicon showed that DX centres associated with the Ge atoms bound two electrons [18], and Mössbauer measurements on Sn-doped GaAs have been interpreted as showing that two or three electrons are localized at each DX centre [19]. Recent interpretations of thermal capture and emission kinetics data of DX centres have been shown to be consistent with a negatively charged DX centre but not with a neutral DX centre [20, 21]. Studies of the change in mobility caused by a DX transformation should also help determine if these defects are charged or neutral, but the interpretation of existing data has not led to definitive conclusions [22–24]. An experiment whose results are in conflict with the negative-U model used magnetic susceptibility measurements to show that DX centres in several approximately 10 μm thick epilayers of $Al_xGa_{1-x}As$ ($x > 0.23$) doped with Si or Te were paramagnetic, implying that only one electron was bound [25]. However, a similar experiment performed on a 200 μm thick epilayer of $Al_{0.3}Ga_{0.7}As:Te$ which had its substrate removed showed that the concentration of paramagnetic centres was an order of magnitude less than the concentration of DX centres [26].

We report the use of a well established technique, far-infrared Fourier transorm spectroscopy (FIRFTS) of local vibrational modes (LVMs), to study the Si DX centre in GaAs. Spectroscopy of local vibrational modes has been extensively used for studying defects in semiconductors [27], and in fact the conclusion that the DX centre is due to an isolated substitutional donor and not a complex was implied by combining LVM spectroscopy and Shubnikov–de Haas effect data [5]. However, the LVM of the Si DX centre in GaAs has not been previously observed because of the following difficulties inherent in such an experiment. If one wishes to perform experiments without applying pressure, alloys must be used. However, the vibrational spectra of alloys are extremely difficult to interpret because variations in the local environment of the defect lead to substantial broadening of LVM peaks. If pressure is employed to avoid the use of alloyed samples, then the sample must be placed in a diamond anvil cell (DAC). This is the only tool available for achieving the high pressures necessary in a cell which allows optical access. Unfortunately, DACs limit the sample size to the order of a few hundred micrometres in diameter and roughly 100 μm in thickness. While it is common practice to do spectroscopy in a DAC on samples where the absorption is due to intrinsic effects (i.e. 10^{22} atoms/cm³), Fourier transform spectroscopy of defects, where typical concentrations are 10^{17}–10^{18} cm⁻³, has not been successfully performed to our knowledge. The problem is one of achieving sufficient signal-to-noise ratio.

Transmission experiments performed in a DAC examining semiconductor defects have been previously reported. Electronic levels in GaAs were studied using magneto-spectroscopy [28], but this technique is quite different from the Fourier transform spectroscopy reported here. Magneto-spectroscopy has the advantage that a laser is used as a monochromatic light source instead of a black body, so achieving reasonable signal-to-noise ratio is less of a problem. It has the disadvantage, however, that the absorption frequency must be tuned using the magnetic field to coincide with the laser wavelength.

We have overcome the difficulties involved in performing FIRFTS in a DAC and present the first observation of an LVM of the Si substitutional donor and DX centre in GaAs under hydrostatic pressure up to 40 kbar [29].

While LVM spectroscopy is normally used to study defect structure, we also combine our spectroscopic data with Hall effect and resistivity data to infer the charge state of the DX centre.

2. Experimental technique

We describe our technique for performing LVM FTIR spectroscopy with a sample mounted in a DAC. All spectra were recorded with a Digilab FTS-80 Fourier transform spectrometer. The DAC, which has been previously described by its developers [30], was mounted in a Janis Super VP cryostat. Diamonds with 750 μm culets were used as anvils, and spring steel was used as the gasket material. The pressure medium used was a 4:1 mixture of methanol and ethanol. This mixture has been shown to be hydrostatic [31] up to 100 kbar, well above the pressures used for this work, and is also transmissive in the wavelength region of interest. The ruby fluorescence method [32, 33] was used to measure the pressure in the cell, but we had no means to perform this measurement while the DAC was at liquid-helium temperature. The following technique was therefore employed to calibrate the pressure in the cell at low temperature. The pressure in the DAC was measured at room temperature, after which it was immersed in liquid nitrogen. The pressure was then measured again taking into account dv/dT of the fluorescence, which is independent of pressure [34]. It was assumed that the pressure did not change between 77 and 4 K since the coefficient of thermal expansion of the cell, which is made of Vascomax 350, is small at low temperatures†. It was found that the pressure in the cell at 77 K was roughly 4 kbar below the pressure measured at 300 K after the cooling process. This decrease in pressure is independent of the pressure in the cell.

A Ge:Be photoconductor, which has a photoconductive onset at 200 cm⁻¹, was used as the detector [35]. This detector is most sensitive for low photon fluxes, and its responsivity and noise equivalent power (NEP) are shown as a function of bias for a particular set of operating conditions in figure 1. From the figure it is clear that for these conditions the detector has high responsivity and an NEP which is not far above photon-limited background NEP, which is roughly 10^{-16} W Hz⁻¹/² . For this experiment, we are in a regime of higher photon flux than the case shown in figure 1, but the detector is still performing close to the photon noise limit. To increase the signal-to-noise ratio, a cone was used to concentrate light on the cell, and the detector was mounted directly behind the cell in an integrating optical cavity.

---

† Although we do not know the thermal coefficient of expansion for this particular material at cryogenic temperatures, we safely assume it to be similar to that of other stainless steels and therefore extremely small. We are confident that our pressure calibration is correct to within ± 1 kbar since an extrapolation of the high pressure frequencies of the Si_{Ga} LVM down to zero pressure implies that the mode should be where it is actually observed at zero pressure.

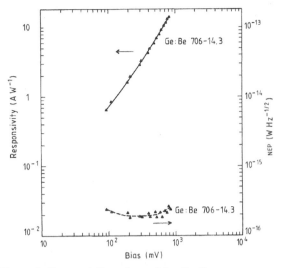

Figure 1. Responsivity and NEP of the Ge:Be photoconductor detectors used in this experiment. Conditions for these data are $\nu = 238$ cm^{-1}, photon rate $= 1.5 \times 10^8$ photons/s and $T = 4$ K. The photon rate in the present work is higher.

The integrated assembly of the aforementioned parts is shown in figure 2. This design has two critical features. First, stray radiation, a small amount of which could dominate the small number of photons reaching the detector through the DAC, is virtually eliminated. Second, alignment is trivial, because it is only necessary to have the infrared beam enter the 12.7 mm diameter entrance to

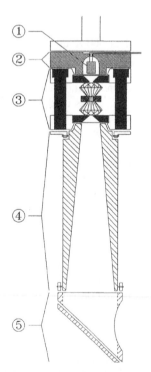

Figure 2. Detail of the cryostat insert for performing FIRFTS with sample mounted in a diamond anvil cell.
1, photoconductor; 2, photoconductor assembly; 3, diamond anvil cell; 4, light concentrating cone; 5, mirror assembly.

the mirror assembly. If the detector is separated from the DAC, positioning of the DAC cell becomes extremely critical and therefore very difficult.

Since a round hole is made in the gasket for the purpose of holding the sample and pressure medium, round samples were made in order to minimize light leakage. The samples were cylinders of GaAs:Si 300 μm in diameter and approximately 100 μm thick. They were prepared by hand lapping the GaAs to the appropriate thickness in a slurry of 3 μm grit and then cutting out the discs using an ultrasonic cutter.

3. Results and discussion

Our experiment can be divided into two main steps. We first identify a new LVM which we observe in hydrostatically stressed GaAs:Si as an LVM of the Si DX centre. We then combine our spectroscopic data with Hall effect and resistivity analysis to determine the charge state of the DX centre.

In order to determine if some change in the LVM occurs upon transforming into a DX centre, it is first necessary to determine where the mode of the untransformed substitutional donor lies under pressure. This is done using a piece of GaAs:Si with $n_{Si} = 6.3 \times 10^{17}$ cm^{-3}, subsequently referred to as sample 1I. This sample had been irradiated with 1 MeV electrons. The irradiation creates electronic levels near the middle of the band gap which absorb all free carriers [36]. As a result, the sample becomes transparent to infrared radiation at all pressures and formation of the DX centre is suppressed. The irradiation does not, however, affect the LVMs of the donor atoms [36]. A spectrum of sample 1I in the DAC taken at $P = 35 \pm 2$ kbar is shown in figure 3. We observe one peak, whose frequency we can observe as a function of pressure. Spectra at different pressures are

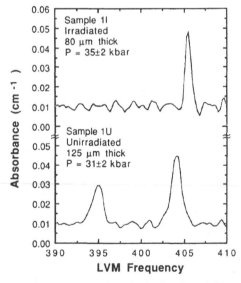

Figure 3. Absorption spectra of samples 1I and 1U at $T = 5$ K.

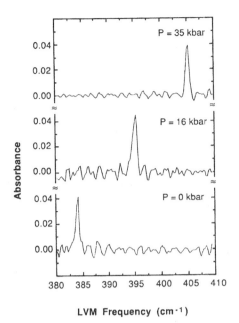

Figure 4. Absorption spectra of sample 1I at three pressures. $T = 5$ K.

shown in figure 4, and the data for all samples we observed are summarized in figure 5. We unambiguously identify this peak as the Si_{Ga} LVM since its frequency at zero pressure matches what has been previously observed for the substitutional donor [37]. The dependence of the LVM frequency on pressure is linear over the range of this study and is given by $d\nu_{Si_{Ga}}/dp = 0.66 \pm 0.03$ cm^{-1} kbar^{-1}.

Figure 3 also shows the spectrum of sample 1U at 30 ± 2 kbar, which is well above the pressure necessary to cause the DX transformation. This sample is identical to sample 1I, having been cut from the same wafer, except that it had not been irradiated. There are two peaks clearly observed in the spectrum, one at 404 cm^{-1} and the other at 395 cm^{-1}. The peak at 404 cm^{-1} is precisely where the Si_{Ga} LVM is expected at 31.5 kbar, but the peak at 395 cm^{-1} is a new feature. The pressure dependence of this new feature is shown in figure 5 and is $d\nu/dp =$

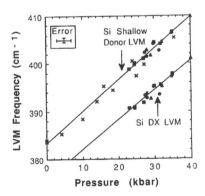

Figure 5. Pressure dependence of the Si_{Ga} and Si_{DX} LVM frequencies. x, ●, ■, and ▲ refer to samples 1I, 1U, and 2 and 3 respectively.

0.61 ± 0.04 cm^{-1} kbar^{-1}, roughly the same as that of the Si_{Ga} peak. There are no data below 23 kbar because the sample is opaque below these pressures due to free-carrier absorption. This new feature will be identified as the LVM of the Si DX centre.

As a first step in making this identification, we confirmed that DX centres were present in the sample by performing the following two experiments. First, the sample was illuminated with white light while at 4 K. The amount of infrared radiation reaching the detector was drastically reduced because of free-carrier absorption resulting from the persistent photoconductivity of the DX centres, and an absorption spectrum of the sample could no longer be taken. Second, the sample was brought to room temperature and recooled at a pressure of only 21 kbar, too low to cause the DX transformation. Once again, no spectrum of the sample could be taken. Bringing the sample back to room temperature and increasing the pressure to 24 kbar resulted in the sample once again being transparent.

In heavily doped GaAs:Si samples, many LVM peaks in addition to the Si_{Ga} peak have been observed with FIRFTS [37]. They can be assigned to $^{28}Si_{As}$, $^{28}Si_{Ga}$–$^{28}Si_{As}$, Si–X, Si–Y, $^{29}Si_{As}$ and $^{30}Si_{As}$. We exclude the possibility that the lower-freqency peak we observe is due to any of the defects listed above based on the results of the following experiment. A bulk piece of sample 1I was investigated by conventional absorption spectroscopy at zero pressure, and the only observable peak was that due to $^{28}Si_{Ga}$. This is consistent with the fact that only one peak was observed when Sample 1I was in the DAC, but provides a more stringent test since the signal-to-noise ratio for conventional spectroscopy is roughly a factor of 20 better than for spectroscopy performed in the DAC. As an additional check, if we extrapolate the frequency of the new peak to zero pressure, it would be found at 376 ± 1.5 cm^{-1}. This value lies between the frequencies of the ^{29}Si and ^{30}Si LVM peaks reported in [37], and the absorbance in these peaks is less than 10% of the absorbance due to $^{28}Si_{Ga}$. In contrast, the new LVM peak we observe has roughly half the absorbance of the $^{28}Si_{Ga}$ LVM peak. Finally we note that our technique is not sensitive to defect concentrations much below 6×10^{17} cm^{-3}, and it is unlikely that the three unirradiated samples we studied would have a defect besides Si at this high a concentration. The above arguments demonstrate conclusively that the lower-frequency LVM is due to a previously unobserved defect related to Si. Since this defect is only formed by the application of pressure in a sample where free electrons are available, we identify this new peak as the LVM of the Si DX centre.

We now explain how combining our spectroscopic data with Hall effect and resistivity data allows us to infer the charge state of the DX centre. We compared the area in the LVM absorption peak of sample 1I with the sum of the areas of the Si_{Ga} and Si_{DX} peaks of sample 1U for several spectra and found that they were equal to within 30%. This implies that the absorption cross section for these two defects is the same to within this accuracy. This in turn implies that the ratio of the area in the LVM absorption peaks $A_{\mathrm{SiGa}}/A_{\mathrm{SiDX}}$ observed for a given sample is equal to $n_{\mathrm{SiGa}}/n_{\mathrm{SiDX}}$ to within 30%, where n is the concentration of the corresponding defect. The theoretical ratio $n_{\mathrm{SiGa}}/n_{\mathrm{SiDX}}$ depends on the charge state which is assumed for the DX centre and the compensation ratio in the sample. The compensation ratio is defined as $n_{\mathrm{A}}/n_{\mathrm{D}}$, where n_{A} is the concentration of minority acceptors and n_{D} is the concentration of majority donors. If the DX centre were a neutral defect, all uncompensated donors should undergo the DX transformation, and the ratio of the concentration of Si shallow donors to Si DX centres would be

$$\frac{n_{\mathrm{SiGa}}}{n_{\mathrm{SiDX}}} = \frac{n_{\mathrm{a}}}{n_{\mathrm{d}} - n_{\mathrm{a}}} = \frac{\theta}{1 - \theta}.$$

If the DX centre were negatively charged, only one half of the uncompensated donors could transform into DX centres, and the concentrations would be related to the compensation ratio by

$$\frac{n_{\mathrm{SiGa}}}{n_{\mathrm{SiDX}}} = \frac{\frac{1}{2}(n_{\mathrm{d}} - n_{\mathrm{a}}) + n_{\mathrm{a}}}{\frac{1}{2}(n_{\mathrm{d}} - n_{\mathrm{a}})} = \frac{1 + \theta}{1 - \theta}.$$

The compensation ratio can be determined from the mobility and free-carrier concentration in the sample [38]. This information can be obtained from a combination of Hall effect and resistivity analysis, and this was done for all the unirradiated samples which have been examined in this study. The results are given in table 1. Our analysis took into account the concentration dependence of the effective mass [39]. We combine the Hall effect analysis with our spectroscopic data, and plot our results in figure 6 along with curves for the predictions of the DX^- and DX^0 models. Our results clearly support the negative-U model for the DX centre.

Since there is not complete agreement between different sources on obtaining compensation ratios from concentration and mobility data, we consider how using theories other than the one chosen here would affect the interpretation of the data. Any model which includes scattering effects other than those considered in [38] would lead to lower compensation ratios, and would therefore shift the data points even further away from the DX^0 curve shown in figure 6. A recent model [40] which claims that screening effects in other models have been overestimated also leads to lower compensation ratios. It therefore appears that using other models would not change our conclusion that the DX centre is negatively charged.

We now discuss the consistency of our results with other experimental and theoretical work. In [3], data

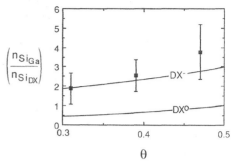

Figure 6. Comparison between experimental results and theoretical predictions for the ratio $n_{\mathrm{SiGa}}/n_{\mathrm{SiDX}}$ as a function of compensation.

describing electron capture at DX centres are used to infer that they should have an LVM at a frequency much lower ($< 80\ \mathrm{cm}^{-1}$) than the frequency of the DX centre LVM we observe. Our work does not exclude the existence of such a mode since our detector was not sensitive to light at these low frequencies. Also, if the negative-U model proposed in [9] and [10] is correct, the Si substitutional donor transforms into a DX centre by breaking a bond with one of the nearest-neigbour arsenic atoms and moving into an interstitial position. It is therefore intriguing that the LVM frequency of the DX centre is so close to that of the unperturbed donor. If the vibration of the Si atom is along the direction of the broken bond, it is reasonable to expect that the LVM frequency will be much lower than that of the Si substitutional donor. However, modes exist in which the donor vibration lies in a plane parallel to the plane containing the three nearest-neighbour arsenic atoms. Rough arguments suggest that the vibrational frequency of at least one such mode would not be drastically different from that of the substitutional donor.

4. Conclusion

In conclusion, the LVM of the Si DX centre in GaAs has been observed for the first time. The shift of the LVM frequency with pressure was found to be $\mathrm{d}\nu_{\mathrm{SiDX}}/\mathrm{d}p = 0.61 \pm 0.04\ \mathrm{cm}^{-1}\ \mathrm{kbar}^{-1}$ for the Si DX centre and $\mathrm{d}\nu_{\mathrm{SiGa}}/\mathrm{d}p = 0.66 \pm 0.03\ \mathrm{cm}^{-1}\ \mathrm{kbar}^{-1}$ for $\mathrm{Si_{Ga}}$. The ratio of the area of the $\mathrm{Si_{Ga}}$ absorption peak to that of the Si DX absorption peak has been combined with Hall effect and resistivity analysis to provide further evidence supporting the negative-U model for the DX centre. With suitable theoretical work it is hoped that the frequency of the Si DX LVM will provide a definitive clue to the microscopic structure of this defect.

Acknowledgments

The authors would like to gratefully acknowledge J Beeman for his help in designing and constructing the photoconductor detector, E Bourret and Grant Elliot for providing samples, R Berg and W J Moore for irradiating

Table 1. Results from Hall effect and resistivity analysis.

Sample	$N_{\mathrm{D}} - N_{\mathrm{A}}$ ($10^{18}\ \mathrm{cm}^{-3}$)	$\mu_{110\mathrm{K}}$ ($10^3\ \mathrm{cm^2\ V^{-1}s^{-1}}$)	θ
1I	0.63		
1U	0.63	2.70	0.31
2	2.9	1.53	0.39
3	2.1	1.41	0.47

samples, J Guitron and J Emes for their help preparing samples, L Falicov for helpful discussions and M Pasternak for providing the design for our diamond anvil cell and advice on how to use it. One of us (EEH) would like to acknowledge the support of the Miller Institute for Basic Research in Science. This work was supported by the Director, Office of Energy Research, Office of Basic Energy Sciences, Materials Science Division, of the US Department of Energy under Contract DE-AC03-76SF00098 and the US NSF under grants DMR-88-06756 and EAR-89-03801.

References

[1] Nelson R J 1977 *Appl. Phys. Lett.* **31** 351
[2] Lang D V and Logan R A 1977 *Phys. Rev. Lett.* **39** 635
[3] Lang D V 1986 *Deep Levels in Semiconductors* ed S T Pantelides (New York: Gordon and Breach) p 489
[4] Mizuta M, Tachikawa M, Kukimoto H and Minomura S 1985 *Japan. J. Appl. Phys.* **24** L143
[5] Eaves L *et al* 1988 *Gallium Arsenide and Related Compounds 1987 (Inst. Phys. Conf. Ser. 91)* ed A Christou and H S Rupprecht (Bristol: Institute of Physics) p 355
[6] Theis T N and Mooney P M 1990 *Mater. Res. Soc. Symp. Proc.* **163** 729
[7] Dobaczewski L, Kaczor P, Langer J M, Parker A R and Poole I 1990 *Proc. 20th Int. Conf. on the Physics of Semiconductors* ed E M Anastassakis and J D Joannopoulos (Singapore: World Scientific) p 497
[8] Dobaczewski L and Kaczor P 1991 *Phys. Rev. Lett.* **66** 68
[9] Chadi D J and Chang K J 1988 *Phys. Rev. Lett.* **61** 873
[10] Chadi D J and Chang K J 1989 *Phys. Rev. B* **39** 10063
[11] Anderson P W 1975 *Phys. Rev. Lett.* **34** 953
[12] Troxell J R and Watkins G D 1980 *Phys. Rev. B* **22** 921
[13] Watkins G D and Corbett J W 1961 *Phys. Rev.* **121** 1001
[14] Skowronski M, Neild S T and Kremer R E 1990 *Appl. Phys. Lett.* **57** 902
[15] Mooney P M, Theis T N and Wright S L 1988 *Appl. Phys. Lett.* **53** 2546
[16] Hayes T M, Williamson D L, Outzourhit A and Small P 1989 *J. Electron Mater.* **18** 207
[17] Kitano T and Mizuta M 1987 *Japan. J. Appl. Phys.* **26** 1806
[18] Fujisawa T, Yoshino J and Kukimoto H 1989 *Japan. J. Appl. Phys.* **29** L388
[19] Gibart P, Williamson D L, Moser J and Basmaji P 1990 *Phys. Rev. Lett.* **65** 1144
[20] Theis T N, Mooney P M and Parker B D 1991 *J. Electron Mater.* **20** 35
[21] Mosser V, Contreras S, Piotrzkowski R, Lorenzini Ph, Robert J L, Rochette J F and Marty A 1991 *Semicond. Sci. Technol.* **6** 505
[22] Maude D K, Eaves L, Foster T J and Portal J C 1989 *Phys. Rev. Lett.* **62** 1922
[23] Chadi D J, Chang K J and Walukiewicz W 1989 *Phys. Rev. Lett.* **62** 1923
[24] O'Reily E P 1989 *Appl. Phys. Lett.* **55** 1409
[25] Katchaturyan K, Awschalom D D, Rozen J R and Weber E R 1989 *Phys. Rev. Lett.* **63** 1311
[26] Katsumoto S, Matsunaga N, Yoshida Y, Sugiyama K and Kobayashi S 1990 *Japan. J. Appl. Phys.* **29** L1572
[27] Barker A S Jr and Sievers A J 1975 *Rev. Mod. Phys.* **47** S1
[28] Dmochowski J E, Wang P D and Stradling R A 1991 *Semicond. Sci. Technol.* **6** 118
[29] Wolk J A, Kruger M B, Heyman J N, Walukiewicz W, Jeanloz R and Haller E E 1991 *Phys. Rev. Lett.* **66** 774
[30] Sterer E, Pasternak M P and Taylor R D 1990 *Rev. Sci. Instrum.* **61** 1117
[31] Piermiani G J, Block S and Barnett J D 1973 *J. Appl. Phys.* **44** 5377
[32] Barnett J D, Block S and Piermiani G J 1973 *Rev. Sci. Instrum.* **44** 1
[33] Piermiani G J, Block S, Barnett J D and Forman R A 1975 *J. Appl. Phys.* **46** 2774
[34] Jahren A H, Kruger M B and Jeanloz R *J. Appl. Phys.* to be submitted
[35] Haegel N M, Haller E E and Luke P N *Int. J. Infrared Millimetre Waves* 4
[36] Theis W M and Spitzer W G 1984 *J. Appl. Phys.* **56** 890
[37] Woodhead J, Newman R C, Tipping A K, Clegg J B, Roberts J A and Gale I 1985 *J. Phys. D: Appl. Phys.* **18** 1575
[38] Walukiewicz W 1990 *Phys. Rev. B* **41** 10218
[39] Raymond A, Robert J L and Bernard C 1979 *J. Phys. C: Solid State Phys.* **12** 2289
[40] Meyer J R and Bartoli F J 1987 *Phys. Rev. B* **36** 5989

Semicond. Sci. Technol. **6** (1991) B84–B87. Printed in the UK

Electron paramagnetic resonance of the shallow Sn donor in GaAs/Al$_{0.68}$Ga$_{0.32}$As:Sn heterostructures

W Wilkening†, U Kaufman† and E Bauser‡

† Fraunhofer-Institut für Angewandte Festkörperphysik, Tullastrasse 72, 7800 Freiburg, Federal Republic of Germany
‡ Max-Planck-Institut für Festkörperforschung, Heisenbergstrasse 1, 7000 Stuttgart 80, Federal Republic of Germany

Abstract. A new, light-induced electron paramagnetic resonance (EPR) signal with apparent tetragonal symmetry has been observed in n-type Al$_{0.68}$Ga$_{0.32}$As:Sn layers grown on SI-GaAs. A comparison with previous magnetic resonance data for the shallow Si donor indicates that the signal corresponds to the shallow, X-valley-associated effective-mass state of the Sn donor. The optical behaviour of the new signal shows that the shallow state is a metastable state of the Sn DX centre. The nearly vanishing EPR intensity for magnetic fields in the layer is attributed to the large spin–valley interaction expected for the shallow Sn donor in indirect AlGaAs.

1. Introduction

In Al$_x$Ga$_{1-x}$As group IV and group VI donors on cation and anion sites, respectively, form bistable defects [1]. According to the present understanding, the relaxed ground state configuration for $x > 0.22$ gives rise to the deep DX level, while the unrelaxed (tetrahedral) metastable configuration corresponds to the conventional hydrogenic shallow donor centres expected for these dopants. Experimental evidence for the shallow donor states originally stemmed from the observation of 1s–2p hydrogenic absorption lines in direct ($x < 0.35$) Al$_x$Ga$_{1-x}$As [2] and of donor photoionization continua in indirect ($x > 0.35$) material [3]. Another very clear-cut spectroscopic fingerprint for shallow donors in indirect AlGaAs came from magnetic resonance studies which can reveal the symmetry and multiplicity of the lowest conduction band valleys involved. Both optically detected magnetic resonance (ODMR) [4–8] and conventional electron paramagnetic resonance (EPR) [9–11] showed that Si introduces a shallow donor state associated with the X valleys in indirect AlGaAs. Photo-EPR provided clear evidence that this state is a metastable state of the Si DX state. An independent valley model was found to account satisfactorily for the Si EPR data, and it was concluded that the effect of spin-valley (SV) interaction on the donor resonance is suppressed by random internal strains and alloy disorder.

In this paper we present the first EPR results for the shallow Sn donor state in indirect Al$_{0.68}$Ga$_{0.32}$As:Sn layers. The EPR characteristics of the Sn resonance are similar to those of the Si shallow donor signal except for one significant difference which is reminiscent of the large spin–valley interaction of Sn as compared with Si. We also point out that the present Sn signal is different from the deep Sn donor resonance reported elsewhere [12]. Therefore the claim that previous assignments to the Si shallow donor may have to be revised must be rejected.

2. Experiment

The layers investigated were grown by liquid phase epitaxy. They were deposited onto both sides of an (001)-oriented, undoped SI-GaAs substrate. The 6 μm thick n-type Al$_{0.68}$Ga$_{0.32}$As:Sn layers are separated from the substrate by 2 μm thick undoped Al$_{0.68}$Ga$_{0.32}$As buffer layers to prevent formation of a two-dimensional electron gas. The Sn concentration in the doped layer is about 7×10^{17} cm^{-3}. EPR measurements were performed at 9.4 GHz.

Figure 1 shows EPR spectra of the GaAs/Al$_{0.68}$Ga$_{0.32}$As:Sn heterostructure before and after illumination with near-infrared light. In the dark, a signal from the sample is not observed, the structure below 345 mT being due to a cavity background signal. For the magnetic field H along the [001] growth direction, the line attributed to the shallow Sn donor resonance is centred at 349 mT ($g_{001} = 1.955$) and is sharpest, 5.7 mT. When the field is rotated towards the [110] direction

0268-1242/91/100B84+04 $03.50 © 1991 IOP Publishing Ltd

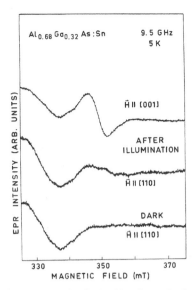

Figure 1. EPR spectra of a GaAs/Al$_{0.68}$Ga$_{0.32}$As:Sn heterostructure in the dark and after illumination. The line near 349 mT is attributed to the shallow Sn donor. The structure below 345 mT is a cavity background signal.

($g_{110} = 1.933$), the line moves to higher fields according to the angular dependence shown in figure 2, broadens and looses intensity. For H parallel to [110] the integrated intensity is lower by a factor of about 5 and the linewidth has increased to about 9 mT. A search for $^{117/119}$Sn hyperfine satellites has been made, in particular in field ranges where they would be expected on the basis of 24 GHz EPR data detected via the magnetic circular dichroism (MCD) [12]. No such satellites were found, although the sensitivity was sufficient to observe them.

3. Interpretation

The reasons to associate the above EPR signal with the Sn shallow donor state in the doped alloy layer are threefold:

(i) such a signal is known neither for undoped Al$_{0.68}$Ga$_{0.32}$As nor for the undoped GaAs substrate.

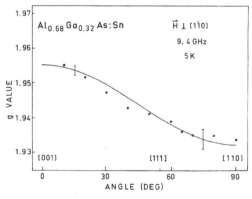

Figure 2. Angular dependence of the signal in figure 1 upon rotation of the magnetic field in a (110) plane.

(ii) The g values, the angular dependence and the linewidth are very similar to those of the Si shallow donor in indirect AlGaAs [4–11]. In particular the single-valley g factors, $g_t = 1.955$ and $g_l = 1.910$, evaluated from g_{001} and g_{110}, are very close to those of the Si shallow donor in Al$_{0.6}$Ga$_{0.4}$As:Si [11].

(iii) The failure to observe $^{117/119}$Sn hyperfine lines indicates a small hyperfine splitting, probably hidden within the linewidth, and therefore a delocalized state.

Thus, the assignment of the EPR line in figure 1 to the shallow Sn donor in uniaxially strained AlGaAs is well founded. Obviously, the present resonance is different from the isotropic Sn deep donor resonance observed by MCD-EPR [12] and from two isotropic EPR lines also attributed to the Sn donor [13].

In zero magnetic field the effective Hamiltonian describing the relevant interactions within the threefold valley-degenerate 1s (T$_2$) shallow donor ground state has the form [10]

$$\mathcal{H} = \delta\mathcal{L}_z^2 + \varepsilon(\mathcal{L}_\xi^2 - \mathcal{L}_\eta^2) + \lambda\mathcal{L}S$$

where $\mathcal{L}$ is a vector operator transforming as the angular momentum operator L. The first term is dominant ($|\delta| \sim 100\,\text{cm}^{-1}$) and describes the axial field splitting (see Figure 3) resulting from the small lattice-mismatch-induced effective tetragonal strain along the [001] growth direction z. The previous results for the Si donor have shown that δ is negative, i.e. that the valley doublet X$_x$, X$_y$ is stabilized. The second term represents random, in-plane strains including crystal fields of lower than

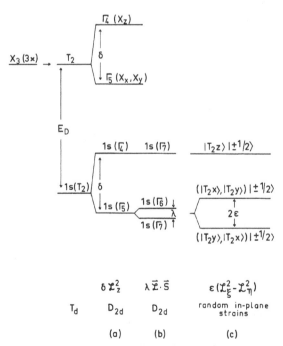

Figure 3. Splitting of the 1s(T$_2$) ground state of a group III site hydrogenic donor tied to the three X$_3$ minima in indirect AlGaAs due to (a) heteroepitaxial uniaxial strain along the [001] growth direction, (b) spin–valley interaction, neglecting random in-plane ε-type strains and (c) in-plane ε-type strains larger than the spin–valley splitting λ.

axial symmetry due to alloy disorder. The parameter ε is not a constant but can vary from one donor site to the other and follows a distribution with mean values of $|\varepsilon| \sim 5$ cm^{-1} [10]. The last term describes the spin–valley interaction which is the analogue of spin–orbit interaction in atoms. The SV coupling constant λ depends strongly on the donor species as is well known for Si and Sn in GaP [14–16]. For Si in indirect AlGaAs $|\lambda|$ has been estimated as about 1 cm^{-1} [10]. For the Sn donor in Al$_{0.68}$Ga$_{0.32}$As we estimate it as $|\lambda| \approx 15$ cm^{-1}.

The donor ground state and its resonance properties (g-factors) depend essentially on the relative magnitude of the parameters ε and λ. This is most easily seen by considering two limiting cases:

(i) For $|\varepsilon| \gg |\lambda|$, the X$_x$ and X$_y$ valleys are independently stabilized, corresponding to the situation shown in figure 3(c), where the first-order effect of the SV interaction is suppressed. In this case the ground state g factors are those of the independent valley model. Along and perpendicular to the growth direction they are given by $g_{001} = g_t \simeq 2$ and $g_{110} = [(g_t^2 + g_l^2)/2]^{1/2} \simeq 2$ respectively. Here g_l and g_t are the longitudinal and the transverse g factor of a single X valley. Changes in linewidth and integrated intensity when H is rotated from [001] to [110] are not expected in this limit. This is exactly the situation previously encountered for the shallow Si donor [10], and this is consistent with the fact that $|\varepsilon| > |\lambda|$ is expected because of the small $|\lambda|$ of Si.

(ii) On the other hand, in the opposite limit where $|\varepsilon| \ll |\lambda|$, a SV state is stabilized as sketched in figure 3(b). Independent of the sign of λ, its g factors are given by $g_{001} = g_t \simeq 2$ and $g_{110} \simeq 0$. Thus, when the field direction is rotated from [001] towards [110], the resonance line will move rapidly to very high field values and will broaden beyond detection. It is this case which is expected for the shallow Sn donor because of its large SV coupling constant, and the significant intensity loss in the [110] spectrum of figure 1 is attributed to this broadening effect. In this picture, the residual intensity of the donor line in the $H \| [110]$ spectrum arises from a small fraction of Sn donor centres for which the condition $|\varepsilon| < |\lambda|$ is not satisfied. This fraction behaves similarly to Si donors since it experiences particularly large ε values.

The Si resonance previously observed in indirect AlGaAs by ODMR [4–8] and EPR [9–11] has been interpreted as an X-valley-associated shallow donor signal, since the data revealed the symmetry and the multiplicity of the energetically lowest valleys in the strained AlGaAs layers. However, other workers [12] have claimed that this interpretation may have to be revised, and have suggested that the resonance in question is due to the neutral DX centre, DX^0. We reject this claim, since the shallow donor model for the Si EPR in [4–11] and for the Sn EPR in this work is compelling. Obviously, the Sn resonance observed by MCD-EPR [12] is different from the Sn donor signal reported here and corresponds to a deep Sn donor. The reason why this signal is not observed with conventional EPR is obvious. Its linewidth is 51 mT, exceeding that of the shallow donor in figure 1 by

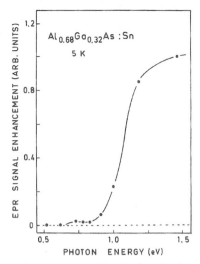

Figure 4. Excitation curve of the shallow Sn donor resonance in figure 1.

an order of magnitude. In the sample studied here its intensity is therefore below the detection limit.

As already mentioned, the Sn resonance in figure 1 is only observed after illumination. At 5 K it decays to about half its intensity within half an hour after the light is switched off. An excitation curve obtained by the initial slope technique is shown in figure 4. The low-energy threshold at $h\nu \approx 0.9$ eV and the general shape of the curve are similar to the excitation curve of the Si shallow donor resonance [9] and suggest that the same mechanism is involved. We therefore interpret the excitation of the shallow Sn EPR as photoionization of the Sn DX centre and subsequent capture of the electron into the hydrogenic Sn donor level. Thus the present data further support the view that the shallow effective-mass state is a metastable state of the deep donor state, the two states having different configurations. An alternative interpretation of the excitation mechanism in terms of compensation appears highly unlikely since one had to postulate residual acceptors in a concentration comparable to that of the Sn dopant.

4. Summary

We have presented the first EPR data for Sn in strained indirect AlGaAs. By comparison with the shallow Si donor EPR we have argued that the observed resonance line is due to the hydrogenic effective-mass of the Sn donor. The significant loss of EPR intensity for magnetic field orientations perpendicular to the growth direction has been traced back to the large spin–valley coupling of the Sn shallow donor. The data for its photoexcitation support that it is a metastable configuration of the deep Sn donor.

Acknowledgments

We are grateful to W Hornischer for help in growing the epitaxial layers and K Sambeth for technical assistance.

References

[1] For recent reviews see Mooney P M 1990 *J. Appl. Phys.* **67** R1; Dobaczewski L and Langer J 1991 *Proc. 4th Int. Conf. on Shallow Impurities in Semiconductors, London, 1990* ed G Davies (Zürich: Trans. Tech. Publications) p 433; Mizuta M 1990 *Proc. Int. Conf. on the Science and Technology of Defect Control in Semiconductors, Yokohama, 1989* ed K Sumino (Amstersdam: North-Holland) p 1043

[2] Theis T N, Kuech T F, Palmateer L F and Mooney P M 1985 *Gallium Arsenide and Related Compounds 1984 (Inst. Phys. Conf. Ser. 74)* ed B de Cremoux (Bristol: Institute of Physics) p 241

[3] Dmochowski J E, Langer M, Raczynska J and Jantsch W 1988 *Phys. Rev. B* **38** 3276

[4] Glaser E, Kennedy T A and Molnar B 1989 *Shallow Impurities in Semiconductors (Inst. Phys. Conf. Ser. 95)* ed B Monemar (Bristol: Institute of Physics) p 233

[5] Glaser E, Kennedy T A, Sillmon R S and Spencer M G 1989 *Phys. Rev. B* **40** 3447

[6] Glaser E, Kennedy T A, Molnar B and Mizuta M 1990 *Mater. Res. Soc. Symp. Proc.* **163** 753

[7] Kennedy T A, Glaser E R, Molnar B and Spencer M G 1990 *Proc. Int. Conf. on the Science and Technology of Defect Control in Semiconductors, Yokohama, 1989* ed K Sumino (Amsterdam: North Holland) p 975

[8] Montie E A, Henning J C M and Cosman E C 1990 *Phys. Rev. B* **42** 11808

[9] Mooney P M, Wilkening W, Kaufmann U and Kuech T F 1989 *Phys. Rev. B* **39** 5554

[10] Kaufmann U, Wilkening W, Mooney P M and Kuech T F 1990 *Phys. Rev. B* **41** 10206

[11] Wilkening W and Kaufmann U 1991 *Proc. 4th Int. Conf. on Shallow Impurities in Semiconductors, London, 1990* ed G Davies (Zürich: Trans Tech. Publications) p 397

[12] Fockele M, Spaeth J-M and Gibart P 1991 *Proc. 4th Int. Conf. on Shallow Impurities in Semiconductors, London, 1990* ed G Davies (Zurich: Trans. Tech. Publications) p 443

[13] von Bardeleben H J, Bourgoin J C, Basmaji P and Gibart P 1989 *Phys. Rev. B* **40** 5892

[14] Dean P J, Faulkner R A and Kimura S 1970 *Phys. Rev. B* **2** 4062

[15] Mehran F, Morgan T N, Titel R S and Blum S E 1972 *Phys. Rev. B* **6** 3917

[16] Dean P J, Schairer W, Lorenz M and Morgan T N 1974 *J. Lumin.* **9** 343

Semicond. Sci. Technol. **6** (1991) B88–B91. Printed in the UK

ODMR investigations of DX centres in Sn- and Si-doped Al$_x$Ga$_{1-x}$As

M Fockele†, J-M Spaeth†, H Overhof† and P Gibart‡

† University of Paderborn, Fachbereich Physik, Warburger Str. 100 A, D-4790
Paderborn, Federal Republic of Germany
‡ Laboratoire de Physique du Solide et Énergie Solaire, Centre National de la
Recherche Scientifique, Sophia Antipolis, 06560 Valbonne, France

Abstract. In a thick (100 μm) Sn-doped Al$_{0.35}$Ga$_{0.65}$As layer and an 11 μm Si-doped
Al$_{0.41}$Ga$_{0.59}$As layer the magnetic circular dichroism (MCDA) of the optical
absorption, the optically detected electron spin resonance (ODEPR) and the
photoconductivity (only in the Sn layer) have been measured simultaneously. All
signals appear together after the photoionization of the DX centre and are directly
correlated to each other. The ODEPR spectra measured in the MCDA bands show the
hyperfine doublets due to the magnetic ^{117}Sn and ^{119}Sn isotopes ($I = 1/2$) in
Sn-doped samples and an ODEPR line ($I = 0$) in Si-doped layers. The 4.7 % abundant
^{29}Si ($I = 1/2$) isotopes could not be observed. We show that the MCDA is due to a
neutral (Si^0, Sn^0) donor, a deep level which is stable at 1.6 K in the dark. This DX^0
state needs thermal energy or light to capture an electron to form the negative-U
DX$^-$ ground state which contains one donor dopant in a large lattice configuration.

1. Introduction

DX centres in Al$_x$Ga$_{1-x}$As are deep donors introduced
by doping with group IV or group VI elements such as Si,
Sn, Se or Te with unusual properties such as a much
larger photoionization threshold energy (of the order of
1 eV) than the thermal ionization energy (0.2 eV for Si)
(Lang 1985, Bourgoin 1990, Mooney 1990, Mooney *et al*
1987, Lang and Logan 1977, Lang *et al* 1979). Important
properties of the DX centres are still controversial (for a
detailed recent review see Mooney (1990)). These include
the microscopic configuration of the relaxed state, whether or not there is a large lattice relaxation (LLR) associated with it and the charge state of the DX ground state
(Lang and Logan 1977, Lang *et al* 1979, Morgan 1987,
Mooney *et al* 1987, Oshiyama and Ohnishi 1986, Hasegawa and Ohno 1986). In this paper we address the
question of the charge state and, associated with it, the
question of the large versus small lattice relaxation.

A substitutional donor in its neutral charge state is
expected to have a paramagnetic ground state. However,
in the dark no electron paramagnetic resonance (EPR)
was found (Mooney *et al* 1989, von Bardeleben *et al*
1989). EPR was observed only after photoexcitation with
light of energy exceeding the photoionization threshold
(Mooney *et al* 1989, von Bardeleben *et al* 1989, 1990).
Chadi and Chang (1988, 1989) suggested that the donor
is not stable in the neutral paramagnetic charge state D^0
but dissociates into D$^+$ and D$^-$ and that the donor is a

negative-U centre, in which, upon capture of a second
electron, the energy is lowered by an LLR. This model
implies a diamagnetic ground state.

We present an investigation of the magnetic circular
dichroism of the optical absorption (MCDA), and of the
optically detected EPR (ODEPR) of Sn- and Si-doped
Al$_x$Ga$_{1-x}$As samples as well as of the photocurrent (PC)
of the Sn-doped sample. Preliminary results were presented elsewhere (Fockele *et al* 1990a,b, 1991).

2. Experiment

The experiments were performed with an Sn-doped
Al$_{0.35}$Ga$_{0.65}$As layer, which was 100 μm thick, removed
from the substrate, and had $N_{Sn} = 3 \times 10^{18}$ cm^{-3}. The
thick layer was grown in a conventional metal organic
vapour phase epitaxy (MOVPE) vertical reactor on semi-
insulating, undoped 500 μm thick GaAs substrate, including a 0.5 μm thick Ga$_{0.5}$Al$_{0.5}$As undoped buffer
layer. MCDA and ODEPR experiments were also performed on an Si-doped ($N_D - N_A = 1 \times 10^{18}$ cm^{-3})
Al$_{0.41}$Ga$_{0.59}$As layer of 11 μm thickness. This sample
was grown by MOVPE on a 500 μm thick semi-insulating
GaAs substrate with a 1 μm thick undoped spacer layer
of Al$_x$Ga$_{1-x}$As.

After cooling the Sn-doped layer in the dark there is
no conductivity nor any MCDA. Upon illumination with
light of energy above 0.9 eV both MCDA and PC appear

0268-1242/91/100B88 + 04 $03.50 © 1991 IOP Publishing Ltd

and saturate with a light power of about 15 μW mm^{-2} for all photon energies. The MCDA obtained is a single band which extends from approximately 0.7 eV (limit of the Ge detector) to 1.5 eV (see figure 1 in Fockele *et al* 1990a). When measuring the saturated MCDA and the PC as a function of the photon energy of the exciting light one obtains the typical photoionization curve of the DX centres with a threshold of 0.9 eV for both the MCDA and the PC (Fockele *et al* 1991). Within experimental error there is a perfect linear relationship between the saturation values of the MCDA and the PC. Furthermore the build up of both signals is correlated. The MCDA spectrum of the Si-doped layer is modified due to the presence of the substrate. Details will be published elsewhere.

The ODEPR spectrum of the Sn-doped layer, measured as a microwave-induced decrease of the MCDA signal (Ahlers *et al* 1983), is isotropic and shows besides a central line a resolved asymmetric hyperfine (HF) doublet (see figure 3 in Fockele *et al* 1990a). The spectrum is explained in the following way: the central line is due to the non-magnetic Sn isotopes (83.46%), the doublet due to the ^{117}Sn and ^{119}Sn isotopes with nuclear spin $I = 1/2$ in the expected intensity ratio due to their natural abundances of 7.61% and 8.58% respectively. The g value is $g_e = 1.97 \pm 0.03$ and the half width is 51 ± 5 mT at low microwave power for the central resonance line. Diagonalization of the spin Hamiltonian yielded an isotropic HF constant of $a/h = 10.1$ GHz for ^{119}Sn (9.6 GHz for ^{117}Sn). The asymmetry of the HF doublet line positions with respect to the central line is caused by the size of the HF interaction with respect to the Zeeman energy (Abragam and Bleaney 1986). A similar HF split isotropic spectrum was recently also observed with conventional EPR in a ^{119}Sn-enriched sample by von Bardeleben *et al* (1991). In figure 1 the ODEPR spectrum of the Si-doped layer is shown. The g factor of the central line is $g = 1.90 \pm 0.03$, the half-width 8 ± 2 mT, i.e. much smaller than the Sn line. An HF doublet due to ^{29}Si could

not be observed. Note that the abundance of ^{29}Si ($I = 1/2$) is only 4.7%. The excitation spectra of the ODEPR lines (MCDA tagged by EPR (Ahlers *et al* 1983); see figure 4 in Fockele *et al* (1990a) for the Sn-doped sample) were shown to be identical to the MCDA spectra.

After switching off the photoexcitation except for the measurement light below the photoionization threshold both the MCDA and the PC decay simultaneously with a time constant of about 15 min at 1.6 K in the Sn-doped samples. In contrast to that the MCDA remains persistent for several hours in the case of Si doping. During the decay the DX centre ground state is repopulated, since after photoionization the PC and the MCDA are recovered. For Sn doping, however, under illumination with light of energy lower than the ionization threshold both the PC and the MCDA decay at 1.6 K within a much shorter time. Hence the recovery of the DX ground state can be, not only thermally achieved, but also optically induced.

The observation of the Sn HF interactions proves directly that we are measuring the EPR spectrum of the dopant. Our ODEPR results, i.e. g factors and linewidths, agree well with the EPR spectra observed previously by conventional EPR by von Bardeleben *et al* (1989) in an Sn-doped sample and with the one measured by Mooney *et al* (1989) in the same Si-doped sample which was used by us. There also EPR lines were observed only after photoionization of the DX centres. An HF structure was observed only in the thick Sn layer (von Bardeleben *et al* 1991).

3. Discussion

The first observed single-line Sn and Si EPR spectra were interpreted by von Bardeleben *et al* (1989) and Mooney *et al* (1989) as being due to a shallow excited hydrogenic state associated with the DX centres. However, our measurements show that this interpretation must be revised. Neither MCDA nor ODEPR spectra are those of a hydrogenic level associated with a conduction band minimum. From a shallow hydrogenic level to the conduction bands an optical transition starting at about 0.73 eV is not possible. For a hole transition to the valence band a much larger photon energy would be needed. Furthermore, the Sn HF interaction of 10.1 GHz corresponds to a spin density of about 13% relative to an Sn 5s orbital. The spin density of a full, unpaired 5s electron was calculated to be 73.9 GHz (^{119}Sn) making use of the local spin density approximation. 13% localization is typical for a deep defect. For instance, the localization of the unpaired electron at the ^{75}As central nucleus in the EL2 defect is 15% (Meyer *et al* 1984).

Figure 2 summarizes schematically the explanation we propose for the understanding of our experimental results. The DX ground state is diamagnetic and negatively charged. The LLR of this state is indicated symbolically by a configuration coordinate (CC). Upon photoionization one creates the paramagnetic charge state DX^0, and it is the EPR spectrum of this state which is

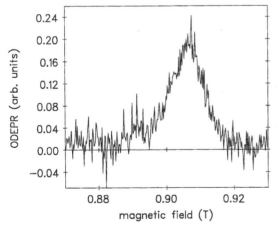

Figure 1. Optically detected electron paramagnetic resonance (ODEPR) spectrum of the Si-doped Al$_{0.41}$Ga$_{0.59}$As layer measured after additional excitation of below-band-gap light at 1.24 eV (15 μW mm^{-2}). The half width of the line is $\Delta B = 8 \pm 2$ mT, $g = 1.90 \pm 0.03$.

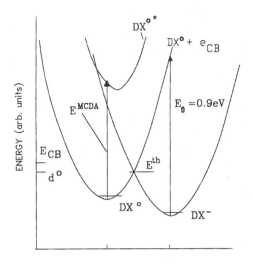

Figure 2. Schematic representation of electronic plus elastic energies in a simple configuration coordinate diagram for $Al_xGa_{1-x}As$. Sketched are the DX^- ground state, the singly ionized centre ($DX^0 + e_{CB}$) and its excited state DX^{0*}. The configuration coordinate symbolically represents the large lattice relaxation. E_0 is the photoionization energy, d^0 is a hydrogenic shallow donor state and E_{CB} the conduction band minimum. E^{th} is the thermal barrier against electron capture by DX^0. E^{MCDA} represents the intracentre transition energy of DX^0.

observed both with our MCDA technique and with the conventional EPR. In this charge state we have another LLR (for which we should perhaps have used a different CC). After photoionization one electron of the DX^- ground state is transferred to the conduction band and shallow donor levels d^0. Therefore the MCDA signal is always correlated to the PC.

At the lowest temperature the PC and the MCDA are rather stable. Therefore the DX^0 state must have a barrier against electron capture. Apparently it is also relatively stable against photoionization, otherwise we would not have been able to observe it optically. A substitutional donor with T_d symmetry is expected to be in a shallow hydrogenic state from which only optical ionization transitions in the far infrared have been reported. Since we have seen an optical transition at about 0.7 eV, the neutral donor must be deep and off centre (LLR). We speculate that this optical absorption band is due to an intracentre transition. For electron capture the DX^0 needs thermal activation to achieve the LLR position in which it can accommodate a second electron to form DX^-. The photoinduced decay of both PC and MCDA indicates that a photoexcitation can also achieve this. We do not know if optical excitation with light above 0.7 eV can also reform the DX^- from DX^0. The steep spectral dependence of the MCDA at low energy below the photoionization threshold may be the result of such a process. A similar photoinduced decay of the PC was also recently observed in Si-doped samples by He *et al* (1989).

The measured observed isotropy of the Sn^0 ODEPR spectrum is unexpected (detailed measurements for the Si-doped sample are yet to be made), since a static off-centre configuration as proposed by Chadi and Chang (1989) should give rise to an anisotropic spectrum. Probably a dynamic pseudo-Jahn–Teller effect averages the anisotropies. If the g factor anisotropy is only small it might be hidden in the broad ODEPR line.

4. Conclusion

In conclusion we have direct experimental evidence that the DX centre ground state is diamagnetic, i.e. a DX^-, supporting the proposed negative-U properties of DX (Chadi and Chang 1989). DX^- can be photoionized into a deep paramagnetic LLR state, which (for Sn^0 and Si^0) is rather stable at low temperature. Both thermal and photoinduced electron capture occur via the DX^0 state exclusively.

Acknowledgments

The financial support by a European Community ESPRIT Contract No 3168 is gratefully acknowledged. We thank P M Mooney and T F Kuech for kindly providing us with the Si-doped $Al_xGa_{1-x}As$ layer.

References

Abragam A and Bleaney B 1986 *Electron Paramagnetic Resonance of Transition Ions* (New York: Dover) pp 163–86

Ahlers F J, Lohse F, Spaeth J-M and Mollenauer L F 1983 *Phys. Rev.* B **28** 1249

Bourgoin J C 1990 *Physics of DX Centers in GaAs Alloys (Solid State Phenomena 10)* ed J C Bourgoin (Vaduz: Sci. Tech. Publications) p 1

Chadi D J and Chang K J 1988 *Phys. Rev. Lett.* **61** 873
——1989 *Phys. Rev.* B **39** 10063

Fockele M, Spaeth J-M and Gibart P 1990a *Mater. Sci. Forum* **65–66** 443
——1990b *Proc. 20th Int. Conf. on the Physics of Semiconductors* ed E M Anastassakis and J D Joannopoulos (Singapore: World Scientific) p 517

Fockele M, Spaeth J-M, Overhof H and Gibart P 1991 *Phys. Rev. Lett.* submitted

Hasegawa H and Ohno H 1986 *Japan. J. Appl. Phys.* **25** L319

He L X, Martin K P and Higgins R J 1989 *Phys. Rev.* B **39** 1808

Lang D V 1985 *Deep Centers in Semiconductors* ed S T Pantelides (New York: Gordon and Breach) p 489

Lang D V and Logan R A 1977 *Phys. Rev. Lett.* **39** 637

Lang D V, Logan R A and Jaros M 1979 *Phys. Rev.* B **19** 1015

Meyer B K, Spaeth J-M and Scheffler M 1984 *Phys. Rev. Lett.* **52** 851

Mooney P M 1990 *Appl. Phys. Rev.* **67** R1

Mooney P M, Caswell N S, and Wright S L 1987 *J. Appl. Phys.* **62** 4786

Mooney P M, Northrop G A, Morgan T N and Grimmeiss H G 1988 *Phys. Rev.* B **37** 8298

Mooney P M, Wilkening W, Kaufmann K and Kuech T F 1989 *Phys. Rev.* B **39** 5554

Morgan T N 1986 *Phys. Rev.* B **34** 2664

Oshiyama A and Ohnishi S 1986 *Phys. Rev.* B **33** 4320

von Bardeleben H J, Bourgoin J C, Basmaji P and Gibart P 1989 *Phys. Rev.* B **40** 5892

von Bardeleben H J, Bourgoin J C, Delerue C and Lannoo M 1991 *Phys. Rev. Lett.* submitted

von Bardeleben H J, Zazoui M, Alaya S and Gibart P 1990 *Phys. Rev.* B **42** 1500

Semicond. Sci. Technol. **6** (1991) B92–B96. Printed in the UK

Magneto-optical properties of the DX centre in $Al_{0.35}Ga_{0.65}As$:Te

R E Peale, Y Mochizuki†, H Sun and G D Watkins

Department of Physics and Sherman Fairchild Lab 161, Lehigh University, Bethlehem, PA 18015, USA

Abstract. Magneto-optical absorption spectra of a 0.4 mm thick, single-crystal $Al_{0.35}Ga_{0.65}As$:Te sample give evidence for two bleachable absorbers, one of which is identified as the DX centre. The bleached-state absorption coefficient and magnetic circular dichroism (MCD), measured from 0.66 to 2.2 μm at 1.7 K, are adequately described by the Drude free-electron model, in terms of which a value for the electron effective mass is obtained. Cooling the sample in darkness leads to transmission transients, from which the absorption coefficient and optical-conversion cross section for the bleachable deep DX ground state are derived. The MCD at the beginning of each transient is identified also with the DX ground state, and its temperature dependence reveals that the bulk of it has a non-paramagnetic origin. We conclude that the paramagnetic contribution to the MCD from the DX ground state is very small, being less than 0.004 % of its peak absorption coefficient. This provides strong support to the diamagnetic ground-state, negative-U model of Chadi and Chang. The origin of the second bleachable absorber (threshold ~ 1.6 eV) has not been established.

1. Introduction

Large lattice relaxations affect the electrical and optical properties of defects in many solids. Such effects are intuitively expected for ionic solids, where Coulombic forces induce large atomic displacements after a change in electronic charge distribution. Examples are F centres in alkali halides [1] and indium impurities in CdF_2 [2].

Large lattice relaxations are also found in the more-nearly covalent, compound semiconductors and even in perfectly covalent, elemental semiconductors. Prototypical examples are the vacancy [3] in Si and the DX centre in AlGaAs and other compound-semiconductor alloys [4, 5]. The recent focus on DX stems in part from its influence on AlGaAs/GaAs modulation-doped field-effect transistors, which are being developed for high-speed circuit applications. More fundamentally, the DX phenomenon is a fascinating, challenging and still unsolved, scientific puzzle.

It is now well established that DX centres in $Al_xGa_{1-x}As$ arise from the isolated dopants which control n-type conductivity [4]. For $x > 0.22$, these group IV or group VI impurities induce both a deep level and a shallow effective-mass level in the gap. The deep level's optical ionization energy is very much larger than its

† Present address: Fundamental Research Laboratories, NEC Corporation, 34, Miyukigaoka, Tsukuba, Ibaraki 305, Japan.

thermal one. The shallow level can be metastably occupied below 100 K, resulting in persistent conductivity. Recapture by the deep level is a thermally activated, multiphonon process; optical capture does not occur. Large lattice relaxation accounts for these features, though its extent and microscopic nature are still in question.

Chadi and Chang [6, 7] have predicted that group VI donors become deep when an adjacent group III host atom breaks away in a $\langle 111 \rangle$ direction from the donor, and this distortion is only stable if the defect binds two electrons. Hence, the DX centre should be a negative-U system [8], and diamagnetism is expected for the deep negatively charged state since the two spins should pair oppositely.

Prominent among the many experiments performed to test the DX for negative-U properties have been attempts therefore to determine the magnetic state of the deep level. Conflicting static magnetic susceptibility measurements have been reported, one concluding that the ground state is paramagnetic [9], another concluding that the observed paramagnetism is an order of magnitude smaller than predicted assuming equal free spin and DX concentrations [10]. The failure [11, 12] to observe spin resonance associated with the deep level is consistent with negative U, but there are many possible reasons for failure to detect a paramagnetic defect (inhomogeneous alloy broadening, rapid spin-lattice relaxa-

0268-1242/91/100B92+05 $03.50 © 1991 IOP Publishing Ltd

tion broadening, etc), and this also must be considered inconclusive.

Our purpose in the present work is to determine the paramagnetic contribution to magnetic circular dichroism (MCD) in the near-infrared absorption band of the deep DX state. This contribution is relatively insensitive to the width and relaxation times of the ground Zeeman-split states and should therefore more nearly reflect the ground state static susceptibility. In addition, being detected in an absorption band specific to the DX centre, the measurement is defect specific. Therefore, if a paramagnetic component is detected, it will convincingly rule out negative-U properties of DX. On the other hand, if none is detected, it will represent a strong argument for DX negative-U properties.

2. Experiment and results

MCD arises from spin–orbit interaction in the excited state involved in the optical absorption transition [13]. Its magnitude should therefore reflect the atomic spin–orbit interaction of the constituent atoms in the core of the defect. For this reason, the heavier atom tellurium was selected as the DX donor in our studies.

Our 0.037 cm thick, LPE, $Al_{0.35}Ga_{0.65}As:Te$ sample had its substrate lapped off prior to measurements. Sample transmittance and MCD were measured at liquid-helium temperatures in an optical-access, superconducting magnet cryostat.

When AlGaAs is cooled in the dark, some DX centres may remain in the metastable shallow state, in principle. Hence, we must first characterize the magneto-optical properties of the metastable state in order to correctly identify those effects arising exclusively from the ground state. For this purpose, the optical-absorption coefficient, α_m, was measured after white-light illumination converted all of the DX centres to their metastable states. Figure 1 presents α_m versus wavelength λ. Fundamental absorption is dominant for $\lambda < 0.65\ \mu m$. The long-wave-

length dependence is approximately λ^2 as predicted by the free-electron, Drude model, which gives [14, 15]

$$\alpha_m = \frac{Ne^2}{m^*nc^3\pi\tau_e}\lambda^2 \qquad (1)$$

where c is the speed of light, n the index of refraction of the material, e the electron charge, τ_e the relaxation time, m^* the effective mass and N the free-carrier concentration. The best fit of the function $\alpha = (const)\lambda^2$ to the data is the smooth curve in figure 1.

Magnetic circular dichroism (MCD) is defined as $\alpha_L - \alpha_R$, where α_L (α_R) is the absorption coefficient for left (right) circularly polarized light propagating along the magnetic field direction. Figure 2 presents the measured MCD versus magnetic field at 0.8 and 1.4 μm after white-light bleaching. A spectrum without the sample shows that the instrumental background circular dichroism is field independent, and we take this as our zero, as shown. The MCD is positive, linear in magnetic field B and increases with wavelength. Measurements at $T = 1.7$ K and 4.2 K give identical results within experimental uncertainty, showing that paramagnetism of the metastable state plays an insignificant role in its MCD.

Figure 3 plots the slopes of the metastable state MCD versus wavelength from 0.9 to 2.0 μm. The data increase approximately as wavelength cubed, as indicated by the best cubic fit curve. The theory of the Faraday effect [16], using the Drude conductivity tensor, gives the expression

$$MCD = \frac{16\pi\sigma_0\omega_c}{nc\tau_e^2\omega^3} \qquad (2)$$

where $\omega_c = eB/m^*c$ is the cyclotron frequency and $\sigma_0 = Ne^2\tau_e/m^*$ is the DC conductivity. Equation (2) is positive, linear in B and cubic in λ, in agreement with our observations.

Dividing equation (2) by equation (1) gives the simple relation

$$\frac{MCD}{\alpha_m} = 4\frac{\omega_c}{\omega}. \qquad (3)$$

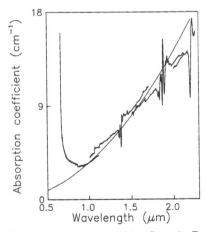

Figure 1. Absorption spectrum of $Al_{0.35}Ga_{0.65}As:Te$ at 1.7 K after white-light illumination.

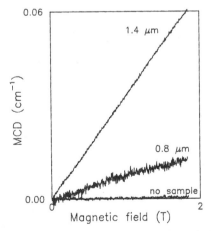

Figure 2. Field dependence of 1.7 K, $Al_{0.35}Ga_{0.65}As:Te$ MCD after white-light illumination.

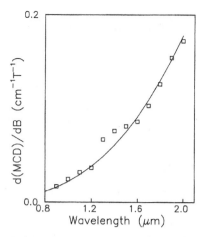

Figure 3. Spectral dependence of 1.7 K, $Al_{0.35}Ga_{0.65}As$:Te MCD after white-light illumination.

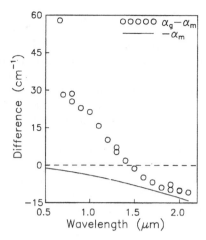

Figure 5. Spectral dependence of $\alpha_g - \alpha_m$. The smooth curve represents $-\alpha_m$ from Figure 1.

Equation (3) together with our data gives $m^* = 0.06\, m_0$, which compares well with the value $0.08\, m_0$ predicted [17] and measured [18] for the Γ minimum when $x = 0.35$. Evidently, the Drude model adequately describes both free-carrier absorption and MCD.

The optical transmission is strongly time dependent after cooling the sample in the dark. Figure 4 presents the transmitted intensity versus time at wavelengths 0.7 μm and 1.6 μm. At time $t = 0$ the shutter is opened, and the 0.7 μm transmission increases from I_0 to a value I_∞ as absorption is bleached. At time 1.6 μm, transmission decreases exponentially to a saturation value I'_∞. Additional white-light illumination further reduces the transmission to its final value (I_∞). The 1.6 μm $I'_\infty(I_\infty)$ levels persist unchanged in darkness for 30 min after terminating the 1.6 μm, or additional white-light, illumination. The significance of these two transient effects will be returned to later.

If bleachable absorbers, initially in their ground states, are completely converted to the metastable (free-

carrier) state, then the difference between ground- and metastable-state absorption coefficients is

$$\alpha_g - \alpha_m = -\frac{1}{d}\ln\left(\frac{I_0}{I_\infty}\right). \tag{4}$$

Figure 5 presents $\alpha_g - \alpha_m$ (open circles) versus λ. For $\lambda \gtrsim 1.5\ \mu$m, $\alpha_m > \alpha_g$, and the data approach the negative of the $\alpha_m \sim \lambda^2$ behaviour of figure 1, revealing that α_g is negligible here.

The open circles of figure 6 are estimates of α_g obtained by adding values of α_m (scaled by a factor of 0.75) determined from the smooth curve of figure 1 to the $\alpha_g - \alpha_m$ data of figure 5. (The data of figure 5, determined from equation (4), required no change of the optical setup and are therefore relatively accurate and reproducible. On the other hand, the scale of the results in figure 1 depends on our estimate of the incident intensity, I_i. This is determined by lifting the sample, which can perturb the delicate optical alignment. The 0.75 scaling factor required to match the long-wavelength tail in figure 5 is

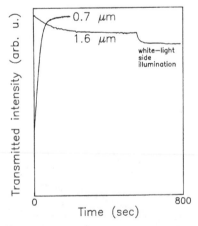

Figure 4. Transmission transients observed after cooling $Al_{0.35}Ga_{0.65}As$:Te to 1.7 K in the dark. The effect of white-light side illumination on the long-wavelength transient is also shown.

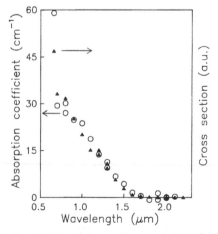

Figure 6. Spectral dependence of ground state absorption coefficient (open circles) and conversion cross section (full triangles) for bleachable absorbers in $Al_{0.35}Ga_{0.65}As$:Te at 1.7 K.

fully consistent with our estimate of the accuracy of the absolute scale for the data of figure 1.) The α_g values are seen to rise to a 30 cm^{-1} plateau between 0.7 and 0.8 μm, and then abruptly rise again to a value of 60 cm^{-1} at 0.66 μm.

This represents to our knowledge the first direct transmission measurement of the optical absorption coefficient of the deep bleachable absorbers in AlGaAs. Previously it has been primarily the cross section σ for optical ground- to metastable-state conversion that has been measured and assumed to be proportional to the absorption coefficient. We can test this, since

$$\sigma = (\tau\varphi)^{-1} \qquad (5)$$

where the conversion rate, τ^{-1}, and a relative measure of the photon flux, φ, are also obtained in our experiment. Figure 6 presents σ (full triangles) versus λ. Clearly, σ and α_g indeed do have the same spectral dependence.

The conversion rate, τ^{-1}, can be estimated accurately from data such as that shown in figure 4 for values ranging from a few seconds to many minutes, giving accurate values of σ over many decades. The results, extending therefore over a much wider spectral range than available for the absorption coefficient, are plotted in figure 7. The full curve is the DX conversion cross section found in Al$_{0.37}$Ga$_{0.63}$As:Te from capacitance transients (from figure 4 of [19]). The points and curve agree well for $\lambda \geq 0.9$ μm, confirming that here our transients reflect the well established DX conversion only. For $\lambda \leq 0.8$ μm, however, the data and curve diverge significantly, revealing the onset of a second, more-rapidly bleaching absorber. The sudden increase in the bleachable absorption coefficient at 0.66 μm (figure 6) also supports this identification of a second bleaching process.

This unexpected and interesting result is currently under study. We have established, for example, that bleaching this shorter-wavelength band produces additional persistent photoconductivity and is the origin

(figure 4) of the increased free-carrier absorption at 1.6 μm after white-light illumination. Preliminary experiments indicate also that recovery from this second bleaching process occurs at a somewhat lower temperature (~ 10 K lower) than does the DX bleaching process. We suspect that it may arise from an unrelated defect —metastable or simply compensating—in high concentration in our particular sample. We cannot rule out, however, that it is DX-related and simply missed by previous studies. We will defer further discussion concerning this source of additional persistent photoconductivity until our current studies are completed, and the results will be presented in a subsequent more detailed publication. For our purposes here, we can ignore this second bleachable stage and proceed with the results of MCD studies on the confirmed DX-related band at $\lambda > 0.9$ μm.

Like the transmission, the MCD is also found to be time dependent after cooling the sample in darkness, but differs from the transmission in being magnetic field dependent. Figure 8 presents examples measured at 0.9 μm for fields of 0 and 1.9 T. At $B = 0$, the MCD is independent of time and serves to define the zero, as shown. At 1.9 T, MCD(t) increases from an initially finite, positive value to a steady-state value MCD(∞), identified as the free-carrier MCD.

Figure 9 presents the initial MCD versus B at 900, 1032 and 1250 nm. The circles (crosses) are 1.7 K (4.2 K) data. The long-dashed (short-dashed) lines are linear fits to the 1.7 (4.2 K) data. The slopes decrease rapidly with increasing wavelength, opposite to the free-carrier MCD spectral dependence, so any free-carrier contribution to the initial MCD is small. Hence, we identify the initial MCD with the DX ground state absorption.

The initial MCD shows no statistically significant temperature dependence. Taking the scatter as an upper limit for the possible paramagnetic contribution to the DX ground-state MCD, a value of $\leq 0.004\%$ of the peak absorption coefficient (30 cm^{-1}) is found. Since fractional MCD (scaled to our temperature and field strength) of paramagnetic defects in semiconductors or insulators

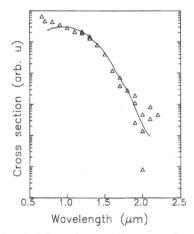

Figure 7. Spectral dependence of the conversion cross section for the bleachable absorbers. The smooth curve, from figure 4 of [19], is σ for DX centres in Al$_{0.37}$Ga$_{0.63}$As:Te.

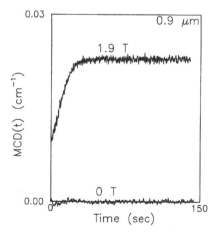

Figure 8. Transient MCD at 0.9 μm after cooling Al$_{0.35}$Ga$_{0.65}$As:Te to 1.7 K in the dark.

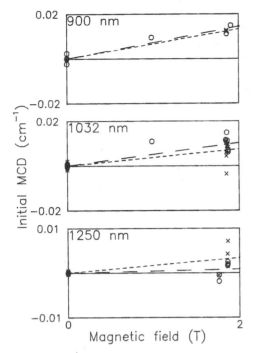

Figure 9. Initial MCD versus magnetic field at 900, 1032 and 1250 nm. Data measured at 1.7 K (4.2 K) are given by open circles (crosses) and fit by long-dashed (short-dashed) lines.

is often of the order of 0.1 to 10%, this result lends strong support to the diamagnetic ground state, negative-U model of Chadi and Chang.

It is important to point out, however, that it is not a *proof*. As pointed out in the introduction, the *detection* of a significant paramagnetic contribution would have served as a proof for a paramagnetic ground state. However, as in EPR, the *failure* to detect paramagnetism is not a proof of its absence. In the case of MCD, its strength is proportional to spin-orbit interaction in the final level of the optical absorption transition. We have no direct information of its magnitude and it could be small and strongly quenched by the low symmetry of the defect. Thus, the very low upper limit that we have established for the paramagnetic MCD component must be considered another strong 'nail in the coffin' for the neutral, paramagnetic, single-electron state model, but the 'coffin' is still not completely closed and should not yet be buried without a few more 'nails'.

Acknowledgements

This work was supported by National Science Foundation Grant No. DMR–89–02572. In addition, YM would like to acknowledge M Mizuta for fruitful discussions, and he is also grateful to F Saito, H Watanabe, M Ogawa and Y Wada for support.

References

[1] Fowler W B (ed) 1968 *Physics of Color Centers* (New York: Academic)
[2] For a brief review see Langer J M 1980 Large defect-lattice relaxation phenomena in solids in *New Developments in Semiconductor Physics* ed F Beleznay, G Ferenczi and J Giber (Berlin: Springer) p 123
[3] Watkins G D 1986 *Deep Centers in Semiconductors* ed S T Pantelides (New York: Gordon and Breach) pp 147–83
[4] Mooney P M 1990 *J. Appl. Phys.* **67** R1
[5] Lang D V 1986 *Deep Centers in Semiconductors* ed S T Pantelides (New York: Gordon and Breach) pp 489–539
[6] Chadi D J and Chang K J 1988 *Phys. Rev. Lett.* **61** 873
[7] Chadi D J and Chang K J 1989 *Phys. Rev. B* **39** 10366
[8] Watkins G D 1984 *Festkörperprobleme* **XXIV** 163
[9] Khachaturyan K A, Awschalom D D, Rozen J R and Weber E R 1989 *Phys. Rev. Lett.* **63** 1311
[10] Katsumoto S, Matsunaga N, Yoshida Y, Sugiyama K and Kobayashi S 1990 *Japan. J. Appl. Phys.* **29** L1572
[11] von Bardeleben H J, Zazoui M and Alaya S 1990 *Phys. Rev. B* **42** 1500
[12] Mooney P M, Wilkening W, Kaufmann U and Keuch T F 1989 *Phys. Rev. B* **39** 5554
[13] Henry C H and Slichter C P 1968 *Physics of Color Centers* ed W Beall Fowler (New York: Academic) pp 384–403
[14] Turner W J and Reese W E 1960 *Phys. Rev.* **117** 1003
[15] Fan H Y and Becker M 1951 *Proc. Reading Conf. on Semiconducting Materials* ed H K Henisch (London: Butterworths Scientific Publications) pp 132–47
[16] Mavroides J G 1972 *Optical Properties of Solids* ed F Abeles (New York: Elsevier) pp 351–528
[17] Harrison J W and Hauser J R 1976 *J. Appl. Phys.* **47** 292
[18] Inoshita T and Iwata N 1990 *Phys. Rev. B* **42** 1296
[19] Lang D V, Logan R A and Jaros M 1979 *Phys. Rev. B* **19** 1015

Semicond. Sci. Technol. 6 (1991) B97–B100. Printed in the UK

Studies of donor states in Si-doped $Al_xGa_{1-x}As$ using optically detected magnetic resonance with uniaxial stress

E R Glaser and T A Kennedy

Naval Research Laboratory, Washington, DC 20375 USA

Abstract. Optically detected magnetic resonance experiments with uniaxial stress along the [1$\bar{1}$0] and [100] directions have been performed on Si-doped epitaxial layers of AlAs and $Al_{0.4}Ga_{0.6}As$ grown on (001) GaAs substrates. These studies enable the symmetry of the donor wavefunction to be probed in detail. The results obtained with $T \parallel$ [100] confirm the independent valley description for Si donor states in AlAs. The lack of a response to uniaxial stress along [1$\bar{1}$0] provides evidence against the contribution of donor states derived from the L-point conduction band minima and against a purely spin–valley coupled state at $x = 0.4$.

1. Introduction

The doping of AlGaAs with donor atoms from either group IV or group VI results in the formation of both shallow and deep levels [1]. It is generally accepted that these levels are associated with the same defect, the DX centre. Electron paramagnetic resonance experiments and magnetic resonance studies detected on photoluminescence have been most successful in probing the nature of the shallow, metastable states associated with the X-point conduction band minima in Si-doped $Al_xGa_{1-x}As$ with $x \geq 0.35$. It is of particular interest to study the nature of the donor levels in $Al_xGa_{1-x}As$ with x near 0.4 since in this regime the conduction band minimum abruptly changes from the Γ point to the X point. In addition, the conduction band edge derived from the L points is nearly degenerate with those derived from the Γ and X points near $x = 0.4$.

Early electron paramagnetic resonance (EPR) studies of Si-doped $Al_xGa_{1-x}As$ layers ($0.55 < x < 0.80$) grown on (001) GaAs substrates revealed a single resonance with an anisotropic g value in the (1$\bar{1}$0) plane [2]. The anisotropy was assigned to 'valley repopulation' among the conduction band valleys associated with the X-point minima caused by the heteroepitaxial strain. The strain sensitivity of the Si donor states was further demonstrated by optically detected magnetic resonance (ODMR) experiments on similar samples studied on and after removal from the parent GaAs substrates [3]. The full X-point symmetry of the shallow state in this Al mole fraction regime was revealed explicitly by angular rotation studies in the (001) plane [4]. The anisotropy and splitting behaviour observed from the rotation studies in the (1$\bar{1}$0) and (001) planes of an Si-doped AlAs/GaAs structure could be understood using an independent valley model in the presence of heteroepitaxial strain. The Si donor ground state in AlAs/GaAs heterostructures was described as an orbital (valley) doublet with tetragonal symmetry about the [001] growth direction.

The nature of the donor states revealed from EPR and ODMR experiments on Si-doped $Al_xGa_{1-x}As$/GaAs heterostructures with $x < 1$ is not as well understood. A monotonic decrease in both the g-value anisotropy and splitting with decreasing Al mole fraction is observed [2–7]. This behaviour has recently been attributed to increasing L–X interband mixing [5, 6] or a spin–valley coupling interaction [3, 4] as x approaches 0.4 from above.

In order to test these models, ODMR experiments have been performed with externally applied uniaxial stress along the [1$\bar{1}$0] and [100] directions located in the (001) growth plane. These uniaxial stress experiments enable the dependences of the g values on aluminium mole fraction (x) and the built-in heteroepitaxial stress to be studied separately. Si-doped $Al_xGa_{1-x}As$ samples with Al mole fraction from 0.4 to 1.0 have been investigated. The results indicate that the symmetry of the donor wavefunction at $x = 0.4$ is still X-like. No evidence for an L-like contribution is found. However, the results provide evidence that the character of the Si donor states in AlGaAs crystals changes as a function of aluminium mole fraction.

0268-1242/91/100B97 + 04 $03.50 © 1991 IOP Publishing Ltd

2. Experimental aspects

The two samples discussed in this paper are epitaxial layers (1.5–5.5 μm) of Si-doped $Al_xGa_{1-x}As$ grown on semi-insulating, 500 μm thick (001) GaAs substrates. One of the samples was grown by molecular beam epitaxy (MBE) with x equal to 0.4. In addition, experiments were performed on an Si-doped AlAs layer grown by organo-metallic vapour phase epitaxy (OMVPE). The samples were Si doped to concentrations of 5×10^{16} and 1×10^{18} cm^{-3}, respectively.

The ODMR experiments were performed with uniaxial compressive stress applied in the plane of the layers either along the [1$\bar{1}$0] or [100] directions (see figure 1). The uniaxial stress apparatus is described elsewhere [7]. The cross section of the samples was 0.75 mm $\times$ 0.5 mm so that pressures up to 200 MPa (2 kbar) could be easily attained. The length was 2 mm to avoid bowing and stress inhomogeneity.

The magnetic resonance was detected synchronously as a change in the total intensity of donor–acceptor recombination which was coherent with the on/off modulation at 170 Hz of 50 mW of microwave power in a K-band (24 GHz) spectrometer. The photoluminescence (PL) was excited with above-band-gap radiation provided by a Kr$^+$ laser at 476 nm. Deep photoluminescence from 1.0–1.8 μm was detected by a Ge photodiode. The samples were studied at 1.6 K in an optical cryostat. The magnetic field (B) was supplied by a 9" pole-face electromagnet with a maximum field of 1.1 T. The measurements were obtained in the Voight geometry.

3. Theoretical background

The theory of the donor ground state in $Al_xGa_{1-x}As$ with $x \geq 0.35$ for group IV impurities substitutional on the III lattice sites (Ga, Al) can be described within the virtual crystal approximation following work by Morgan [8]. The conduction band constant energy surfaces are ellipsoidal in momentum space about the X-point minima with long axes along the $\langle 001 \rangle$ cube-edge directions. In the hydrogenic effective-mass approximation, the wavefunction of the donor ground state is derived from (i) Bloch functions for the X_x, X_y and X_z valleys, and (ii) 1s-like envelope functions that satisfy the effective-

mass equation. From symmetry arguments, the central cell potential does not mix the three hydrogenic effective-mass states for group IV donors on the group III site [8]. Hence, the ground state of an electron bound to a group IV donor is an orbital (valley) triplet. This degeneracy can be lifted by any internal or external stress along a direction which removes the equivalence of the three X-point conduction band minima.

Due to the multiplicity of valleys associated with the X point and the finite spin (1/2) of the donor electron, a spin–valley coupling must also be considered since this interaction ($\lambda L \cdot S$) can mix the X_x, X_y and X_z valleys. This interaction can also lift the three-fold degeneracy of the ground state. However, the presence of internal random strains in the crystal can render the magnetic resonance of a spin–valley coupled donor unobservable. For example, it was necessary to apply large uniaxial stresses in appropriate directions in order to observe EPR signals in Si- and Sn-doped bulk GaP samples [9].

Non-degenerate states, such as the Γ minimum or a deep A_1 state, are unaffected by uniaxial stress. The heteroepitaxial stress and the externally applied stresses do affect the valleys associated with the X- and L-point conduction band edges. The built-in biaxial compression along [100] and [010], due to the lattice mismatch of the $Al_xGa_{1-x}As$ and GaAs lattice constants at 1.6 K, produces an elongation of the AlGaAs lattice constant in the [001] direction through the Poisson effect. This tensile strain raises the energy of the X_z valley relative to the X_x and X_y valleys. The splitting is $\sim$14 meV for AlAs on GaAs [7]. Uniaxial stress applied along [1$\bar{1}$0] further enhances the splitting between the X_z and X_x, X_y valleys. However, uniaxial stress along [100] can lift the remaining degeneracy of the X valleys as it lowers the symmetry from tetragonal to orthorhombic. Since the long axes of the L-point conduction band constant energy ellipsoids are located along the $\langle 111 \rangle$ axes in momentum space, neither the biaxial compression nor an externally applied stress along a cube-edge direction can remove the four-fold degeneracy of the L-point valleys. However, the equivalency of the four L-point valleys is removed with stress applied in the [1$\bar{1}$0] direction. The maximum stress (200 MPa) applied in this study was approximately equal to the value of heteroepitaxial stress in AlAs on GaAs [10].

4. Results and discussion

4.1. Si donors in AlAs

ODMR spectra for the Si-doped AlAs/GaAs heterostructure with uniaxial stress (T) applied in the [100] direction and the magnetic field (B) parallel to the other in-plane cube-edge direction are shown in figure 2. Two resonances with approximately equal amplitudes are observed for zero *applied* stress with g = 1.917 $\pm$ 0.001 and 1.976 $\pm$ 0.001. However, the intensity of the high-field resonance decreases while the intensity of the low-field resonance increases with equal rate as a function of

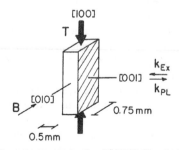

Figure 1. Sample geometry for $T \parallel$ [100]. The hatched face represents the AlGaAs epitaxial layer.

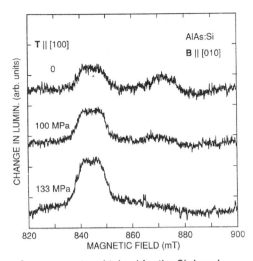

Figure 2. ODMR spectra obtained for the Si-doped AlAs/GaAs heterostructure with uniaxial stress (*T*) along [100]. These data demonstrate valley repopulation.

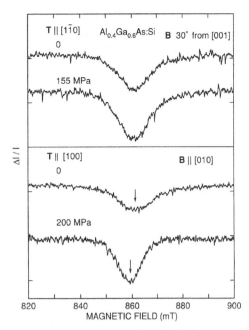

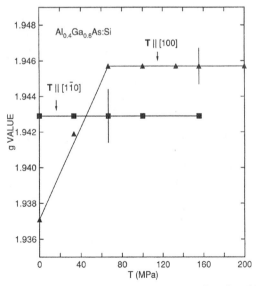

Figure 3. ODMR spectra obtained for the Si-doped Al$_{0.4}$Ga$_{0.6}$As/GaAs sample with **T**∥[110] (top half) and **T**∥[100] (bottom half).

increasing uniaxial stress along [100]. In addition, the *g* value and linewidth of the low-field line do not change up to 200 MPa.

The resonance features obtained for the Si-doped AlAs layer in the absence of external stress (**T** = 0) can be understood using an independent valley model in the presence of the built-in heteroepitaxial strain [4, 7]. Two well separated resonances of about equal amplitude are observed with **B**∥[010] because the field is simultaneously oriented parallel to the short axis (*g* value denoted by $g_\perp$) of the X$_x$ valley (low-field resonance) and parallel to the long axis (*g* value denoted by $g_\parallel$) of the X$_y$ valley (high-field resonance). The X$_z$ valley is not populated, and hence does not participate in the recombination.

The results obtained with **T**∥[100] confirm the independent valley description. For these stress conditions, donor states derived from the X$_x$ valley are lowered in energy with increasing |**T**| relative to the states derived from the X$_y$ valley. Thus, the resonance associated with the X$_y$ valley (high-field line) diminishes with increasing |**T**| while the resonance associated with the X$_x$ valley (low-field line) simultaneously grows in intensity. The donor states derived from the X$_y$ valley are depopulated for uniaxial stress along [100] greater than ∼120 MPa. These results demonstrate a pure valley repopulation. The symmetry of the Si donor state in AlAs is lowered from tetragonal to orthorhombic with **T**∥[100]. The results obtained in the absence of external stress [4, 7] and the present uniaxial stress results both provide evidence for negligible valley-orbit interactions, consistent with the symmetry arguments predicted by Morgan [8], and negligible spin-orbit (i.e. small λ) coupling for Si donors in AlAs [6, 7].

4.2. Si donors in Al$_{0.4}$Ga$_{0.6}$As

ODMR spectra for the Si-doped Al$_{0.4}$Ga$_{0.6}$As/GaAs heterostructure with uniaxial stress applied in the [1$\bar{1}$0] and

[100] directions are shown in figure 3. A compilation of the donor *g* values as a function of the magnitude of the stress for the two stress geometries is given in figure 4. There is no shift of the *g* values and little change in the linewidth of the resonance up to 200 MPa applied in the [1$\bar{1}$0] direction. However, an immediate change in the character of the resonance is observed with stress along [100]. The *g* value of the resonance with **B**∥[010] shifts from 1.937 ± 0.001 for |**T**| = 0 to 1.946 ± 0.001 for |**T**| ∼

Figure 4. ODMR *g*-values for Si donors in Al$_{0.4}$Ga$_{0.6}$As with **T** applied in the [100] (triangles) and [110] (squares) directions. The magnetic field is along [010] for **T**∥[100] and 30° from [001] for **T**∥[110]. Full lines are guides to the eye.

60 MPa. The g value does not shift further for larger values of stress. This limiting g value is equal within error to that obtained with $B \| [001]$ in the absence of externally applied stress [4–7]. Also, the linewidth of the resonance narrows with increasing stress from ~ 15.8 mT to a limiting value of ~ 10 mT. In addition, the integrated intensity of the resonance does not change as a function of stress, similar to the behaviour observed for the sum of the integrated intensities of the two donor resonances in the Si-doped AlAs sample with $B \| [010]$ discussed earlier (see figure 2).

The lack of a response to uniaxial stress along $[1\bar{1}0]$ provides evidence against the contribution of donor states derived from the L-point conduction band minima in Si-doped $Al_xGa_{1-x}As$ at $x = 0.4$. The results obtained with uniaxial stress applied along [100] indicate that the donor wavefunction is still X-like for this Al mole fraction regime.

Additional mechanisms that can potentially influence the nature of the Si donor ground state $Al_xGa_{1-x}As$ with $x < 1$ include a finite spin–valley coupling interaction and alloy disorder. Contrary to an earlier analysis [3, 4], the results of the uniaxial stress experiments with $T \| [100]$ provide evidence against a purely spin–valley coupled state at $x = 0.4$. In particular, the $g_\perp$ value observed for the Si-doped AlAs sample ($g_\perp = 1.976$) should be recovered with $T \| [100]$ and $B \| [010]$ at some finite stress. Instead, the limiting g value is 1.946.

Alloy disorder may also alter the nature of the Si donor ground state in $Al_xGa_{1-x}As$ with $x < 1$ [2]. The short-range component of the disorder can mix states of different k vector [11], and thus lead to a coupling interaction between states derived from the X_x and X_y valleys or between states derived from the Γ- and X-point conduction band minima. The local strains from the disorder are built into the crystal and hence the donors become less sensitive to the heteroepitaxial and externally induced uniform strains. The potential fluctuations due to the disorder may lead to a deepening of the Si shallow hydrogenic effective-mass states in analogy to effects seen for P and As donor states in hydrogenated amorphous Si and Ge [12]. Further work is needed to determine the strength of these interactions.

Recently, absorption and photoluminescence experiments on Si-doped GaAs under hydrostatic pressure have provided evidence for a strongly localized donor state with A_1 symmetry [13, 14]. These experiments suggest that there exist three different states of the same donor impurity: the hydrogenic effective-mass shallow state, the 'deep' A_1 state and the DX-like state. The present ODMR results on Si-doped $Al_xGa_{1-x}As$ samples with x near 0.4 may reflect a contribution of these localized, but optically active, A_1 states to the make-up of the donor states probed.

In summary, ODMR experiments performed on Si-doped AlAs and $Al_{0.4}Ga_{0.6}As$ samples with externally applied uniaxial stress along the $[1\bar{1}0]$ and [100] directions provide evidence against the contribution of donor states derived from the L-point conduction band minima and against a purely spin–valley coupled state at $x = 0.4$. The results obtained with $T \| [100]$ confirm the independent valley model for Si donor states in AlAs.

Acknowledgment

This work was supported in part by the US Office of Naval Research.

References

[1] For a review see Mooney P M 1990 *J. Appl. Phys.* **67** 3 R1
[2] Wartewig S, Böttcher R and Kuhn G 1975 *Phys. Status Solidi* b **70** K23
[3] Glaser E, Kennedy T A and Molnar B 1989 *Shallow Impurities in Semiconductors 1988 (Inst. Phys. Conf. Ser. 95)* ed B Monemar (Bristol: Institute of Physics) p 233
[4] Glaser E, Kennedy T A, Sillmon R S and Spencer M G 1989 *Phys. Rev.* B **40** 3447
[5] Montie E A, Henning J C M and Cosman E C 1990 *Phys. Rev.* B **42** 11808 and references therein
[6] Kaufmann U, Wilkening W, Mooney P M and Kuech T F 1990 *Phys. Rev.* B **41** 10206;
Wilkening W and Kaufmann U 1991 *Proc. 4th Int. Conf. on Shallow Impurities in Semiconductors, London 1990* ed G Davies (Zurich: Trans Tech Publications) p397
[7] Glaser E R, Kennedy T A, Molnar B, Sillmon R S, Spencer M G, Mizuta M and Kuech T F 1991 *Phys. Rev.* B in press
[8] Morgan T N 1986 *Phys. Rev.* B **34** 2664 and references therein
[9] Mehran F, Morgan T N, Title R N and Blum S E 1972 *Phys. Rev.* **6** 217;
Title R S and Morgan T N 1970 *Bull. Am. Phys. Soc.* **15** 267
[10] Rozgonyi G A, Petroff P M and Panish M B 1974 *J. Cryst. Growth* **27** 106
[11] Klein M V, Sturge M D and Cohen E 1982 *Phys. Rev.* B **25** 4331
[12] Stutzmann M and Street R A 1985 *Phys. Rev. Lett.* **54** 1836
[13] Dmochowski J E, Wang P D and Stradling R A 1990 *Proc. 20th Int. Conf. on the Physics of Semiconductors* ed E M Anastassakis and J D Joannopoulos (Singapore: World Scientific) p 658
[14] Liu X, Samuelson L, Pistol M E-, Gerling M and Nilsson S 1990 *Phys. Rev.* B **42** 11791

Semicond. Sci. Technol. **6** (1991) B101–B104. Printed in the UK

Magnetic resonance of Sn-doped Al$_x$Ga$_{1-x}$As detected on photoluminescence

T A Kennedy†, E R Glaser† and T F Kuech‡§

† Naval Research Laboratory, Washington DC 20375, USA
‡ IBM Thomas J Watson Research Center, Yorktown Heights, NY 10598, USA

Abstract. Photoluminescence and optically detected magnetic resonance (ODMR) have been studied for three Sn-doped Al$_x$Ga$_{1-x}$As samples with mole fractions from 0.45 to 1.0. The near-band-edge photoluminescence is weak and broad but strong deep bands occur for each sample. The ODMR is dominated by broad, luminescence-enhancing signals with $g = 2.00$. The results are quite different from those for Si or other dopants. These experiments reveal that Sn is producing an optically active deep level in these samples.

1. Introduction

Sn doping of Al$_x$Ga$_{1-x}$As produces donors when the Sn atoms are incorporated on group III sites in the lattice. These donors, like Si and the group VI dopants S, Se and Te, exhibit anomalous effects which have been called 'DX' properties. At least two states are associated with each donor. Firstly, there is a shallow, hydrogenic effective-mass (HEM) state which has the character of the Γ conduction band for low Al mole fraction (x) and the character of the X conduction band for high x. Secondly, there is a deep state with a barrier for the capture and emission of electrons. These properties have been determined using deep-level transient spectroscopy [1].

Other experiments on Sn-doped samples reveal the donor states described above and, in some cases, different states. Photoluminescence of Sn-doped Al$_x$Ga$_{1-x}$As grown by molecular beam epitaxy (MBE) has the emission characteristic of a shallow donor [2]. However, photoluminescence studies of the Sn acceptor in GaAs reveal that it is a deep state [3]. A ^{119}Sn Mössbauer study in Al$_x$Ga$_{1-x}$As found evidence of both substitutional shallow donors and a clustered species [4]. Optically detected magnetic resonance (ODMR) detected in absorption [5] and electron paramagnetic resonance (EPR) [6] experiments have detected a state with a large hyperfine interaction from the magnetic Sn nuclei. The experiments were done on samples with x from 0.3 to 0.4 and the large hyperfine interaction denotes a deep state.

§ Present address: Department of Chemical Engineering, University of Wisconsin, Madison, WI 53706, USA.

Our group at NRL has been studying the properties of donors in Al$_x$Ga$_{1-x}$As using magnetic resonance detected on photoluminescence. Our work [7], along with that of other groups [8, 9], has shown that there is an HEM state associated with the X minimum for samples with $x > 0.4$. This state is weakly perturbed by alloy disorder, band mixing and spin–valley coupling to different degrees at different x values [7-9]. Samples doped with S, Se and Te show the X-related HEM state but modified by a central-cell effect [10, 11]. This difference between a group IV donor (Si) and group VI donors is consistent with a prediction by Morgan [12].

In this work, the magnetic resonance detected on photoluminescence is extended to Sn-doped samples grown by organo-metallic vapour phase epitaxy (OMVPE). Samples with mole fractions from 0.4 to 1.0 have been studied. Both the photoluminescence and the ODMR experiments reveal a deep state—in contrast to the states revealed for the other chemical species.

2. Photoluminescence Results

The samples were grown by OMVPE with Al mole fractions of 0.45, 0.6 and 1.0 and an Sn doping of 5×10^{17} cm^{-3}. Each sample had a 1 μm undoped layer beneath the doped layer. Layer thicknesses were 4 μm for $x = 0.45$, 3 μm for $x = 0.6$ and 7 μm for the AlAs. The AlAs sample was capped with 0.2 μm of Al$_{0.5}$Ga$_{0.5}$As. For comparison, some results are included for an Si-doped sample grown by MBE. This sample had a mole

0268-1242/91/10B101+04 $03.50 © 1991 IOP Publishing Ltd

fraction of 0.41 and a donor concentration of 5×10^{16} cm^{-3}.

Photoluminescence experiments were performed at 1.6 K with a 0.22 m, double-grating spectrometer. Excitation was provided by a Kr laser operating on the 476 nm line. The power density was 1 W cm^{-2}. An Si photodiode was used to detect wavelengths shorter than 1 μm and a cooled Ge photodiode for longer wavelengths.

Photoluminescence in the band edge region was weak and broad for these Sn-doped samples (see figure 1). Using the Si photodiode, emission in the band edge region was only detected for the sample with $x = 0.45$. In contrast, an Si-doped sample showed sharper, stronger emission. The poor results in these samples are rather puzzling. As stated in the introduction, Sn-doped samples grown by MBE exhibit good band edge emission [2]. However, there is an independent report of rather broad emission from an Sn-doped sample grown by OMVPE [13]. The degradation seems to depend on the Sn doping since undoped Al$_x$Ga$_{1-x}$As grown by OMVPE shows good band edge emission for all x up to 0.80 [14].

The Sn-doped samples exhibit strong deep photoluminescence bands at different energies (see figure 2). In this regard, these samples are similar to the Al$_x$Ga$_{1-x}$As doped with other chemical species. For the $x = 0.6$ sample, the strong peak around 1.5 μm is identical to what is often seen in samples grown by OMVPE [8, 10, 11]. The AlAs sample has a distinct peak around 1.1 μm (1.1 eV in energy), which will be shown to be connected to a strong, distinct ODMR. As in most of the previous studies, the ODMR was detected on the deep bands.

3. ODMR detected on photoluminescence

ODMR was detected as a change in the intensity of the photoluminescence in a Voight-geometry spectrometer.

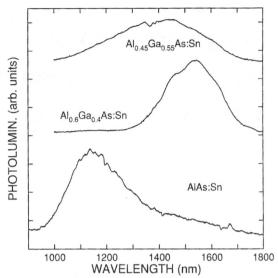

Figure 2. Deep photoluminescence for three Sn-doped Al$_x$Ga$_{1-x}$As samples with different mole fractions. Different bands dominate in each sample.

Fifty milliwatts of 24 GHz microwave power was on/off modulated at 170 Hz. The sample temperature was 1.6 K with magnetic fields up to 1.1 T available from an electromagnet. The photoexcitation was the same as for the photoluminescence experiments. Light was detected by the Ge photodiode preceded by a 1.0 μm long-pass filter.

The ODMR results for the three Sn-doped samples differ from each other and from the other dopants (see figure 3). All three Sn-doped samples do show resonances

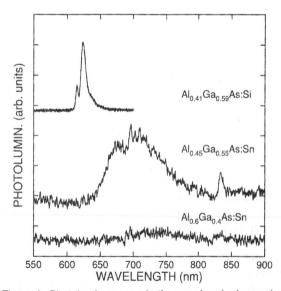

Figure 1. Photoluminescence in the near-band-edge region for an Si-doped and two Sn-doped samples. The emission in the Si-doped sample is much stronger and sharper than that for the Sn-doped samples.

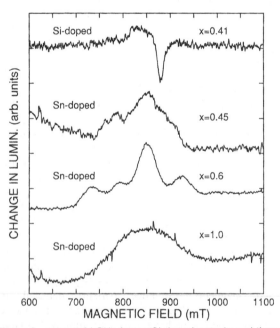

Figure 3. ODMR at 24 GHz for an Si-doped sample and three Sn-doped samples. The Si-doped sample exhibits a sharp negative signal around 874 mT corresponding to $g = 1.94$. The Sn-doped samples exhibit differing positive signals. All have maxima around 850 mT corresponding to $g = 2.00$. The signal in the $x = 0.6$ sample is superposed on a four-line spectrum probably due to Ga interstitials.

with g values equal to the free electron, i.e. 2.00. The spectra for $x = 0.45$ and 0.6 are isotropic and broad. For $x = 0.45$, the spectrum may have partially resolved hyperfine structure. Resolved hyperfine structures from Sn nuclei have been reported for slightly lower x values [5, 6]. The spectrum for $x = 0.6$ seems best interpreted as a single line at $g = 2.00$ on top of a hyperfine-split spectrum from a Ga interstitial [15]. All three spectra are quite distinct from the sharp, anisotropic spectra with an average g-value of 1.94 observed for Si-doped samples [7-9, 11] (see figure 3).

The AlAs:Sn sample exhibited two lines whose intensity varied with crystal orientation (see figure 4). A strong positive line was observed with $B\|[001]$ with $g = 2.00$ and a linewidth of 150 mT. This line decreased in intensity as the field was rotated to [110]. Converse anisotropy was observed for a negative ODMR with $g = 2.09$ and a linewidth of 20 mT. Strong anisotropies in intensity have not been reported before for dopants in Al$_x$Ga$_{1-x}$As.

The anisotropy observed for the AlAs:Sn sample was about the [001] axis, the growth axis of the layers. Since the AlAs was grown on a GaAs substrate, the lattice mismatch of 1.4×10^{-3} produced a tensile strain in the [001] direction. In order to further probe the strain sensitivity of the resonances, some preliminary experiments have been performed with externally applied uniaxial stress. For the $x = 0.6$ sample, stresses up to 200 MPa were applied along the [110] direction, enhancing the built-in [001] elongation in that sample. The spectra were insensitive to the stress. For the AlAs sample, stresses up to 200 MPa were applied along [100], which along with the built-in strain led to an ortho-

rhombic symmetry. For $B\|[010]$, the spectrum was insensitive to the added stress. There are some differences between the [001] spectra in the usual geometry and those in the stress geometry. Smaller samples were used for the stress measurements and hence that geometry contains less microwave electric field.

4. Discussion

Photoluminescence and ODMR of these Sn-doped samples contrast sharply with results for Si-doped samples. Good band edge emission is not observed. The ODMR is at $g = 2.00$, the free-electron value, rather than centred around 1.94, the value characteristic of HEM donors associated with the X minima [7-10]. The linewidths for Sn are much broader than for Si. The line positions are insensitive to strain. These factors indicate a deep, symmetric (a$_1$) state and the results are consistent with the observations of a large hyperfine interaction in samples with slightly lower x [5, 6]. A deep, symmetric state has also been found for Ge in GaP [16].

The results for AlAs:Sn are distinct from the data for other mole fractions. The two lines with anisotropic intensities are difficult to explain. Two partial explanations are offered. Firstly, the anisotropy in intensity may indicate a deep donor to shallow acceptor recombination process with the anisotropy arising from the strained valence band. The selection rules producing the intensity-ODMR would break down as B moves away from the symmetry axis (the [001]). This effect has been reported in (hexagonal) CdS [17]. Secondly, the two resonances may be either the donor and the acceptor or the closer pairs and the more distant pairs. Similar two-line spectra have been observed in heavily doped GaP [18]. More work is necessary to achieve a complete analysis of the AlAs:Sn results.

What can be deduced about the atomic structure of the Sn-related state? The deep symmetric state seems inconsistent with the Sn being on the group III site and unrelaxed. However, at present it is difficult to choose among the other possibilities: a relaxed isolated donor, an Sn complex on either lattice site, or Sn occupying predominately the As site.

Two further experiments are indicated to address the questions raised. First, different growth and doping methods should be checked to see whether a different Sn state is produced. We are trying ion implantation of Sn into undoped Al$_x$Ga$_{1-x}$As. Previously, implantation of Si and S produced the HEM states [7, 8, 10, 11]. Second, ENDOR measurements of the nearest neighbours would serve to demonstrate the symmetry of the electronic spin resonances now observed. We are currently developing an ENDOR spectrometer and will apply it to this problem.

In summary, photoluminescence and ODMR experiments on a set of Sn-doped Al$_x$Ga$_{1-x}$As samples reveal a deep state rather than the hydrogenic effective-mass state revealed for Si and other dopants.

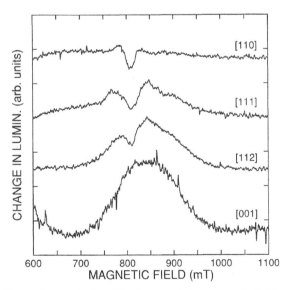

Figure 4. ODMR in the AlAs:Sn sample for magnetic field directions from the [110] to the [001]. Two resonances are evident with intensities which vary with angle. A negative signal is observed at 820 mT ($g = 2.09$) which is strongest for $B\|[110]$. A positive signal is observed at 850 mT ($g = 2.00$) which is strongest for $B\|[001]$.

Acknowledgement

This work was supported in part by the Office of Naval Research.

References

[1] Mooney P M 1990 *J. Appl. Phys.* **67** R1
[2] Wicks G, Wang W I, Wood C E C, Eastman L F and Rathbun L 1981 *J. Appl. Phys.* **52** 5792
[3] Schairer W, Bimberg D, Kottler W, Cho K and Schmidt M 1976 *Phys. Rev. B* **13** 3452
[4] Gibart P, Williamson D L, El Jani B and Basmaji P 1988 *Phys. Rev. B* **38** 1885
[5] Fockele M, Spaeth J-M and Gibart P 1990 *Proc. 20th Int. Conf. on the Physics of Semiconductors* ed E M Anastassakis and J D Joannopoulos (Singapore: World Scientific) p 517
[6] von Bardeleben H J, Bourgoin J C, Delerue C, Lannoo M and Gibart P *Preprint*
[7] Glaser E, Kennedy T A, Sillmon R S and Spencer M G 1989 *Phys. Rev. B* **40** 3447
[8] Montie E A, Henning J C M and Cosman E C 1990 *Phys. Rev. B* **42** 11808
[9] Kaufmann U, Wilkening W, Mooney P M and Kuech T F 1990 *Phys. Rev. B* **41** 10206
[10] Glaser E, Kennedy T A, Molnar B and Mizuta M 1990 *Mater. Res. Soc. Symp. Proc.* **163** 753
[11] Glaser E, Kennedy T A, Molnar B, Sillmon R S, Spencer M G, Mizuta M and Kuech T F *Phys. Rev.* submitted
[12] Morgan T N 1968 *Phys. Rev. Lett.* **21** 819
[13] El Jani B, Koehler K, N'Guessan K, Bel Hadj A and Gibart P 1988 *J. Appl. Phys.* **63** 4518
[14] Kuech T F, Wolford D J, Veuhoff E, Deline V, Mooney P M, Potemski R and Bradley J 1987 *J. Appl. Phys.* **62** 632
[15] Kennedy T A, Magno R and Spencer M G 1988 *Phys. Rev. B* **37** 6325
[16] Godlewski M, Goldys E, Heymink Liesert B J and Sienkiewicz A 1990 *20th Proc. Conf. on the Physics of Semiconductors* ed E M Anastassakis and J D Joannopoulos (Singapore: World Scientific) p 646 and references therein
[17] Patel J L, Nicholls J E and Davies J J 1981 *J. Phys. C: Solid State Phys.* **14** 1339
[18] Godlewski M and Monemar B 1988 *J. Appl. Phys.* **64** 200

Semicond. Sci. Technol. **6** (1991) B105–B110. Printed in the UK

Magnetic resonance studies of group IV and VI donors in $Ga_{1-x}Al_xAs$

H J von Bardeleben

Groupe de Physique des Solides, Université Paris VII, Centre National de la Recherche Scientifique, Tour 23, 2 place Jussieu, 75251 Paris Cedex 05, France

Abstract. The magnetic resonance results for the group IV (Si, Sn) and group VI (S, Se, Te) donors in GaAlAs alloys are reviewed. For the donors Si, S, Se, Te the DX deep donor state has apparently not been observed; the paramagnetic state generated by photoexcitation has been attributed to the X conduction band derived effective-mass (EM) state. On the contrary recent results for Sn, where the central hyperfine interaction could be resolved, demonstrate clearly that for Sn the paramagnetic donor state is not EM-like; it is attributed to the neutral charge state of the DX centre. A comparison of the photoexcitation process and the associated persistent photoconductivity gives a strong indication for a negative-U DX ground state. It is shown that recent deep-level transient spectroscopy results further support this assignment. However, it is unclear why in some cases (Si, S, Se, Te) apparently only the EM states are observed and not the deep donor state, whereas in the case of Sn the situation is reversed.

1. Introduction

The coexistence of shallow effective-mass states and deep DX states is one of the characteristics of group IV donors Si, Ge, Sn and group VI donors S, Se, Te in $Ga_{1-x}Al_xAs$ alloys. Depending on the alloy composition either only the effective-mass state gives rise to a gap level ($x \lesssim 0.2$, atmospheric pressure) or for $x \gtrsim 0.2$ both of them can be observed simultaneously. The properties of the deep state, DX, which becomes the ground state for $x \gtrsim 0.2$ have been studied in detail in the past few years and the results have been reviewed recently [1, 2]. The basic properties of the group IV and VI donor-related DX states are very similar: a common existence region $x \gtrsim 0.2$, a thermal barrier for electron capture, being of comparable magnitude to the activation energy for electron emission E_{th}, and a high photoionization threshold E_{opt} with $E_{opt} \gtrsim 3E_{th}$; nevertheless, important chemical shifts of these properties have been reported indicating different local atomic configurations [3, 4].

Different models have been proposed for the origin of the deep DX state: small lattice relaxation SLR models [5, 6], which relate the variation of the electron emission and capture barriers to the conduction band structure as well as large lattice relaxation (LLR) models implying Jahn–Teller distortions of the $T_2(X)$ states [7] or strong configurational instabilities leading to bond breaking and [111] off-centre displacements with a negative-U ground state [8]. In spite of the large number of experimental results the microscopic configuration of the DX ground state is still a matter of debate. Magnetic resonance techniques are particularly well adapted to the study of the electronic structure of the defect ground state and they have, of course, been applied to the DX centre in $Ga_{1-x}Al_xAs$: static magnetic susceptibility [9, 10], electron paramagnetic resonance (EPR) [1, 11–15], optically detected magnetic resonance (ODMR) [1, 16, 17] and magnetic circular dichroism (MCD) measurements [18,19]. With the exception of the negative-U model of Chadi and co-workers, [8, 20], which predicts a diamagnetic two-electron ground state, all other models imply a paramagnetic $S = 1/2$ ground state. And, indeed, paramagnetic donor states have been observed for the Si [11–13, 16, 17], Sn [14, 18, 19], S [21], Se [21] and Te [15] donors in $Ga_{1-x}Al_xAs$ (figure 1); however, none of them seemed to correspond to the DX ground state. The EPR and ODMR results were interpreted in the model of X conduction band (CB) related effective-mass states [22], which can be populated by low-temperature photoexcitation or are already occupied after cooling the sample in the dark due to the incomplete freeze-out of the carriers on the DX ground state for high AlAs mole fractions. Very recently the observation of a strong central hyperfine interaction in the case of the Sn donor in $Ga_{1-x}Al_xAs$ [18, 19, 23] has demonstrated that the effective-mass (EM) model is inadequate for this donor; for the first time a deep donor state, which is attributed to the neutral DX state, is seen by the magnetic resonance technique. Further, we will show that an analysis of the photoexcitation process of the paramagnetic state, as well

0268-1242/91/10B105+06 $03.50 © 1991 IOP Publishing Ltd

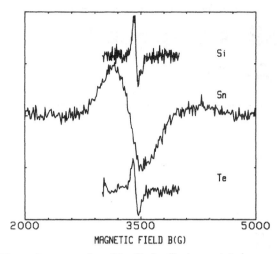

Figure 1. EPR spectra of the Si, Sn, Te donor state in $Ga_{1-x}Al_xAs$ with $x \approx 0.6$; $T = 4$ K, $B \| [001]$.

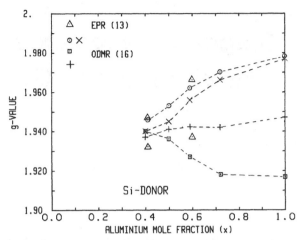

Figure 2. Compilation of the g values of the Si donor as a function of AlAs mole fraction with B rotated in the (110) (DX) and (001) (0, +) planes.

as of its thermal stability, gives strong indications for a negative-U character of the DX ground state. Experimental evidence for the observation of the paramagnetic state by other experimental techniques such as deep-level transient spectroscopy (DLTS) is presented.

2. Magnetic resonance results in Si- and Te-doped $Ga_{1-x}Al_xAs$: the effective-mass model

As bulk crystals of $Ga_{1-x}Al_xAs$ are generally not available, most of the magnetic resonance studies had to be performed on GaAlAs/GaAs heterostructures, where the GaAlAs layer is biaxially strained due to the lattice mismatch with a maximum strain of 1.4×10^{-4} for $x = 1.0$. The first results on the Si donor in GaAlAs showed the importance of this built-in strain for the paramagnetic donor spectrum. In 1973 Böttcher et al [11] published the only study concerning bulk crystals for alloy compositions $x = 0.6 \cdots 0.8$. An isotropic single-line spectrum with $g = 1.963$ could be observed without application of uniaxial stress, contrary to what might have been deduced from the group IV donor properties in GaP [24]. When these measurements were repeated on Si-doped GaAlAs/GaAs heterostructures ($x = 0.56$, 0.70, 0.80) the EPR spectrum was found to be anisotropic. Its angular variation for a rotation of the magnetic field in the (110) plane could be described by $g^2(\theta) = g_{\|[001]} \cos^2 \theta + g_{\perp[001]} \sin^2 \theta$. These authors [11, 12] attributed the EPR spectrum to the X CB-derived effective-mass ground state, which is a $T_2(X)$ state for the group IV donor [22]. These preliminary results were more recently completed by ODMR studies of the Si donor, covering the entire indirect gap composition range $0.37 \leqslant x \leqslant 1.0$ [16, 17] as well as by EPR studies for $x = 0.41$ and $x = 0.60$ [13]. The particular symmetry of the resonance spectrum — a rotation of the magnetic field in the (001) growth plane revealed a splitting of the single-line spectrum into two components of equal intensities with a

fourfold symmetry around the [001] growth axis — seems to support the original assignment by Böttcher et al.

An overview of the reported g values for rotations of the magnetic field in the (110) and (001) planes is shown in figure 2. Both the anisotropy of the g factor $g_{\|[100]} - g_{\|[001]}$ and the absolute g values are composition dependent. The interpretation of these results in the EM model (figure 3) requires the consideration of the hetero-epitaxial strain along the z growth axis which destroys the degeneracy of the T_{2x}, T_{2y} and T_{2z} EM states. Further, the presence of random strains has to be evoked to explain the quenching of the spin–valley coupling [25] and the equal occupation of the T_{2x}, T_{2y} ground states. Whereas this model describes quantitatively the case of AlAs/GaAs, a quantitative comparison of the alloy dependence of the g tensor using the known variation of the strain, effective masses and band gap with x has not yet been presented.

Montie et al [17] have extended the effective-mass model for the Si donor to $0.37 < x < 1.0$. They invoke a mixing of the X, L, Γ effective-mass states $T_2(X_3)$, $T_2(L_1)$, $A_1(\Gamma)$ which requires a lowered C_2 symmetry, the origin of which is not specified.

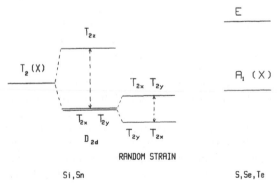

Figure 3. X CB EM ground states of the group IV and VI donors.

Table 1.

	Te	Se	S
Symmetry	A_1	A_1	A_1
g_0	1.952	1.955	1.957
$(A_1 - E)$ splitting (meV)		20	19
References	[15]	[21]	[21]

EPR [15] and ODMR [21] results on group VI donors S, Se, Te (figure 1) have also been reported (table 1). EM theory predicts in this case an $A_1(X)$ ground state, insensitive to random strain, giving rise to an isotropic g factor slightly perturbed by the $E(1s)$ state. The experimental results for S, Se, Te seem to be in agreement with these predictions [22, 24]. Contrary to the case of Si, the line splitting for a rotation of the magnetic field in the (001) growth plane is no longer observed. However, an angular variation in the (110) plane with an anisotropy of $g_{//[001]} - g_{//[110]} \lesssim 0.01$ is still present for Se and S [21]. The EPR studies of the Te donor state show for $x = 0.30$ [15] and $x = 0.65$ [26] a single-line EPR spectrum, which is also isotropic for a rotation of the magnetic field in the (110) plane. Hetero-epitaxial strain can nevertheless admix the first excited E state into the A_1 ground state.

Thus for both group IV and VI donors an EPR spectrum is observed which has the characteristics of the X CB EM state. Surprisingly, in none of these studies is the deep donor state, expected to be paramagnetic before or after photoionization, detected.

The observation of the X CB EM states after photoionization of the DX ground state in indirect-gap material ($x > 0.37$), where the X CB is lowest, is in agreement with the thermal barrier for electron capture of the DX centre. However, when we pass to direct-gap material ($x < 0.37$) with the lowest conduction band changing to Γ, we expect the EPR spectrum of the EM state to change drastically. Indeed, no ODMR or EPR result for the Si donor has been reported in this case. EPR results of the Te donor show, however, an unexpected smooth variation of the g factor and linewidth at $x \gtrsim 0.37$ and the EPR spectrum can be observed down to alloy compositions $x = 0.30$ [26]. At this composition the $A_1(X)$ state lies ≈ 90 meV above the Γ CB minimum [27]. Thus in order to explain the results in the X CB EM model we must assume a thermal ionization energy E_A for $A_1(X)$ higher than 90 meV, or, if we accept the theoretical value of $E_A \approx 40$ meV, we have to assume a shift of the quasi-Fermi level after low-temperature photoexcitation to $E_\Gamma + 50$ meV, which is reasonable in these highly doped ($> 3 \times 10^{18}$ cm^{-3}) samples.

3. Magnetic resonance in Sn-doped $Ga_{1-x}Al_xAs$: inadequacy of the EM model

Whereas no magnetic resonance result has been reported for Si in direct-gap material, the group IV donor Sn does give rise to a paramagnetic state [14, 18, 19, 23]. Its EPR

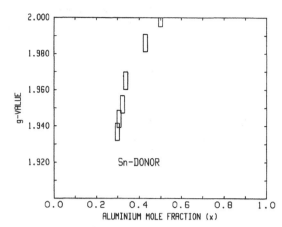

Figure 4. g-factor variation as a function of alloy composition for the Sn donor.

spectrum is isotropic and is only observed after photo-excitation with a threshold of 0.8 eV. In one case ($x = 0.31$) of a particularly highly doped sample a lineshape analysis for different optical excitation conditions allowed decomposition of the EPR spectrum into two single-line spectra with $g_1 = 1.92$, $\Delta B_{pp} = 520$ G and $g_2 = 1.95$, $\Delta B_{pp} = 200$ G [14]. The two spectra were at first tentatively attributed to the T_{2x}, T_{2y} and T_{2z} X CB EM states. Further EPR studies on samples with $x = 0.30$, 0.33, 0.35, 0.43, 0.50 showed in all cases one isotropic single-line spectrum, the g factor of which varies quite differently from those of the other group IV donor Si (figure 4). It changes smoothly at the Γ-X cross-over and for $x = 0.5$ is already very close to $g = 2.000$ [23]. In the highest-doped samples ($\gtrsim 3 \times 10^{18}$ cm^{-3}) an additional line is observed at the high-field side (figure 5) whose intensity ratio central line/secondary line ≈ 15 indicates its hyperfine origin (^{119}Sn, ^{117}Sn; 7.6%, 8.6%). This assignment has been confirmed by the study of a 75% ^{119}Sn-enriched sample (figure 5) [23]. The results confirm those of a previous MCD K-band study of an $x = 0.35$ sample [18, 19]. These authors attributed the

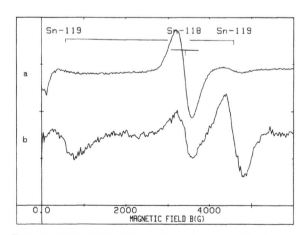

Figure 5. EPR spectra of the Sn donor in non-enriched (a) and isotopically enriched (75% ^{119}Sn) (b) Sn-doped $Ga_{0.6}Al_{0.4}As$.

MCD absorption band with a 0.7 eV threshold to the photoionization of the deep DX state and concluded that the threshold of ~ 0.7 eV is incompatible with a shallow EM state. A comparison of the high central hyperfine interaction constant of $A \approx 10$ GHz with the atomic values of ^{119}Sn allows us [23] to calculate the electron localization on the 5s central donor orbital: 23 %. Whereas such a high value is typical for deep antibonding A_1 states such as Sn in II–VI compounds [28], it is completely incompatible with an EM state for which a localization $\ll 1 \%$ can be expected [23]. Thus a different model for the paramagnetic Sn donor state is required.

We attributed the origin of this deep paramagnetic state to the singly occupied A_1^{ab} antibonding state of the substitutional donor [23]. Recently, it has been shown [29–32] that the A_1^{ab} state is resonant with the CB in GaAs, but becomes localized into the gap in $Ga_{1-x}Al_xAs$ for $x \gtrsim 0.25$. The calculated alloy variation of the A_1^{ab} state is close to that of the DX ground state. Its first excited state, the T_2^{ab} state, has been calculated to lie ~ 1 eV above A_1^{ab} and will thus stay resonant with the lowest CB for all alloy compositions. An attribution of the MCD absorption band to the A_1^{ab}–T_1^{ab} internal transition should be considered. It might be argued that as the MCD spectrum has not been calibrated, this deep donor state is not related to the DX centre but to a minor Sn-related defect complex. Whereas the MCD technique does not allow us to determine the spin concentration as long as the absorption coefficient has not been correlated with a spin concentration by an independent technique, this can be done in an EPR experiment. We have found that within a factor of 2 the spin concentration is equal to the nominal doping concentration. From this we conclude that we observe the paramagnetic one-electron state of the DX centre. The deep paramagnetic state with a resolved central hyperfine interaction could be measured up to alloy compositions $x = 0.5$. The additional EPR spectrum which corresponds to the X CB EM state, is in general not detected for $x < 0.37$.

4. Photoexcitation process, thermal stability

Independent of the direct- or indirect-gap character of the alloy, the paramagnetic state of the Sn donor ($x \leqslant 0.5$) and the Te donor ($x \leqslant 0.65$) is never observed before photoexcitation. The Si donor seems to be an exception as in the case of $x \geqslant 0.61$ it is already observed in the dark [11–13]. The photoexcitation spectrum of the paramagnetic state has been determined for all three donors: Si ($x = 0.60$) [13], Sn ($x = 0.31$) [14], Te ($x = 0.30$) [15] (figure 6). The results show the agreement of the threshold energy for this process with the threshold energy measured by capacitance techniques for the photoionization of the deep DX centre. In the case of Te the optical cross section for each process has been determined quantitatively. From this we obtain the important information that the formation of the paramagnetic state proceeds by ionization of the DX centre. As we also know that the paramagnetic state after photoionization is the

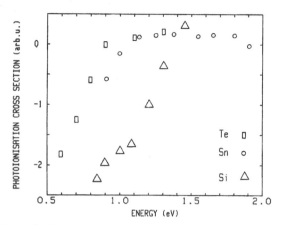

Figure 6. Photoexcitation spectra for the paramagnetic donor state of Te, Sn, Si.

neutral, one-electron $S = 1/2$ state the initial state of DX must be the diamagnetic $S = 0$ state as predicted by the negative-U model. A further test of this model is given by the already reported transport measurements on the Sn donor in $Ga_{1-x}Al_xAs$ ($x < 0.31$) [33]. Shubnikov–de Haas results show that low-temperature photoexcitation leads to the formation of a persistent photoinduced carrier concentration of $\approx 10^{18}$ cm^{-3}, which is close to the photoexcited spin concentration.

These results demonstrate clearly the negative-U character of the Sn DX centre in GaAlAs; they require the following photoexcitation process:

$$DX^- + d^+ \rightarrow DX^0 + d^+ + e^- \rightarrow DX^0 + d^0.$$

They imply, on the other hand, that both final donor states, DX^0 as well as the hydrogenic d^0 state, are expected to coexist and should in principle be detectable by EPR spectroscopy.

The reverse process of the excitation of the paramagnetic state, its thermal annihilation, has equally been studied for the three dopants Si [13], Sn [14, 23], Te [15, 26, 34]. The recovery of the DX^- ground state could proceed in two ways: thermal excitation of the bound electron (d^0) with the CB and recapture via a thermal barrier into the DX ground state

$$DX^0 + d^0 \rightarrow DX^0 + d^+ + e^- \rightarrow DX^- + d^+$$

or direct tunnelling

$$DX^0 + d^0 \rightarrow DX^- + d^+$$

between the DX^0 and d^0 states. The EPR results on the Sn and Te donor states (figure 7) show for both a thermal recovery with two different time constants. The Sn donor, for example, anneals in the 4–20 K range with a time constant of 5 min; this time constant is temperature independent and indicates thus a tunnelling process [35]. At higher temperature, 30–40 K, a second thermally activated slower annealing process is superposed. A particularity for the analysis of the thermal transients in these highly doped materials ($x < 0.37$) is that, due to the low density of states of the Γ conduction band, the

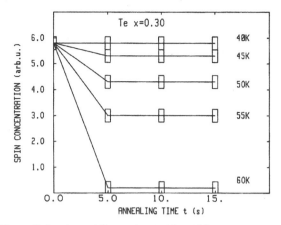

Figure 7. Isochronal thermal annealing of the paramagnetic donor state in Te-doped $Ga_{0.70}Al_{0.30}As$.

thermal barrier for electron capture will depend on the position of the quasi-Fermi level [34], which changes during the capture process. The thermally activated annealing process shows an alloy dependence and varies with the chemical nature of the donor in agreement with the electron capture barriers determined by DLTS spectroscopy $E_c(Sn) < E_c(Te) < E_c(Si)$.

5. Coexistence of two deep donor states DX^-, DX^0

The EPR results for the Si, Sn and Te donors in $Ga_{1-x}Al_xAs$ with alloy composition $x \leqslant 0.5$ have shown that the DX ground state can be transformed by photoionization into a paramagnetic state. For the Sn DX centre the high electron localization demonstrates clearly that the paramagnetic state is not an effective-mass state but the deep DX^0 state. The DX^0 state is not populated under thermal equilibrium conditions but can nevertheless be observed at low temperature after optical or electrical stimulation. Recent DLTS results confirm this model and give additional information. The multiple-peak structure of the DLTS spectra of the Sn [34], Si [36] and Te [37] donors has been frequently reported for alloy compositions $0.3 < x < 0.6$. In the case of Si this structure has generally been interpreted in a local environment model of an interstitially displaced donor. However, the EPR results demonstrate that in addition to the DX^- ground state a second, shallower, deep donor state DX^0 does exist in the gap. Its corresponding electron emission and capture processes should be observable by capacitance transient spectroscopy. The observability of the electron emission from the DX^0 state will depend on its capture barrier for the second electron. If this barrier is low, the capture of a second electron will proceed before emission of the first one and the DX^0 level will generally not be observed in a classical DLTS experiment. However, if the capture barrier is high enough, as might be expected for the Si DX centre, which has the highest electron capture barrier of the group IV and VI donors, the $DX^{0/+}$ state should be observed. Results in agreement with this model have recently been published

[38, 39]. The emission and capture energies of this shallow level, which can be attributed to the $DX^{0/+}$ level are, for $x = 0.38$, 200 meV and 170 meV respectively. Similar results have been reported for the Te DX centre as well as for the Sn DX centre, where current transient spectroscopy, which allows the detection of much faster transients with time constants in the nanosecond domain, revealed a shallower level with a reduced activation energy of ~ 70 meV as compared with the DX level [40]. As the $DX^{0/+}$ level is shallower than the DX ground state, the two states will be expected to have different existence ranges, which are schematically shown in figure 8.

The photo-EPR results make it clear that the deep DX^0 state should be of importance in all experiments not performed under thermal equilibrium conditions. Electron capture into the DX state as well as electron emission from this state are expected to proceed sequentially via the $DX \leftrightarrow DX^0 \leftrightarrow d^+$ states. It is unclear up to now why neither the deep DX^0 state of the Si donor nor the EM state of the Sn donor has yet been observed in ODMR or EPR experiments. An explanation would be that the two metastable neutral donor states d^0 and DX^0 can transform into each other even at 4 K, Si and Sn being the the two extreme cases:

$$d^0 \rightleftharpoons DX^0$$
$$\text{(Si)} \qquad \text{(Sn)}$$

6. Conclusion

The magnetic resonance results of all five donors (Si, Sn, S, Se, Te) show the presence of one paramagnetic state, which — with the exception of Si — is only observed after photoionization of the DX ground state. In the only case where a central hyperfine interaction has been resolved, the corresponding electron localization demonstrates the inadequacy of the EM model. It has recently been shown that the symmetry of the A_1 antibonding state can be expected to be lowered to C_{2v} in the case of a strong pseudo-Jahn-Teller interaction, which allows an

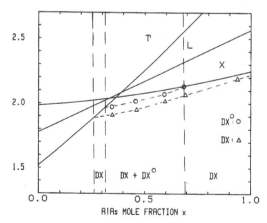

Figure 8. Conduction band structure and DX^- and DX^0 level variations in the $Ga_{1-x}Al_xAs$ alloy system.

alternative interpretation of the Si results [23]. Further magnetic resonance studies on isotopically enriched samples seem to be necessary to establish the localized or delocalized character of the paramagnetic donor states observed for Si, S, Se, Te. Nevertheless, in principle both the neutral DX state and the EM state are expected to be occupied after low-temperature photo-excitation and should thus be observable by magnetic resonance techniques. The EPR results present evidence for a negative-U character of the DX ground state as proposed by the model of Chadi and Chang [8]. However, the observation of an isotropic g tensor for the Sn DX^0 state seems to be at odds with the proposed C_{3v} symmetry of the DX ground state.

Acknowledgments

This work has been financially supported by the Basic ESPRIT contact no 3168.

References

[1] Bourgoin J C (ed) 1990 *Physics of DX Centers in GaAs Alloys (Solid State Phenomena 10)* (Vaduz: Sci. Tech. Publications)
[2] Mooney P 1990 *J. Appl. Phys.* **67** R1
[3] Lang D V, Logan R A and Jaros M 1979 *Phys. Rev.* B **19** 1019
[4] Tachikawa M, Mizuta M and Kukimoto H 1984 *Japan. J. Appl. Phys.* **23** 1594
[5] Henning J C and Ansems J P M 1987 *Semicond. Sci. Technol.* **2** 1
[6] Bourgoin J C and Mauger A 1988 *Appl. Phys. Lett.* **53** 749
[7] Morgan T N 1986 *Phys. Rev.* B **43** 2664
[8] Chadi D J and Chang K J 1988 *Phys. Rev. Lett.* **61** 873
[9] Khachaturyan K A, Awschalom D D, Rosen J R and Weber E R 1989 *Phys. Rev. Lett.* **63** 1311
[10] Katsumoto S, Matsunaga N, Yoshida Y, Sugiyama K and Kobayashi S 1990 *Japan. J. Appl. Phys.* **29** L1572
[11] Böttcher R, Wartewig S, Bindemann R, Kuhn G and Fischer P 1973 *Phys. Status Solidi* b **58** K23
[12] Wartewig S, Böttcher R and Kuhn G 1975 *Phys. Status Solidi* b **70** K23
[13] Mooney P M, Wilkening W, Kaufmann U and Kuech T F 1989 *Phys. Rev.* B **39** 5554
[14] von Bardeleben H J, Bourgoin J C, Basmaji P and Gibart P 1989 *Phys. Rev.* B **40** 5892
[15] von Bardeleben H J, Zazoui M, Alaya S and Gibart P 1990 *Phys. Rev.* B **42** 1500
[16] Glaser E, Kennedy T A, Sillmon R S and Spencer M G 1989 *Phys. Rev.* B **40** 3447
[17] Montie E A, Henning J C H and Cosman E C 1990 *Phys. Rev.* B **42** 11808
[18] Fockele M, Spaeth J M and Gibart P 1990 *Mater. Sci. Forum* **65** 443
[19] Fockele M, Spaeth J M and Gibart P 1990 *Proc. 20th Int. Conf. on Physics in Semiconductors* ed E M Anastassakis and J D Joanopoulos (Singapore: World Scientific) p 517
[20] Zhang S B and Chadi D J 1990 *Phys. Rev.* B **42** 7174
[21] Glaser E, Kennedy T A, Molnar B and Mizuta M 1990 *Mater. Res. Soc. Symp. Proc.* **163** ed D J Wolford, J Bernholc and E E Haller (Pittsburgh, PA: Materials Research Society) p 363
[24] Mehran F, Morgan T, Title R S and Blum S E 1972 *Phys. Rev.* B **6** 3917
[25] Kaufmann U, Wilkening W, Mooney P M and Kuech T F 1990 *Phys. Rev.* B **41** 10206
[26] von Bardeleben H J and Zazoui M unpublished
[27] Guzzi M and Staehli J L 1990 *Physics of DX Centers in GaAs Alloys (Solid State Phenomena 10)* ed J C Bourgoin (Vaduz: Sci Tech. Publications) p 25
[28] Brunthaler G, Jantsch W, Kaufmann U and Schneider J 1985 *Phys. Rev.* B **31** 1239
[29] Yamaguchi E 1987 *J. Phys. Soc. Japan* **56** 2835
[30] Yamaguchi E, Shiraishi K and Ohno T 1990 *Proc. 20th Int. Conf. on the Physics of Semiconductors* ed E M Anastassakis and J D Joannopoulos (Singapore: World Scientific) p 501
[31] Yamaguchi E, Shiraishi K and Ohno T unpublished
[32] Foulon Y, Lannoo M and Allen G 1990 *Physics of DX Centers in GaAs Alloys (Solid State Phenomena 10)* ed J C Bourgoin (Vaduz: Sci. Tech. Publications) p 195
[33] Lavielle D, Sallese J M, Goutiers B, Dmowski L, Basmaji P, Portal J C and Gibart P 1990 *Gallium Arsenide and Related Compounds 1989 (Inst. Phys. Conf. Ser. 106)* ed T Ikoma and H Watanabe (Bristol: Institute of Physics) p 363
[34] Zazoui M, Feng S L, von Bardeleben H J and Bourgoin J C 1990 *Mater. Sci. Forum* **65** 455
[35] von Bardeleben H J and Sheinkman M unpublished
[36] Mohapatra Y N and Kumar V 1990 *J. Appl. Phys.* **68** 3431
[37] Fudamoto M, Tahira K, Tashiro S, Morimoto J and Miyakawa T 1989 *Japan. J. Appl. Phys.* **28** 2038
[38] Jia Y B, Li M F, Zhou J, Gao J L, Kong M Y, Yu P Y and Chan K T 1989 *J. Appl. Phys.* **66** 5632
[39] Seguy P, Yu P Y, Li M, Leon R and Chan K T 1990 *Appl. Phys. Lett.* **57** 2469
[40] Balland B, Vincent G, Bois D and Hirtz P 1979 *Appl. Phys. Lett.* **34** 108

Semicond. Sci. Technol. **6** (1991) B111–B120. Printed in the UK

Metastable defects in silicon: hints for DX and EL2?

George D Watkins

Sherman Fairchild Laboratory 161, Department of Physics, Lehigh University, Bethlehem, PA 18015, USA

Abstract. A review is given of defects that display metastability in silicon, with emphasis on those that have been identified and the various mechanisms that they reveal for the phenomenon. Pair defects described include interstitial-iron-substitutional-group-III-acceptors and ones formed by interstitial carbon with substitutional group V donors or substitutional carbon. Interstitial hydrogen, boron and silicon and substitutional nitrogen and oxygen are taken as examples of isolated single-atom defects that display on-centre to off-centre instabilities. It is argued that this single-atom instability can be understood in terms of a predictable Jahn–Teller effect and that this concept may provide useful insight into the DX and EL2 phenomena in the III–V materials and their alloys.

1. Introduction

In silicon, quite a few defects have now been discovered that display metastability. Several of these have been identified and the microscopic lattice rearrangements involved determined. This is therefore a good system to look at in order to explore the variety of processes that can indeed occur and to understand the physical driving forces that give rise to the phenomenon. Hopefully this in turn may serve as a guide in helping to understand the phenomenon in other systems, in particular the III–V compound semiconductors and their alloys, the major topic of this conference.

In this paper, therefore, we will review what we believe we have learned about metastable defects in silicon. We will first consider pair defects where the metastability is easily understood as arising from changes in the lattice separation and location of the individual constituents. We will then turn to the isolated single-atom defects—first interstitial, then substitutional. Here we will combine experimental information with theory. In so doing, a pattern begins to emerge that is instructive and its relevance to DX and EL2 in the III–V compound semiconductors will be explored. Finally, other interesting metastable systems which have been found in silicon which are not yet identified will also be briefly summarized.

2. Pair defects

When a defect is made up of two constituents, there are clearly an infinite number of arrangements possible.

One of these will be the lowest energy configuration, often the one with the closest separation and all others will be excited ones. In the case where one of the constituents is an interstitial atom, which characteristically has a low activation energy for migration in the lattice ($\leqslant 1$ eV), hopping from one configuration to another becomes possible at ambient temperatures and metastability effects can be anticipated.

2.1. Donor–acceptor pairs

This is conceptually the simplest system to understand. Binding here arises from the Coulomb attraction between a positively charged ionized donor and a negatively charged ionized acceptor. As the charge state of one of the constituents changes, the Coulomb interaction changes accordingly with predictable changes in the relative stability of the various pair separations and their electrical level positions. The equilibrium concentrations of the various defect configurations therefore depend upon the charge state of the pair.

The best established example of donor–acceptor metastability is provided by the interstitial-iron-substitutional-group-III-acceptor pairs [1–8]. We consider here the example of the Fe_iAl_s pair which has been studied by both deep level transient capacitance spectroscopy (DLTS) [2,3] and electron paramagnetic resonance (EPR) [4,5]. In DLTS, two hole emission peaks are observed, the deeper one at $E_v + 0.20$ eV being strongly favored when cooled under zero bias ($Fe_i^{2+}Al_s^-$ charge state) and a shallower one at $E_v + 0.13$ eV becoming more prominent when cooled under zero bias ($Fe_i^+Al_s^-$ charge state). Study of the energetics and the

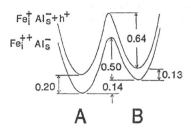

Figure 1. cc diagram for the nearest (A) and next-nearest (B) Fe_iAl_s pairs in Si. (Energies in eV.)

conversion kinetics of the two has led to the configurational coordinate (CC) diagram of figure 1. EPR studies, on the other hand, have revealed two distinct $Fe_i^+Al_s^-$ pairs, one with $\langle 111 \rangle$ symmetry indicative of a closest pair ($r = 2.35$ Å) and the other with C_{2v} $\langle 100 \rangle$ symmetry indicative of the next-nearest pair (2.72 Å). The relative energies of the configurations of figure 1 match very closely the predicted Coulomb energy differences for the two separations for each of the two charge states and the barrier for conversion is as expected being comparable to the diffusional migration energy for interstitial iron (~ 0.85 eV) [9]. Taken together, these studies establish convincingly that it is the nearest (A) and next-nearest pairs (B) that are being observed in the DLTS studies and that the simple Coulomb interaction model contains all of the essential physics of the system.

The other Fe_i–acceptor systems reveal departures from this simple behaviour in that the relative stability of the nearest versus next-nearest pairs appears to depend upon the size of the acceptor atom, reversing in the case of In [3]. This points out that additional effects such as strain must also be taken into account for detailed agreement [10].

In the case of each of these simple pairs, at most two configurations have actually been detected. A much more dramatic example of donor–acceptor metastability has recently been discovered for *interstitial-carbon-acceptor-substitutional-group-V-donor pairs* [10–17]. For the C_iAs_s and C_iSb_s pairs, four distinct bound pairs have been detected by DLTS and for C_iP_s, five! [16,17] The defects can be cycled reversibly between the various configurations by different sequences of diode bias, temperature and minority carrier injection and relatively complete CC diagrams have now been established for each [17]. Again a Coulomb model of C_i^0, C_i^-, or $C_i^=$ paired off with the positively charged donor of various separations appears to account for the general features, the barriers for most of the conversions being comparable to the diffusional activation energy in this case of C_i (~ 0.7–0.8 eV) [18]. Recent EPR studies reported for two of the C_iP_s configurations [19] provide promise of further establishing the microscopic configurations.

2.2. The C_iC_s pair

The most thoroughly studied metastable defect in silicon is the C_iC_s pair [20–22], formed when interstitial

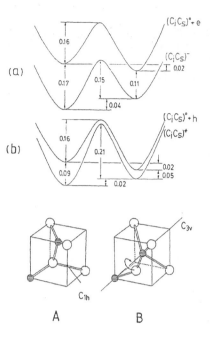

Figure 2. Models and cc diagrams for the C_iC_s pair in Si: (a) acceptor state, (b) donor state. The smaller cross-hatched atoms are carbon. (Energies in eV.)

carbon, produced by electron irradiation, migrates and is captured by substitutional carbon. Combining DLTS, EPR, PL (photoluminescence) and ODMR (optical detection of magnetic resonance) results, a fairly complete picture of the defect has emerged [22]. Two nearly equally energetic configurations exist for the defect and models deduced from these studies for the microscopic structure of each are shown in figure 2. It is a bistable defect, configuration A being the stable one for both the ionized donor (+) and acceptor (−) states and B the stable one for the neutral state. For each of the charge states, the energies for the two configurations and the barrier for conversion between them have been determined, as summarized in the CC diagrams also shown in figure 2.

In this case there is no long range Coulomb attraction between the constituents since C_s is electrically inactive (neutral). The binding of the pair therefore is probably best thought of as arising from the strain fields of each. This is particularly evident in configuration A, where the arrangement can be recognized as a substitutional (small) carbon atom (the fourfold coordinated one in the lower left-hand corner) in a position to relieve the compressional strain on the silicon end of a $\langle 100 \rangle$-oriented silicon–carbon 'interstitialcy', which is the normal configuration of isolated interstitial carbon [18,23]. The nearly energetic B configuration in turn must reflect the chemical near equivalence of carbon and silicon, allowing a trade-off of chemical versus strain energy for the two configurations. Both A and B have the same number of completed bonds and the low barrier for interconversion results because the two differ only by a simple molecular bond switching. The electrical activity of configuration A arises from

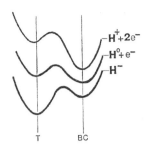

Figure 3. Schematic cc diagram for H in Si.

the single available non-bonding orbital on each of the three-fold coordinated silicon and carbon atoms which supply levels in the gap. In configuration B, the activity arises from the two non-bonding orbitals on the bond-centred silicon atom. There are therefore the same number of non-bonding available orbitals also for the two configurations.

(Evidence of a very similar bond switching is also apparent in the C_iAs_s and C_iSb_s pairs described in the previous section, where two of the configurations for each also form a bistable pair with a conversion barrier of only ~ 0.2 eV [15,16].).

3. Isolated interstitial atoms

3.1. Hydrogen

The experimental evidence from muonium spin resonance (μSR) [24–26] studies (muonium can be considered a short-lived light isotope of hydrogen) and EPR of 77 K H-implanted samples [27] reveals that the stable configuration for neutral interstitial hydrogen is a bond-centred one (BC) where the hydrogen is nestled between two nearest neighbour silicon atoms. Also detected in the μSR studies is a metastable position for H^0 which reflects the high symmetry tetrahedral (T) vacant interstitial site in the lattice [24].

Isolated interstitial hydrogen therefore appears to provide another simple textbook example of metastability where the detailed microscopic configurations of the neutral state have been established. In figure 3, we illustrate this in a schematic CC diagram where we have also sketched in a consensus of recent theoretical predictions [28–32] for the H^+ and H^- charged states. In most of the calculations so far (not [32]), the BC configuration has been predicted to be the stable one for H^0, consistent with experiment, and in all of the calculations it is the predicted configuration for the ionized H^+ charge state. The calculations of Van de Walle et al [31] and Chang and Chadi [32], on the other hand, have also explored the H^- surface and conclude that for it the stable configuration converts to the T site, as shown. If correct, hydrogen is actually a *bistable* defect, similar to the C_iC_s pair, where the stable configuration changes with charge state.

The relative energies illustrated in figure 3 for each charge state have been chosen to reflect the results of Van de Walle et al [31]. In particular, their results suggest that hydrogen is a negative-U system, the calculated $H^- \rightarrow H^0 + e^-$ ionization energy being ~ 0.4 eV greater than that of $H^0 \rightarrow H^+ + e^-$ and that H^0 is therefore not a thermodynamically stable charge state of the system. I have taken liberties with their results, however, in inserting barriers between the two configurations for each charge state. Their results give none and they suggest that the clear metastability evidenced in the μ^0 (T) site could rather result from rapid tunneling motion of the light mass muon, which inhibits the trapping into the BC site but would therefore probably not apply to hydrogen. And so controversies remain. Recent DLTS studies [33] of a defect believed to be interstitial hydrogen in 77 K hydrogen-implanted silicon (and which displays metastability) show promise of resolving some of these questions.

3.2. Interstitial boron

From combined DLTS [34,35] and EPR [36] studies, interstitial boron has been established to have negative-U properties between its B_i^+, B_i^0, and B_i^- charge states and the configuration for the neutral paramagnetic B_i^0 state has been deduced to be distorted slightly off bond-centre as illustrated in figure 4(b). In addition to its negative-U properties it displays athermal reorientation and migration under optical excitation and electrical minority carrier injection, strongly suggestive of substantial lattice rearrangements versus charge state. Combining bits of experimental evidence from these studies led to suggested models for the other charge states as shown in figure 4(b).

Very recent theoretical calculations [37] have thrown new light on the defect. An off-bond-centred configuration is apparently indeed found to be the lowest in energy for B_i^0 and negative-U properties are predicted. On the other hand, the B_i^- configuration is concluded to remain off-bond-centred and that for B_i^+ is predicted instead to be what is in effect a $Si_i^{2+}B_s^-$ pair formed as the silicon atom is displaced backward into the nearby interstitial T site and the boron atom takes its place. These are also shown in figure 4(c). This alternative configuration for B_i^+ was not anticipated in the early EPR studies even though, in hindsight, it actually appears to explain better the various experiments done at the time to probe its structure. (It continues to explain why the $B_i^0 \rightarrow B_i^+ + e^- \rightarrow B_i^0$ cycle retains the $\langle 111 \rangle$ defect alignment but loses memory of the particular off-axis distortion, but at the same time explains better the sense of the $\langle 111 \rangle$ defect alignment under applied stress [36]). Our best guess therefore at this juncture is that this new B_i^+ configuration predicted by theory and shown in figure 4(c) is probably the correct one. We will reserve judgement on the B_i^- configuration at this stage, however, because the configurations of figure 4(c) alone do not appear to supply a mechanism for the experimentally observed recombination-enhanced migration.

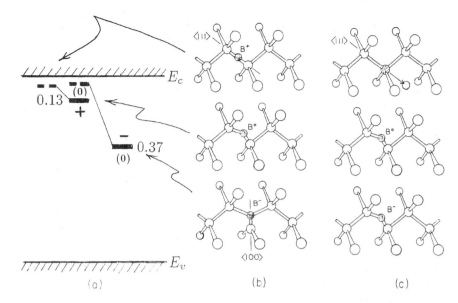

Figure 4. (*a*) Level positions (eV from the conduction band) and models suggested, (*b*) originally from experiment [36] and (*c*) recently from theory [37], for interstitial B in Si.

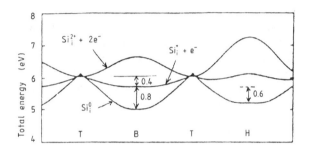

Figure 5. cc diagram for interstitial Si (after [39]). (Energies in eV.)

3.3. Interstitial silicon

At present, we have no direct experimental information about the properties of interstitial silicon except that it is highly mobile and attempts to freeze it in for studies, even in a 4.2 K electron radiation damage experiment, have been unsuccessful [38]. We do have hints, however, from the structure found for defects formed from the trapping of interstitials, such as for example, the C_iC_s pair and B_i already discussed. A close inspection of figures 2 and 4 reveals that although we have labeled these for convenience as C_iC_s and B_i, which atom is actually the interstitial or whether it is shared between more than one is highly variable and depends upon charge state. In just these two examples we see that we can have a twofold or a threefold shared silicon interstitialcy or a case where a silicon atom is predicted to be the 'true' interstitial in the T site. This charge state dependence of various similarly energetic bonded and non-bonded interstitial configurations supplies an important hint as to possible bi- and metastable properties for the isolated interstitial as well as clues to the

mechanism for its high diffusional mobility.

Two independent theory groups have recently modelled the silicon interstitial [39, 40] and their results are remarkably consistent with our clues above. In the Si_i^{2+} state, the T site is predicted to be the most stable, being ~ 0.5–0.7 eV more stable than the BC site. In the Si_i^+ and Si_i^0 states, the T site is no longer the minimum energy position and the interstitial moves to one of several possible other positions which appear to be very close in energy. These include the BC site, another bonding configuration having a split $\langle 110 \rangle$ character and the hexagonal site (the open lattice site halfway between two tetrahedral positions). These form a set of rather flat energy surface paths through the crystal explaining perhaps the low activation energy for migration. In addition, alternating cycling back to the Si_i^{2+} T site with hole capture and return with electron capture provides an effective athermal migration mechanism which can explain its long range migration under electron irradiation conditions at 4.2 K.

Both groups predict negative-U behaviour and in figure 5 we summarize their results in the form of a cc diagram. (The values given are from the results of Car *et al* [39] who predicted the BC configuration lowest for Si_i^+ and Si_i^0). Again, as for interstitial hydrogen, they predict no barriers between the configurations and, in this case, we have not included them in figure 5. Whether they exist or not is of course crucial as to whether metastability effects actually occur, with the possibility that the defect can be frozen into its higher energy configuration. For our purposes, this is a subtle point, however, the critical point which contains the *physics* of metastability being the large lattice rearrangements versus charge state and consequent effects upon the electronic properties of the defect.

3.4. The Jahn–Teller effect

The examples of interstitial silicon and boron suggest a simple physical mechanism for the large lattice relaxational changes. It is the Jahn–Teller effect. Si^{2+} or B^+ as free ions have the configuration ns^2 and, placed in the interstitial T site of a crystal, are non-degenerate and therefore stable against a symmetry lowering distortion. (In the predicted configuration for B_i^+ of figure 4(c) the arguments are the same, the on-centre stability reflecting in that case Si^{2+}). On the other hand, Si^+ or B^0 (s^2p^1) and Si^0 or B^- (s^2p^2) are Jahn–Teller unstable and gain energy by off-centre distortions, entering into the bonds of their neighbours.

Following this simple logic, Al_i^{2+} (s^1) and Al_i^+ (s^2) are predicted to be stable in the T site, a fact confirmed for Al_i^{2+} by EPR [38, 41, 42] and for Al_i^+ from the existence of sharp zero-phonon structure for the excitation spectrum of Al_i^+ [43]. The observation on the other hand of recombination-enhanced migration for the defect provides strong evidence of large lattice relaxational changes available for Al_i^0 (s^2p^1) (and Al_i^- (s^2p^2) also if it is found to exist).

Interstitial hydrogen must be treated as a special case because it has no inner electron shell core and its bare proton can enter into high electron density regions forming the very special 'hydrogen bond'. A certain logic still remains in the above context, however, this bonding occurring only for H^+ ($1s^0$) and H^0 ($1s^1$) and not for the filled shell H^- ($1s^2$).

4. Isolated substitutional impurities

4.1. Substitutional nitrogen

Substitutional neutral nitrogen is observed by EPR to be statically distorted off-centre in a ⟨111⟩ direction away from one of its four nearest silicon neighbours [45, 46]. No direct evidence exists for freezing in a metastable on-centre configuration but an unusual temperature dependence found for the nitrogen hyperfine interaction has been cited as evidence that such a configuration does exist for the neutral defect [46]. In this regard it is interesting to recall the theoretical results of an early semi-empirical quantum mechanical treatment on a small cluster to simulate the defect [47]. The results are shown in figure 6. In addition to predicting the ⟨111⟩ distortion, a small local on-centre minimum was also found.

4.2. Substitutional oxygen

One might have expected substitutional oxygen to be a deep double donor and on-centre, as is found for the other chalcogens S, Se, Te. Instead, it is off-centre in a ⟨100⟩ direction, bonding to only two of the four silicon neighbours [48, 49]. This result was also obtained in the early cluster calculations [47], as shown in figure 7. In this configuration, only a single acceptor level $(-/0)$ is experimentally observed in the gap, at $E_c - 0.16$ eV. A moment's thought reveals that this implies that it

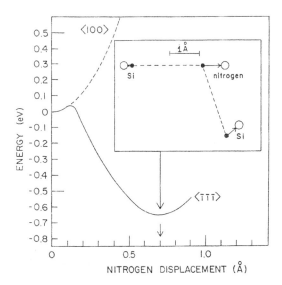

Figure 6. Theoretical prediction of off-centre distortion for substitutional N^0 in Si (from [47]).

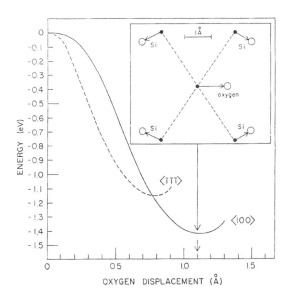

Figure 7. Theoretical prediction of off-centre distortion for substitutional O^0 in Si (from [47]).

is basically a negative-U centre, the single donor state $(0/+)$ in its off-centre position being deep in the valence band, this electron being bound by an energy much greater than the electron would be in an on-centre $(0/+)$ state. This is illustrated schematically in figure 8, which reveals further that in actual fact the off-centre $(0/+)$ level must be deep enough in the valence band so that the $(0/2+)$ 'occupation' level (halfway between the on-centre $(+/2+)$ level and the off-centre $(0/+)$ level) is also in the valence band, in order to explain the apparent lack of any electrical activity beyond its $(-/0)$ acceptor level.

Viewed this way, substitutional oxygen can be considered to contain all of the *physics* of metastability but without the manifestation, simply because the ef-

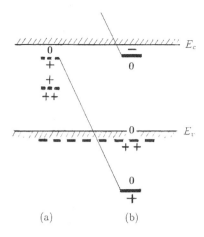

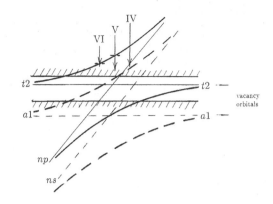

Figure 8. Level positions for substitutional oxygen, (a) if on-centre and (b) in its observed off-centre configuration.

Figure 9. The electrical level structure of a substitutional atom viewed as the result of interaction between its s and p valence orbitals and the a1 and t2 orbitals of the vacancy into which it is inserted.

fect is too strong. This suggests a simple experiment. If hydrostatic pressure were applied tending to favor the on-centre configuration, could the (0/2+) occupation level be raised above the valence band to be detected? This experiment was actually tried several years ago precisely to probe this question and the results were negative and therefore unfortunately were not reported [50]. Let me take this opportunity to correct that omission. In this experiment, the room temperature free carrier absorption at 10.6 μm (CO_2 laser) was monitored in a diamond anvil cell for a heavily doped p-type sample ($B \sim 5 \times 10^{18}$ cm^{-3}) which had been irradiated with 2×10^{18} cm^{-2} electrons (2.5 MeV) to produce $\sim 2 \times 10^{17}$ cm^{-3} substitutional oxygen defects. The free carrier concentration had been reduced to $\sim 1 \times 10^{18}$ cm^{-3} by the irradiation and a reduction therefore of $\sim 40\%$ was anticipated if the occupation level passed upward through the room temperature Fermi level at $\sim E_v + 0.08$ eV and each on-centre defect removed two holes. Unfortunately no such abrupt change was observed up to our maximum reliable pressure, 50 kbar, nature apparently conspiring to cheat us of what would otherwise have been an exciting observation.

4.3. The pseudo-Jahn–Teller effect

In these early theoretical treatments, the physical driving force for these distortions was also explored and it was concluded that they found a natural explanation in terms of what has come to be termed the 'pseudo' Jahn–Teller effect. To generalize this argument, it is instructive to consider the electronic structure of a substitutional impurity as arising from the interaction between its s and p valence orbitals and the corresponding a1 and t2 orbitals of the isolated vacancy into which it is inserted [51]. This is illustrated schematically in figure 9. For a neutral group IV atom (C, Si, Ge, etc), the near chemical equivalence of its atomic valence orbitals to the dangling silicon orbitals of the vacancy causes strong level repulsions between them (p with t2, s with a1) pushing filled bonding orbitals into the

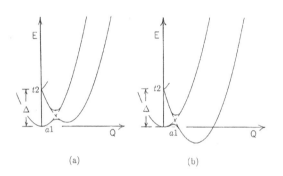

Figure 10. Pseudo-Jahn–Teller effect for a group V atom in Si ($n = 1$). Off-centre distortion is (a) metastable, (b) stable.

valence band and empty anti-bonding orbitals into the conduction band and clearing the gap of states (the vacancy is 'healed'). However, as we increase the nuclear charge Z to the group V or group VI atoms, the anti-bonding orbitals are repelled upward less, the a1 level entering the gap—shallow for group V atoms, but deep for group VI. With an a1^1 configuration for the neutral group V atom and a1^2 for the neutral group VI impurity, no Jahn–Teller distortion would be expected, explaining the observed on-centre character of most group V and group VI donor atoms in silicon.

A potential falacy of this argument, however, is the presence of the localized t2 anti-bonding orbital, still in the conduction band, but lurking nearby, close in energy, having been lowered also by the extra positive nuclear charge of the impurity. It is Jahn–Teller unstable and if the Jahn–Teller stabilization energy becomes comparable to the promotion energy required to excite the $n = 1$ (for nitrogen) or $n = 2$ (for oxygen) electrons into it, a stable or metastable off-centre minimum may occur.

This is illustrated in figure 10 for a group V atom,

as in the case of nitrogen, where the single particle energies conveniently also serve for the total energies of the corresponding configurations. An off-centre ⟨111⟩ distortion splits the t2 state, a single a1 orbital being lowered by $-VQ$, where V is the Jahn–Teller coupling coefficient and Q, the distortion coordinate. Adding $kQ^2/2$ for the elastic energy, we see that if $V^2/2k$, the Jahn–Teller stabilization energy for the t2 configuration, exceeds Δ, the a1–t2 energy separation, the off-centre distortion becomes the lowest energy configuration as in figure 10(b). On the other hand, if $V^2/2k < \Delta < V^2/k$, a secondary metastable off-centre minimum can occur, as in figure 10(a). (These simple arguments have not included off-diagonal Jahn–Teller coupling between the a1 and t2 orbitals, which prevents level crossings of the a1 levels and alters these energy limits somewhat. Its effects are also indicated schematically in figure 10. The limits above, derived for the purely diagonal coupling case, still serve as useful guides, however).

Generalizing the above to include group VI donors ($n = 2$), (simply replacing Δ by $n\Delta$, V by nV, but retaining the same elastic energy) gives

$$\frac{n^2 V^2}{2k} < n\Delta < \frac{n^2 V^2}{k} \tag{1}$$

for metastable off-centre and

$$\frac{n^2 V^2}{2k} > n\Delta \tag{2}$$

for stable off-centre.

These serve to illustrate several instructive features. First, we note that the pseudo-Jahn–Teller effect naturally tends to provide local minima and barriers between them, essential for the display of metastable effects. Second, the much larger effects for oxygen than nitrogen is the result of $n = 2$, the lowering of the Jahn–Teller energy varying as n^2, but with the promotion energy varying only as n. (As revealed in figure 7, the oxygen initially distorts as predicted in a ⟨111⟩ direction. The fact that the final configuration is actually a ⟨100⟩ distortion must result from interactions beyond this simple linear treatment, as has been discussed elsewhere [47]). Finally, the fact that the first row elements N and O have stable off-centre configurations while no such evidence exists for the heavier group-V and group-VI substitutional impurities finds a simple answer in the smaller force constant k expected for the compact first row atoms with weaker on-centre electronic overlap with neighbours at the Si–Si bond distance of the lattice.

This raises an intriguing question. Considering the large effect for oxygen, are we sure that there aren't some vestiges of metastability for the other $n = 2$ chalcogen double donors? Have we really looked? One interesting observation is that the zero-phonon electronic excitation absorption spectra are intense and as expected for on-centre double donors [52], but so far the corresponding luminescence has not been reported. Is it possible that competitive non-radiative recombination occurs at the donors via Jahn–Teller a1¹t2¹ and t2² metastable surfaces?

5. Relevance to DX and EL2

The experimental evidence is strong that DX arises from an isolated substitutional single donor in the relevant compound semiconductor lattice and that its metastable properties arise from a large lattice relaxational change [53]. In the case of EL2 in GaAs, there is still controversy concerning its microscopic identity, but recent theoretical calculations have concluded that an isolated substitutional As_{Ga} antisite double donor should display similar metastable properties [54–56].

Let us therefore consider both together as simple pseudo-Jahn–Teller systems of a substitutional single donor (DX) or double donor (As_{Ga} = EL2?) in analogy to nitrogen and oxygen in silicon and see what insight this might supply. The arguments are the same. The presence of the extra nuclear charge pulls down an a1 level into the gap but with an unoccupied t2 level lurking above.

Consider first the isolated As_{Ga} antisite in GaAs. It is a double donor and, like for oxygen in silicon, $n = 2$ in (1) and (2). Unlike oxygen, however, for which (2) applies strongly, the experimental and theoretical evidence indicate that the on-centre position is the stable one and the ⟨111⟩ off-centre configuration is metastable, equation (1) apparently applying. This reduction in relative strength of the Jahn–Teller energy presumably reflects in part the increased force constant k for the larger As donor atom.

DX, on the other hand, has $n = 1$ for the neutral state, like nitrogen in silicon. A glance at figure 6 or figure 10(b) reveals that if these energy surfaces are indeed accurate, then nitrogen would be expected to display DX properties, with persistent photoconductivity as the on-centre shallow donor state is frozen in at low temperatures. However, the fact that the light nitrogen atom (with small k) satisfies (2) with $n = 1$ perhaps argues little for a single donor such as Si in AlGaAs, for example, for which k should be significantly larger. We note instead, as a guide, that even with $n = 2$ for the double donor As_{Ga} in GaAs, the off-centre minimum is the metastable one. Therefore it would seem reasonable to guess that the neutral $n = 1$ single donor might not display metastability.

This brings us to the negative-U concept for DX, which has recently been predicted by theory [57–59]. In our simple treatment, this corresponds to a pseudo-Jahn–Teller distortion for the $n = 2$ negative charge state. Here equation (2) must be modified to include the added repulsion energy U between the two electrons that must also be overcome by the Jahn–Teller stabilization energy giving, if (2) is not satisfied for $n = 1$, the requirement

$$\frac{2V^2}{k} > 2\Delta + U \tag{3}$$

for a stable off-centre negatively charged $n = 2$ state.

Clearly if U is not too great ($U < 2\Delta$), equation (3) could be satisfied while, for $n = 1$, equation (2) is not. The critical question is how large is U? In the case

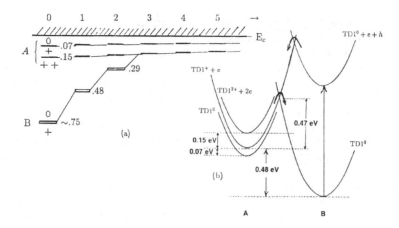

Figure 11. (a) The first TD in Si are bistable with a deep neutral state (B) in negative-U ordering below the shallow effective mass double donor states (A). (b) cc diagram for TD1.

of nitrogen in silicon, our one concrete experimental example, there is no evidence of a stable negative charge state in negative-U ordering and we are led to conclude that there U is sufficiently large to prevent deep second electron capture. On the other hand, the magnitude of U clearly depends critically on the degree of localization of the two electrons and might vary considerably from one system to the other. At this stage, therefore, it is probably best to keep an open mind and not attempt to generalize.

Instead, let us take a different approach. Assume that the stable off-centre DX configuration is the two-electron negative state. Does it make sense that As_{Ga}^0 with $U = 0$ is only metastable off-centre while DX^- with $U > 0$ could be stable off-centre? At first thought it appears grossly inconsistent with (2) and (3). It may be reasonable, however, in the sense that the first donor level (0/+) for on-centre As_{Ga} is very deep ($E_c - 0.742$ eV) [60], while that for on-centre DX is shallow ($\sim E_c - 0.03$ eV). The off-centre As_{Ga} ($-$/0) level could actually be deeper than the off-centre DX level but just doesn't quite compete with its on-centre configuration. Consistent with this idea, the behaviour of the off-centre EL2 configuration versus alloy composition in GaAsP appears to behave in a very similar fashion as the deep DX centre [61].

Viewed this way, As_{Ga} and DX are the same phenomena—an off-centre ⟨111⟩ pseudo-Jahn–Teller distortion of a two-electron system, which is easier in the case of DX because its on-centre level is shallow.

6. Other examples in silicon

6.1. The early thermal donors [62]

Prolonged heat treatment of Czochralski silicon at ~ 450 °C produces a series of much studied effective-mass-like double donors (TD). Recently, it has been discovered that the first few in the series are bistable [63]. This is illustrated in figure 11, where a deep stable single donor level becomes shallower through the progression

TD0, TD1, TD2, until TD3 where it crosses the shallow effective-mass states. Here and beyond it is the shallow state that is stable displaying the normal double donor properties.

Unfortunately, the microscopic structure of these defects and the mechanism for their metastability have not been established. It is clear, however, that they are essentially DX-like centres, being donors with shallow-deep instabilities and with the deep configuration 'tuneable' such that, depending upon the TD, it ranges from being the stable state when neutral, to the metastable state, similar to the tuning in AlGaAs and GaAsP versus alloy composition. We note also that in this case it is a negative-U system, the ionization energy from the neutral state exceeding that from the positive state.

Although we do not know precisely what is going on at the atomic level, this serves to confirm that bistable negative-U behaviour can indeed occur, particularly in this case again where the competition is with a shallow effective-mass-like state.

6.2. Miscellaneous

Several other defects, not yet identified, have also been established to have metastable properties in silicon. These include both irradiation-produced centres [64–67] and those produced by chemical doping [68–69]. Some of these display interconversion barriers of only ~ 0.1–0.2 eV, similar to that found for the C_iC_s and C_i–group-V-donor pairs, suggesting that these may also incorporate C_i as a constituent which is undergoing a molecular bond switching.

7. Summary

A wide variety of defects have been found in silicon that display metastable properties and several have been identified. Those identified include: (i) donor–acceptor pairs where one of the constituents is a mobile interstitial, (ii) a C_iC_s pair where the metastability is associated with a molecular bond switching and (iii) isolated interstitial and substitutional atoms where the configurational change is between an on-centre and off-centre position. In the latter case, the mechanism for the off-centre distortion can be understood in terms of the Jahn–Teller effect arising from electronic degeneracy associated with the defect. In the case of the interstitial, the degeneracy arises from its partially filled atomic p orbitals. For the substitutional impurity, it can be considered to arise from partial occupancy of its p-like molecular states (t2) which are formed with the 'vacancy' orbitals of the four nearest silicon neighbours. In the particular case cited of the nitrogen and oxygen substitutional impurities, it is the degeneracy of an excited t2 state that drives the distortion, a phenomenon called the pseudo-Jahn–Teller effect.

The general conditions for pseudo-Jahn–Teller metastability of a substitutional impurity have been discussed and applied to the As_{Ga} antisite (EL2?) and DX in GaAs and its alloys. Viewed this way, the two can be considered the identical phenomenon—an off-centre $\langle 111 \rangle$ pseudo-Jahn–Teller distortion of a two-electron system. The distorted configuration for As_{Ga}^{0} is the metastable configuration because the competing on-centre configuration is deep. The off-centre configuration for DX^{-} is the stable one even though for it the added Coulomb repulsion term U must be overcome. It is possible because the competing on-centre configuration is shallow.

This idea of regarding off-centre distortions of defects as Jahn–Teller and pseudo-Jahn–Teller in origin is certainly not new and has been called upon to successfully understand and predict the phenomenon in semiconductors for many years [47, 70–72]. It appears clear, for example, that Dabrowski et al in their recent work on As_{Ga} [54, 55] recognize a pseudo-Jahn–Teller origin of the distortion via the excited t2 state, as presented in their one-electron energy diagrams and their treatment of the photoexcitations. Morgan has also earlier suggested Jahn–Teller instability for DX via an excited L-derived t2 state [73]. One might be tempted, in fact, to say that it is trivially obvious—if it were not for the fact that this concept has caused considerable controversy in the recent literature [74–76].

Note added in proof. And if it were not for the very recent conclusion presented at this meeting [77–78] that oxygen remains off-centre in GaAs independent of its charge state, inconsistent with the prediction of the simple pseudo-Jahn–Teller model for isolated substitutional oxygen that we have presented here).

Acknowledgments

This review was made possible by partial support from National Science Foundation grant DMR-89-02572 and Office of Naval Research Electronics and Solid State Program grant N00014-90-J-1264. It was written during a stay at King's College London, made possible by a Visiting Fellowship Research Grant from the Science and Engineering Research Council, which is gratefully acknowledged.

References

[1]   Ludwig G H and Woodbury H H 1962 *Solid State Physics* vol 13 ed F Seitz, D Turnbull and H Ehrenreich (New York: Academic) p 223
[2]   Chantre A and Bois D 1985 *Phys. Rev.* B **31** 7979
[3]   Chantre A and Kimerling L C 1986 *Mater. Sci. Forum* **10–12** 387
[4]   Van Kooten J J, Weller G A and Ammerlaan C A J 1984 *Phys. Rev.* B **30** 4564
[5]   Gehlhoff W, Irmscher K and Rehse U 1989 *Mater. Sci. Forum* **38–41** 373
[6]   Gehlhoff W, Irmscher K and Kreisl J 1988 *New Developments in Semiconductor Physics* ed G Ferenczi and F Beleznay (Berlin: Springer) p 262
[7]   Omling P, Emanuelsson P, Gehlhoff W and Grimmeiss H G 1989 *Solid State Commun.* **70** 807
[8]   Gehlhoff W, Emanuelsson P, Omling P and Grimmeiss H G 1990 *Phys. Rev.* B **41** 8560
[9]   Kimerling L C and Benton J L 1983 *Physica* B **116** 297
[10] Kimerling L C, Asom M T, Benton J L, Drevinski P J and Caefer C E 1989 *Mater. Sci. Forum* **38–41** 141
[11] Song L W, Benson B W and Watkins G D 1986 *Phys. Rev.* B **33** 1452
[12] Chantre A and Kimerling L C 1986 *Appl. Phys. Lett.* **48** 1000
[13] Asom M T, Benton J L, Sauer R and Kimerling L C 1987 *Appl. Phys. Lett.* **51** 256
[14] Song L W, Benson B W and Watkins G D 1987 *Appl. Phys. Lett.* **51** 1155
[15] Benson B W, Gurer E and Watkins G D 1989 *Mater. Sci. Forum* **38–41** 391
[16] Gurer E and Benson B W 1990 *Impurities, Defects and Diffusion in Semiconductors: Bulk and Layered Structures (MRS Symp. Proc. 163)* ed D J Wolford, J Bernholc and E E Haller (Pittsburgh, PA: Materials Research Society) p 295
[17] Gurer E 1990 *PhD Dissertation* Lehigh University to be published
[18] Song L W and Watkins G D 1990 *Phys. Rev.* B **42** 5759
[19] Zhan X D and Watkins G D 1991 *Appl. Phys. Lett.* **58** 2144
[20] Brower K L 1974 *Phys. Rev.* B **9** 2607
[21] Song L W, Zhan X D, Benson B W and Watkins G D 1988 *Phys. Rev. Lett.* **60** 460
[22] Song L W, Zhan X D, Benson B W and Watkins G D 1990 *Phys. Rev.* B **42** 5765
[23] Watkins G D and Brower K L 1976 *Phys. Rev. Lett.* **36** 1329
[24] Patterson B D 1988 *Rev. Mod. Phys.* **60** 69
[25] Kiefl R F, Celio M, Estle T L, Kreitzman S R, Luke G M, Riseman T M and Ansaldo E J 1988 *Phys. Rev. Lett.* **60** 224
[26] Estle T L, Keifl R F, Celio M, Kreitzman S R, Luke G M, Riseman T M and Ansaldo E J 1988 *Bull. Am. Phys. Soc.* **33** 699
[27] Gorelkinskii Y V and Nevinnyi N N 1987 *Pis'ma Zh. Tekh. Fiz.* **13** 105 (*Sov. Tech. Phys. Lett.* **13** 45)
[28] Estreicher S 1987 *Phys. Rev.* B **36** 9122
[29] Deleo G G, Dorogi M J and Fowler W B 1988 *Phys. Rev.* B **38** 7520
[30] Deak P, Snyder L C and Corbett J W 1988 *Phys. Rev.* B **37** 6887
[31] Van de Walle C G, Denteneer P J H, Bar-Yam Y and Pantelides S T 1989 *Phys. Rev.* B **39** 10791
[32] Chang K J and Chadi D J 1989 *Phys. Rev.* B **40** 11644

[33] Holm B, Bonde Nielsen K and Bech Nielsen B 1991 *Phys. Rev. Lett.* **66** 2360

[34] Harris R D, Newton J L and Watkins G D 1987 *Phys. Rev. B* **36** 1094

[35] Troxell J R and Watkins G D 1980 *Phys. Rev. B* **22** 921

[36] Watkins G D 1975 *Phys. Rev. B* **12** 5824

[37] Tarnow E 1990 *Preprint*

[38] Watkins G D 1964 *Radiation Damage in Semiconductors* ed P Baruch (Paris: Dunod) p 97

[39] Car R, Kelly P J, Oshiyama A and Pantelides S T 1985 *Proc. 13th Int. Conf. on Defects in Semiconductors* ed L C Kimerling and J M Parsey Jr (Warrendale, PA: American Institute of Mechanical Engineers) p 269

[40] Bar-Yam Y and Joannopoulos J D 1985 *Proc. 13th Int. Conf. on Defects in Semiconductors* ed L C Kimerling and J M Parsey Jr (Warrendale, PA: American Institute of Mechanical Engineers) p 261

[41] Brower K L 1970 *Phys. Rev. B* **1** 1908

[42] Niklas J R, Spaeth J M and Watkins G D 1985 *Microscopic Identification of Electronic Defects in Semiconductors (MRS Symp. Proc. 46)* ed N M Johnson, S G Bishop and G D Watkins (Pittsburgh, PA: Materials Research Society) p 234

[43] Latushko Y I and Petrov V V 1989 *Mater. Sci. Forum* **38–41** 1171

[44] Troxell J R, Chatterjee A P, Watkins G D and Kimerling L C 1979 *Phys. Rev. B* **19** 5336

[45] Brower K L 1982 *Phys. Rev. B* **26** 6040

[46] Murakami K, Kuribayashi H and Masuda K 1988 *Phys. Rev. B* **38** 1589

[47] DeLeo G G, Fowler W B and Watkins G D 1984 *Phys. Rev. B* **29** 3193

[48] Watkins G D and Corbett J W 1961 *Phys. Rev.* **121** 1001

[49] Corbett J W, Watkins G D, Chrenko R M and McDonald R S 1961 *Phys. Rev.* **121** 1015

[50] Watkins G D and Keilmann F 1984 unpublished

[51] Watkins G D 1983 *Physica B* **117–118** 9

[52] Janzen E, Grossmann G, Stedman R and Grimmeiss H G 1985 *Phys. Rev. B* **31** 8000

[53] Mooney P M 1990 *J. Appl. Phys.* **67** R1

[54] Dabrowski J and Scheffler M 1988 *Phys. Rev. Lett.* **60** 2183

[55] Dabrowski J and Scheffler M 1989 *Phys. Rev. B* **40** 10391

[56] Chadi D J and Chang K J 1988 *Phys. Rev. Lett.* **60** 2187

[57] Chadi D J and Chang K J 1989 *Phys. Rev. B* **39** 10063

[58] Zhang S B and Chadi D J 1990 *Phys. Rev. B* **42** 7174

[59] Dabrowski J, Scheffler M and Strehlow R 1990 *Proc. 20th Int. Conf. on the Physics of Semiconductors* ed E M Anastassakis and J D Joannopoulos (Singapore: World Scientific) p 489

[60] Nissen M K, Steiner T, Beckett D J S and Thewalt M L W 1990 *Phys. Rev. Lett.* **65** 2282

[61] Samuelson L and Omling P 1986 *Phys. Rev. B* **34** 5603

[62] Wagner P and Hage J 1989 *Appl. Phys. A* **49** 123 and references therein

[63] Marenko L F, Markwich V P and Murin L I 1985 *Sov. Phys.–Semicond.* **19** 1192

[64] Bains S K and Banbury P 1985 *J. Phys. C: Solid State Phys.* **18** L109

[65] Chantre A 1985 *Phys. Rev. B* **32** 3687

[66] Awadelkarim O O and Monemar B 1988 *Phys. Rev. B* **38** 10116

[67] Svensson J H and Monemar B 1989 *Phys. Rev. B* **40** 1410

[68] Watkins S P, Thewalt M L W and Steiner T 1984 *Phys. Rev. B* **29** 5727

[69] Singh M, Lightowlers E C and Davies G 1989 *Mater. Sci. Eng. B* **4** 303

[70] Watkins G D 1973 *Radiation Damage and Defects in Semiconductors (Inst. Phys. Conf. Ser. 16)* (Bristol: Institute of Physics) p 228

[71] Lannoo M 1982 *Phys. Rev. B* **25** 2987

[72] Watkins G D 1991 *Electronic Structure and Properties of Semiconductors (Materials Science and Technology 4)* ed W Schroter (Weinheim: VCH) ch 3 at press

[73] Morgan T N 1986 *Phys. Rev. B* **34** 2664

[74] Pantelides S T, Harrison W A and Yndurain F 1986 *Phys. Rev. B* **34** 6038

[75] DeLeo G G, Watkins G D and Fowler W B 1988 *Phys. Rev. B* **37** 1013

[76] Pantelides S T, Harrison W A and Yndurain F 1988 *Phys. Rev. B* **37** 1016

[77] Alt H Ch 1991 *Semicond. Sci. Technol.* this issue

[78] Alt H Ch 1990 *Phys. Rev. Lett.* **65** 3421

Semicond. Sci. Technol. **6** (1991) B121–B129. Printed in the UK

Negative-*U* properties of off-centre substitutional oxygen in gallium arsenide

H Ch Alt

Siemens Research Laboratories for Materials Science and Electronics,
Otto-Hahn-Ring 6, D–8000 München 83, Federal Republic of Germany

Abstract. Recent infrared absorption studies of semi-insulating GaAs have revealed an electrically active, oxygen-related defect with remarkable spectroscopic and microscopic properties. This defect, the structural analogue of the oxygen vacancy centre in Si, occurs in three charge states, the zero-, one- and two-electron states. The experimental fingerprint for each charge state is the local mode frequency which shows a characteristic charge-state-induced shift. Dependent on the Fermi potential, at thermal equilibrium only the local modes corresponding to the zero- or the two-electron state are experimentally observable. The metastable one-electron state disproportionates spontaneously into the zero- and the two-electron states. The energy positions of the associated gap levels are at $E_c - 0.14$ eV and at $E_c - 0.58$ eV for the first and the second electron, respectively. These assignments are derived from the thermally activated decay of the local mode lines and, independently, from the threshold energies of the optical transitions $\sigma_p^0(1)$ and $\sigma_n^0(2)$. Through a comparative deep-level transient spectroscopy and infrared absorption study on neutron-transmutation-doped, n-type samples the second electron level is identified as the well known EL3 level.

1. Introduction

Starting from the observation of two localized vibrational modes (LVM) in the infrared absorption spectrum of semi-insulating (SI) GaAs [1], recent experimental work has brought about an abundance of information on one of the most fascinating impurity-related defects in a compound semiconductor [2–4]. It is now clear that this defect, the off-centre substitutional oxygen impurity, has the electrical characteristics of a negative-*U* centre—a defect which can bind two electrons where the second electron is bound more strongly than the first. There are only a few examples of experimental verification of such a defect behaviour in semiconductors. The classical work on the vacancy and the interstitial boron in Si ranks among the most prominent [5]. Concerning compound semiconductors, there is some evidence that a radiation-induced defect in InP (the M centre) is negative *U* [6]. For the DX centres in $Al_xGa_{1-x}As$ alloys negative-*U* behaviour has been suggested in a recent theoretical model [7]; however, experimental data are still controversially discussed [8].

The role of oxygen in GaAs has been the subject of many speculations for at least 30 years. The first question arose about the influence of oxygen on the SI behaviour. Speculation lasted for about 20 years until finally, based on accurate mass spectrometric work, it became clear that oxygen is not the main deep trap at $E_c - 0.75$ eV [9]. The reasons for this long period of ambiguity was that, contrary to the situation in Si or GaP, neither in optical emission nor in absorption could any spectroscopic evidence of oxygen be found. Also at that time, it became obvious that the oxygen concentration in pure GaAs is well below 1×10^{16} cm^{-3}.

The first spectroscopic signature of oxygen was detected in 1976 [10], when a local vibrational mode absorption at 840 cm^{-1} in oxygen-rich GaAs was attributed to interstitial oxygen. No electrical activity seemed to be correlated with this defect. The off-centre substitutional oxygen, by contrast, is electrically active and occurs in as-grown GaAs in the concentration range up to some 10^{15} cm^{-3}. Thus, with respect to the SI property, it may possess a deleterious influence on stable compensation. However, there is also a possible application in form of an intentional introduction of this defect by ion implantation for electrical isolation of devices [11].

2. Experiment

The majority of samples investigated came from nominally undoped GaAs crystals grown after the liquid encapsulated Czochralski (LEC) method (high pressure and low pressure) from pyrolithic boron nitride crucibles.

0268-1242/91/10B121+09 $03.50 © 1991 IOP Publishing Ltd

Most of these samples were SI with the Fermi level at 0.6–0.7 eV below the conduction band minimum. Some samples were only 'highly resistive' with the Fermi level at about E_c–0.35 eV. The latter samples were grown using a wet (high H_2O content) boric oxide encapsulant which promotes the incorporation of oxygen in the growing crystal. A few samples, grown by the horizontal Bridgman (HB) technique from quartz crucibles, were slightly n-type conducting. In all of these samples oxygen is a residual impurity.

For deep-level transient spectroscopy (DLTS) studies samples were doped by neutron transmutation (NTD). Doses of slow neutrons ranged from 0.67 up to 50×10^{16} cm^{-2}. Annealing of the radiation damage was performed at 750 °C for 30 min under an arsine atmosphere.

The infrared (IR) absorption measurements were carried out using a high-resolution Fourier transform spectrometer (Bruker IFS 113 v). The temperature range was between 7 and 300 K. The cryostat used had a 90° window for illumination of the samples with additional secondary light. In order to avoid any influence of the measurement light on the samples, the spectra of the LVM lines were taken with a long-pass IR filter in front of the sample ($\lambda_{on} = 10 \ \mu m$).

3. Defect model

A typical low-temperature IR absorption spectrum of an oxygen-rich SI GaAs sample in the spectral region of interest is shown in figure 1. Most interesting in this context are the two bands at 730 and 714 cm^{-1} (731 and 715 cm^{-1} at $T = 10$ K) which are denoted as A and B [1]. Other bands in this spectrum are due to the LVM of interstitial oxygen (845 cm^{-1}), the carbon acceptor impurity (582 cm^{-1}), and a so far unidentified defect (605 cm^{-1}). When the bands A and B are investigated with higher spectral resolution, a remarkable triplet fine structure becomes visible (figure 1).

The separation between the individual lines is 0.48 cm^{-1} and the relative intensities are $\sim 3:4:1$ from

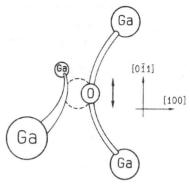

Figure 2. Structural model of off-centre substitutional oxygen in GaAs. Bonding preferentially with two Ga atoms, the oxygen atom moves off-centre in a [100] direction.

high to low energy. From this it is inferred that the splitting is due to the Ga isotopes 69 and 71, occurring in relative natural abundances of 60% and 40%, respectively. Three lines and the experimentally observed intensities are expected for a defect bonding with two Ga atoms in the GaAs matrix. In conjunction with the relatively high vibrational frequency it was speculated [12] that the defect involved might have the molecular structure Ga–O–Ga in analogy to the oxygen vacancy centre in Si [13]. The crucial experiment was done by doping a crystal with ^{18}O [14]. This crystal exhibited only band B at 715 cm^{-1} and, in addition, a band at 679 cm^{-1}. The ratio of these two LVM frequencies is close to the value expected from a simple harmonic oscillator approximation $(M(^{18}O)/M(^{16}O))^{1/2} = 1.06$. The vibrating entity, therefore, contains as its essential constituent one oxygen atom. It is, from all of this, well established that the defect giving rise to the LVM bands at 730 and 715 cm^{-1} is the analogue of the oxygen vacancy centre in Si (figure 2). Stress-splitting studies of the 730 cm^{-1} band are compatible with the assumed defect symmetry C_{2v} [15].

4. Photosensitivity, charge states

The most fascinating property of the bands A and B is their photosensitivity. To be more specific, after cooling a SI sample in the dark, only band A is observed. When the sample is now illuminated with additional light of an energy between 0.8 eV and the band gap a conversion from A to B occurs. At low temperatures this state is stable at least for hours. We first investigated the spectral dependence of the conversion rate $dI(B)/dt \ \Phi(v)$, where $I(B)$ is the intensity of band B and $\Phi(v)$ the photon flux. The low-energy threshold of the conversion is 0.8 eV and the rate increases with increasing energy with a pronounced step at 1.1 eV. The shape of this curve is very similar to the spectral dependence of the photoionization cross section $\sigma_n^0(EL2)$ of the EL2 centre which is well known from the literature [16].

Hence we suggested that the conversion from A to B is due to an optically induced charge transfer process

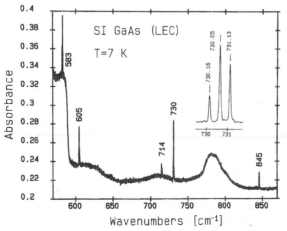

Figure 1. Low-temperature infrared absorption spectrum of oxygen-rich SI GaAs. Inset: fine structure of band B.

from the EL2 centre to the oxygen defect via the conduction band. In a SI sample the Fermi potential is pinned by the EL2 level and, therefore, this level is partially occupied by electrons. A charge transfer requires that the oxygen level is initially unoccupied, which means that it must lie in the upper half of the band gap. The following rate equations describe this process

$$dn/dt = \sigma_n^0(EL2)\Phi N(EL2^0) - c_n(EL2)nN(EL2^+)$$
$$- c_n(O)nN(O^{(0)}) \qquad (1)$$

$$dN(O^{(1)})/dt = c_n(O)nN(O^{(0)}). \qquad (2)$$

Here n is the density of free electrons in the conduction band, c_n is the capture coefficient for electrons, N the concentration of the defects in their different states (neutral ^{0} and positive $^+$ for EL2, unoccupied $^{(0)}$ and occupied $^{(1)}$ for the oxygen defect). It is now assumed that, in first approximation, capture by the oxygen level is negligible compared with capture by EL2. This will be justified below on the basis of more information on the magnitude of the capture coefficients. As a consequence, a steady-state electron density in the conduction band is obtained

$$n = \frac{\sigma_n^0(EL2)\Phi}{c_n(EL2)} \frac{N(EL2^0)}{N(EL2^+)}. \qquad (3)$$

Substituting this expression into equation (2), the change of the oxygen occupancy, which is equivalent to the conversion rate, is given by

$$dN(O^{(1)})/dt = dI(B)/dt$$
$$= c_n(O)N(O^{(0)}) \frac{\sigma_n^0(EL2)\Phi}{c_n(EL2)} \frac{N(EL2^0)}{N(EL2^+)}. \qquad (4)$$

This result explains the observed similarity between the conversion rate and the photoionization cross section $\sigma_n^0(EL2)$ of EL2 in a natural way.

However, there is another experimental result which confirms the proposed charge transfer process. This was obtained by measuring the infrared absorption spectrum of EL2 in the near-infrared (6000–12000 cm^{-1}) before and after the conversion from A to B. The absorption spectrum consists of contributions due to EL2$^+$ and EL2^0 [16]. As, according to the charge transfer model, the occupancy of the EL2 level must have decreased after the conversion from A to B, this decrease should be measurable as a change of the relative contributions of EL2$^+$ and EL2^0 to the absorption spectrum. The change is given by

$$\Delta\alpha(EL2) = (\sigma_p^0(EL2) - \sigma_n^0(EL2))\Delta N(EL2^0)$$
$$= C(\sigma_p^0(EL2) - \sigma_n^0(EL2))N(O)$$

and, therefore,

$$N(O) = \Delta\alpha(EL2)/C(\sigma_p^0(EL2) - \sigma_n^0(EL2))^{-1}. \qquad (5)$$

Here, $\sigma_p^0(EL2)$ is the hole photoionization cross section of EL2, $\Delta N(EL2^0)$ is the change of the neutral EL2 density. $\Delta N(EL2^0)$ is proportional to the density of oxygen defects. We will see later that the proportionality constant C is equal to 2. The experimental result is shown

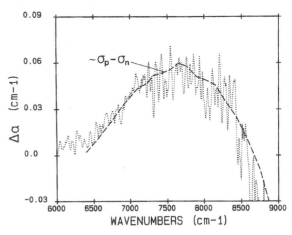

Figure 3. Change of the absorption coefficient in the region of the EL2$^+$ absorption after conversion of band A to band B. The difference $\sigma_p^0 - \sigma_n^0$, normalized to the same peak height, is shown for comparison (broken curve).

in figure 3. The observed change in the near-infrared absorption spectrum follows exactly the spectral dependence of $\sigma_p^0(EL2) - \sigma_n^0(EL2)$. The charge transfer can be monitored by the change of the EL2 absorption spectrum. Equation (5) allows us to estimate the concentration of oxygen defects. It is found that the maximum concentration in the samples investigated is in the lower 10^{15} cm^{-3} range.

This set of experiments shows very impressively that the observation of the oxygen properties is dependent on the presence of the EL2 defect as the dominant deep donor. EL2 pins the Fermi potential and provides a reservoir of optically accessible electrons. Moreover, the knowledge of electrical and optical properties of EL2 is essential to determine the properties of the oxygen defect from the experimental data.

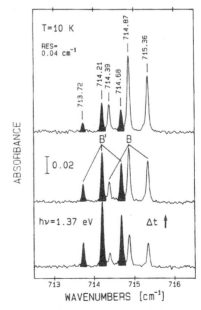

Figure 4. Fine structure of band 'B': B' and B.

Beginning with the first studies of the optically induced conversion from A to B, we observed that at some intermediate stages band B does not show the familiar triplet structure, but a splitting into four lines. Detailed studies revealed that the band 'B' actually consists of two bands which are denoted by B' and B. This is shown in figure 4. Starting the conversion process, band B' appears first. With prolonged illumination time this band disappears again and band B becomes more and more dominant. This new band B' is shifted by ~0.7 cm^{-1} to lower energy, but exhibits exactly the same fine-structure splitting. Thus it is inferred that this band represents a third state of the defect. The detailed kinetics of the conversion A → B' → B are shown in figure 5. Here the intermediate character of band B' comes out clearly. At any time the integrated absorption of A, B' and B is approximately constant. Provided that the oscillator strength of the oxygen defect is the same in each state, this observation

supports the picture that the three bands correspond to three different states of the defect.

From the time dependence of the transients it is immediately obvious that simple exponential laws should give an adequate description of the underlying rate equations. It is proposed that the oxygen defect can bind two electrons where A, B' and B are the zero-, one- and two-electron states. In extension of equations (1) and (2) we make the following 'ansatz'

$$dN(A)/dt = -c_n(A)nN(A)$$
$$dN(B')/dt = +c_n(A)nN(A) - c_n(B')nN(B') \quad (6)$$
$$dN(B)/dt = +c_n(B')nN(B')$$

where $N(i)$ is the defect density in the state i and $c_n(i)$ the corresponding capture coefficient. In general this system of rate equations is non-linear and, therefore, not tractable analytically. However, in analogy to the discussion of equations (1) and (2), we assume that the capture of electrons by the oxygen centre is only a by-path in the recombination which can be neglected in first approximation. This again leads to the condition $n = $ const. Moreover, any photoexcitation processes between the different states $N(i)$ and the bands are neglected. At this point, this can be justified only by the success of the analytical solution of (6) in fitting the experimentally observed conversion kinetics (figure 5). The ratio $c_n(A)/c_n(B')$ is about 2 which means that the capture cross sections of the two states are comparable. An estimate of the steady-state electron density n during the experiments leads to $\sigma_n(A) \sim 6 \times 10^{-20}$ cm^2 at 77 K. This capture cross section is an order of magnitude smaller than $\sigma_n(EL2)$ [17] which is fully consistent with the assumption above that capture by the oxygen defect is much less probable than recombination at EL2.

Until now only the photoinduced conversion from A to B' to B has been discussed. This is indeed the process which is first observed in the energy range $0.8 \le hv \le 1.5$ eV. However, after long illumination times at low temperatures in the range $1.0 \le hv \le 1.3$ eV the reverse process occurs, the conversion B → B' → A. The kinetics of this process is shown in figure 6. Parallel to this process the EL2 absorption band has been monitored. It is found that the decrease of B starts when the transformation of the EL2 defect to its metastable state EL2* [18] is nearly accomplished. If the conversion A → B' → B is interpreted in terms of electron capture, the reverse process must be due to hole capture. What occurs is the following. Illumination with $1.0 \le hv \le 1.3$ eV first quenches the normal state of EL2. As this transformation goes via the neutral charge state of EL2, for the ionized fraction EL2$^+$ the photoionization of an electron from the valence band is the necessary precursor process. At the end of the photoquenching, this leads to an equivalent density of holes in the valence band which are either free or bound to shallow acceptor states like carbon. This is the reservoir of holes accessible by the occupied oxygen levels B and B'. As shown in figure 6, again a reasonable fit is possible from rate equations analogous to (6). As the assumption of a constant hole

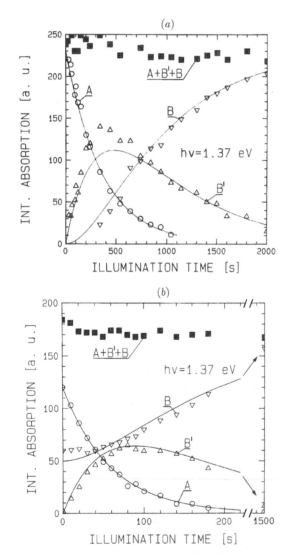

Figure 5. Kinetics of the transition A → B' → B for illumination with 1.37 eV photons: (a) SI sample, (b) 0.4 eV sample. Full curves are model calculations based on capture of photoexcited conduction band electrons.

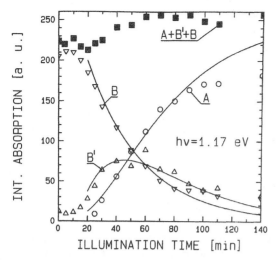

Figure 6. Kinetics of the reappearance of band A during quenching of EL2 (illumination with 1.17 eV photons). Full curves are model calculations based on hole capture. Note that the illumination intensity is about a factor of 100 higher than in figure 5.

density is certainly only a crude approximation for this process, deviations are expected to be larger. Finally, when EL2 is completely quenched, the optical interaction between the oxygen states and the bands must also be taken into account. The essential point is that B′ again is the intermediate state for the conversion process.

There is a third argument in favour of the assignment of A, B′ and B to three different charge states. We carefully measured the change of the EL2 absorption band at different stages of the photoinduced conversion from A to B in a way similar to figure 3. If band B′ did not correspond to a third charge state but simply to another configuration of the defect in the same charge state as band B, no change in the EL2 absorption band should be observable after the vanishing of band A (for example after $t = 1000$ s in figure 5). However, this is in contradiction to the experimental result. The ionized density $N(\mathrm{EL2^+})$ increases continuously until all B′ has converted to B. A linear fit is possible between $N(\mathrm{EL2^+})$ and the fractional occupancy of A, B′ and B if it is assumed that these three states correspond to the zero-, one- and two-electron states.

5. Optical verification of the negative-U behaviour

The experiments presented so far were performed with SI samples. After cooling in the dark these samples show only band A. In comparison, n-type conducting samples show only band B and are not photosensitive. A third type of sample which may be characterized as '0.4 eV' samples, because in temperature-dependent Hall-effect measurements the activation energy is ∼0.4 eV, exhibit both bands, A and B. In these samples the complete conversion to B is possibly similar to SI samples (see figure 5). Again band B′ acts as an intermediate state.

Thus, it is found experimentally that, whatever the position of the Fermi potential in the upper half of the band gap, only bands A and B correspond to thermal equilibrium states. This behaviour necessarily leads to the speculation that off-centre substitutional oxygen forms a negative-U centre.

Such a centre can bind two electrons, where the second electron is bound more strongly than the first. This could be proved directly if it were possible to observe the famous disproportionation phenomenon. In our case, disproportionation can be described by the reaction

$$\mathrm{B'} \rightarrow \tfrac{1}{2}\mathrm{A} + \tfrac{1}{2}\mathrm{B}. \tag{7}$$

This means that the metastable state B′ decays spontaneously into the thermodynamically stable states A and B.

The corresponding set of LVM spectra illustrating the disproportionation is shown in figure 7. The first spectrum was taken after cooling a SI sample to 10 K in the dark. Only band A was observed. A part of this band was photons (see also figure 5). In the next step the sample was warmed up to 95 K, kept at this temperature for 30 min, and then returned to 10 K. Now band B′ disappears again, accompanied by the appearance of a small band B and an increase of band A. Monitoring the integrated absorption at each stage, it comes out that band B′ decays half into A and half into B in exact correspondence to equation (7). In a microscopic picture this means that the one-electron state emits its electron thermally to the conduction band, from where it is captured by another oxygen defect in the state B′. As a net effect, any two defects B′ give one defect A and one

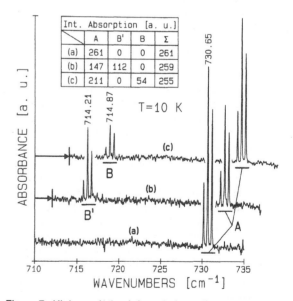

Figure 7. High-resolution infrared absorption spectra illustrating disproportionation of the metastable one-electron state: (a) after cooling the sample to 10 K in the dark, (b) after short illumination with 1.37 eV photons and (c) after warming to 95 K and cooling to 10 K again. Inset: integrated absorption of bands A, B′, B and all bands together (Σ).

defect B. It is found that at temperatures <90 K a fraction of the thermally emitted electrons are lost, probably due to recapture at the EL2 centre.

The thermal decay of the LVM band B' provides a means to derive the activation energy for the emission process from the one-electron state. The corresponding rate equation is

$$dN(B')/dt = -e_n(B')N(B') + c_n(A)nN(A) - c_n(B')nN(B') \tag{8}$$

where $e_n(B')$ is the emission rate to the conduction band. As can be seen from equation (8), the emission transient is always non-exponential due to the capture processes at A and B'. In order to minimize this effect, we first bleached the EL2 defect at 10 K by a strong illumination with 1.17 eV photons. In the next step the metastable state B' was prepared by a short illumination with 1.45 eV photons. This illumination photoexcites electrons from the valence band to the unoccupied oxygen level (see below). Bleaching of EL2 opens an efficient recombination channel for the thermally emitted electrons in the form of recombination with residual holes. This keeps the density of conduction band electrons n low and, therefore, reduces the influence of the second and third term on the right-hand side of equation (8). Using this technique it is possible to derive from the initial slope of the decay characteristics of B' (where $N(B') \gg N(A)$) a time constant corresponding to $(e_n(B'))^{-1}$. From an Arrhenius plot of the emission rate at different temperatures, including the usual T^2 correction, an activation energy of 0.14 ± 0.02 eV is found.

After the complete conversion from A → B at low temperatures, band B is stable up to temperatures of ~ 180 K. In the temperature range 180 K $< T <$ 200 K a decay is observed with decay times suitable for Fourier transfers IR spectroscopy ($\sim$ min). In the picture of the emission from the two-electron state of a negative-U centre, the observed emission process is interpreted as follows. The second electron is thermally emitted to the conduction band, followed immediately by the first due to the smaller binding energy. Thus, this is a two-electron emission process where the emission barrier is given by the binding energy of the second electron. Recapture of the thermally emitted electrons is possible only by the EL2 centre in its positive charge state EL2$^+$. Thus the appropriate rate equation is simply given by $dN(B)/dt = -e_n(B)N(B)$. This unique behaviour is reflected in the experimentally observed purely exponential decay. The evaluation of the data in terms of $\ln(e_n(B)/T^2)$ versus $1/T$ gives an activation energy of 0.58 ± 0.03 eV. Independently from us, disproportionation and the emission have also been studied by another group [19]. Their results are similar to ours.

The latter emission process can be verified by an independent method. If, as postulated above, the thermally emitted electrons are captured by the EL2 centre, then the ionized fraction of this defect, $N(EL2^+)$, must decrease. This can be monitored again by measuring the EL2 absorption band before and after the decay of band

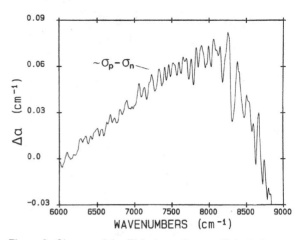

Figure 8. Change of the EL2 absorption coefficient after conversion of band B to band A by thermal emission of electrons.

B (figure 8). The spectral shape of this difference spectrum is similar to the spectrum shown in figure 3 and, moreover, even the magnitude of the absorption change is comparable (both spectra are taken from the same sample). This is exactly what is expected. In other words, the thermally induced conversion from B to A at $T >$ 180 K is the back-reaction of the optically induced conversion from A to B at $T <$ 80 K.

In this section we describe experiments on photoexcited transitions between the localized oxygen states and the band states. In order to access optically the one-electron state B', it is necessary first to transform EL2 to its metastable state, EL2*, which is electrically and optically inactive. Otherwise the interference with the EL2 level would cover the optical properties of the oxygen defect. As described above, at the end of the quenching experiment only band A is observed (see figure 6). Now it is possible to photoexcite electrons from the valence band directly into the unoccupied oxygen state. At 10 K, a well-defined threshold energy of 1.37 eV is observed for this process which is described by the optical cross section $\sigma_p^0(1)$. This energy corresponds to a zero-phonon process and, therefore, locates the one-electron level at $\sim E_c - 0.15$ eV. This is in excellent agreement with the activation energy derived from the thermal decay. Thus, it is concluded that there is no large capture barrier for this state. After the optical filling of the one-electron state the depopulation by photoexcitation to the conduction band could also be observed experimentally down to an energy of 0.35 eV. The expected low-energy threshold of ~ 0.15 eV was out of the range accessible by the illumination optics.

The optical properties of the two-electron level were studied at a temperature of 155 K. This temperature is high enough to avoid quenching of the EL2 defect and low enough to suppress thermal emission. However, if the photoexcitation of the second electron to the conduction band is achieved, the first electron follows immediately due to thermal instability at that temperature. A relatively sharp threshold energy of 0.65 eV is observed for the optical cross section $\sigma_n^0(2)$. Therefore, from this experiment the second electron level is located at ~ 0.60–0.65 eV below the conduction band. In conclusion, it can be stated that the optical excitation experiments confirm the level positions of $\sim E_c - 0.15$ eV and $\sim E_c - 0.60$ eV for the first and the second electron, respectively, and as a direct consequence, the negative-U ordering.

6. Hall-effect and DLTS measurements

As already mentioned, there were a few samples showing both thermodynamically stable bands, A and B, in the dark. From the discussion above it is straightforward to conclude that the Fermi potential in these samples is between the two oxygen levels. In temperature-dependent Hall-effect measurements these samples always give an activation energy for the conduction band electron density of exactly 0.43 eV (figure 9). There are in principle two possibilities. The first one is that there is an additional defect level between the two oxygen levels. The other one is that the Fermi level is pinned by the oxygen negative-U centre itself. We discard the first possibility because the activation energy of 0.43 eV and the appearance of the A/B system are always coupled and observed in crystals grown by completely different techniques (LEC and HB).

If this assumption is correct, then the Hall activation energy gives additional information on the binding energies. It is known from literature that a negative-U system has a very peculiar behaviour in Fermi statistics [20]. Neither the one- nor the two-electron level can pin the Fermi potential. This role is played by the virtual occupancy level being located exactly halfway between the first and the second electron levels at $E_c - (E_1 + E_2)/2$ where E_1 and E_2 are the binding energies for the first and the second electron, respectively. As long as the free-electron density n is much smaller than the oxygen defect concentration (low-temperature approximation), it can be shown analytically that

$$n = \{(g_0/g_2)[2N(O)/(qN(O) - N(S)) - 1]^{-1}\}^{1/2}$$
$$\times N_c \exp[-(1/k_B T)(E_1 + E_2)/2]. \tag{9}$$

Here g_0 and g_2 are the degeneracy factors of the zero- and the two-electron states. $N(O)$ is the oxygen defect density, $N(S)$ the density of compensating shallow states (e.g. the carbon acceptor). q is the charge state of the unoccupied oxygen defect and N_c the density of states of conduction band electrons.

The simulation of the electron density according to (9) is also shown in figure 9. The parameters used are $g_0/g_2 = 1$, $qN(O) - N(S) = 1.5 \times 10^{15}$ cm^{-3} and the experimentally determined activation energies. The fit gives a very reasonable description of the experimental Hall data. The slope is slightly smaller, but taking into account the experimental error of ± 0.03 eV for the activation energies and the unknown temperature dependence of the oxygen level positions relative to the bands, the agreement is quite satisfactory. The Hall-effect data thus support independently the negative-U characteristics. To our knowledge this is the first time that the theoretically proposed occupancy level of a negative-U centre has been found in experiment.

It is only natural to ask whether the oxygen levels found in these studies are related with so far unidentified defect levels in GaAs [21]. This question could be answered by a comparative study between IR absorption and DLTS [22]. Neutron transmutation doping of oxygen-rich GaAs samples (see section 2) was used for two different purposes. First, as these samples contained the off-centre substitutional oxygen defect in a concentration of about $(1$–$2) \times 10^{15}$ cm^{-3}, increasing the doping level from 0 to 1×10^{16} cm^{-3} should induce the conversion from A to B due to the shift of the Fermi level from a near-mid-gap position to the conduction band. This is shown in figure 10. It should be pointed out that neutron irradiation alone neither destroyed nor changed the band A. Thus, this first part of the experiment confirms the model developed so far.

Higher doses of neutrons were used to make the samples n-type conducting (up to $n = 5 \times 10^{16}$ cm^{-3}). These samples were investigated by DLTS. Aside from a strong EL2 peak, two additional peaks with activation energies of 0.58 eV and 0.29 eV were found. These peaks can be attributed to the well known defect levels EL3 and EL7 or EL8, respectively. The coincidence of the activation energies of the EL3 level and the second electron level of oxygen as measured by LVM spectroscopy is striking. In figure 11 the usual plot $\log e_n/T^2$ versus $1000/T$ is shown for these levels. Both data sets can be linked by a straight line with a fair accuracy. This means that both levels also have the same capture cross section. It is therefore concluded that the second electron level of the oxygen defect and the EL3 level are identical. The concentration of this defect in the sample investigated is $\sim 5 \times 10^{14}$ cm^{-3} as determined by DLTS and $(1$–$1.5) \times$

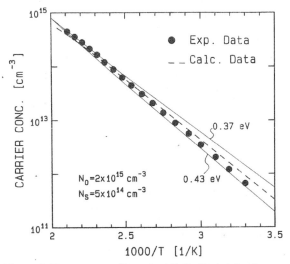

Figure 9. Temperature-dependent Hall-effect data of a 0.4 eV sample.

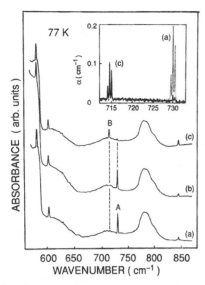

Figure 10. 77 K IR spectra of samples used for neutron-transmutation doping: (a) as-grown material, (b) neutron-irradiated but unannealed sample, (c) same sample after annealing. Inset: high-resolution spectra of samples from (a) and (c).

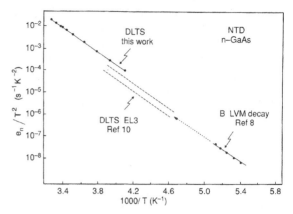

Figure 11. Emission rate versus reciprocal temperature plot for the EL3 level and the thermal decay of band B. Also shown are DLTS data for the EL3 level from the literature [17].

10^{15} cm^{-3} from the LVM absorption strength. This discrepancy of a factor of 2–3 is not in conflict with this conclusion in view of the systematic error of about a factor of 2 in both techniques. There is no emission peak in the DLTS spectra which could be assigned to the first electron level of the oxygen defect. This is expected because also in DLTS the thermal emission of the second electron effects an immediate release of the first electron. Thus the emission rate is governed by the emission of the second electron and each emission process releases two electrons. Without going into too much detail it can be stated that the assignment of the EL3 level to oxygen is not surprising. There are several reports in the literature where the EL3 level has been found in oxygen-doped crystals.

7. Conclusions

High-resolution local vibrational mode spectroscopy has brought about a large amount of information on the optical and electrical properties of off-centre substitutional oxygen in GaAs. The unique feature of this defect is that all the different states can be observed in one sample by their specific LVM bands using appropriate thermal cycling and optical excitation at low temperatures. The interaction with the dominant deep centre EL2 in terms of the different charge transfer processes is essential for the observation of many properties. It is now clear that the defect has three charge states with zero, one and two electrons. There is no doubt that this defect has negative-U properties. With the second electron bound more strongly than the first by ~ 0.45 eV, the one-electron state is a metastable state which can be observed only under non-equilibrium conditions at low temperatures.

Electrical measurements like temperature-dependent Hall-effect measurements and DLTS give a complementary insight into the defect properties. For the first time the occupancy level of a negative-U system, lying halfway between the first and the second electron level, is directly observed as the activation energy of free carriers. DLTS data revealed that the second electron level is identical with the well known EL3 level.

The open question with respect to the microscopic structure of the defect is the origin of the negative-U behaviour. The common picture is a large lattice distortion or relaxation which overcompensates the Coulomb repulsion energy. However, from a careful analysis of the LVM bands no indication for a distortion of the Ga–O–Ga molecule can be derived. Considering this molecule as an XY_2-type molecule in the Herzberg sense [23], it is possible to derive the effective force constant and the bridging angle 2θ from the ^{18}O/^{16}O and ^{71}Ga/^{69}Ga isotope shifts. It comes out that both parameters are constant within $\pm 2\%$ for all three charge states ($2\theta \sim 120°$). Therefore any significant contribution of the Ga–O–Ga molecule to the lattice distortion can be ruled out. This leads to the suggestion that the other part of the defect, consisting of the two Ga atoms binding together (figure 2), plays the essential role with respect to the negative-U ordering of the gap levels. This view is supported by the fact that the change of the LVM frequency is relatively small, amounting to 2.2% after capture of the first electron and to only 0.1% after capture of the second electron. This means that the electrons are located essentially on the Ga–Ga bond. Presumably, the defect may be separated into two relatively independent parts. One part, the Ga–O–Ga molecule, is the IR active part, whereas the other part, the two-site defect Ga–Ga, is responsible for the electrical properties of the defect. Valuable information on this latter part of the defect could come from spin resonance measurements, which also could solve the question of the total charge of the defect in each state. However, these measurements are inherently difficult due to the so far low concentration of the defect.

Acknowledgments

The author is grateful to U Kaufmann and J Schneider from the Fraunhofer Institut IAF in Freiburg for helpful discussions and cooperation for the neutron transmutation doping experiment. This work has been supported by the BMFT (contract no 2716 C).

References

[1] Song C, Ge W, Jiang D and Hsu C 1987 *Appl. Phys. Lett.* **50** 1666–8
[2] Alt H Ch 1989 *Appl. Phys. Lett.* **54** 1445–7
[3] Alt H Ch 1989 *Appl. Phys. Lett.* **55** 2736–8
[4] Alt H Ch 1990 *Phys. Rev. Lett.* **65** 3421–4
[5] Watkins G D 1984 *Festkoerperprobleme: Advances in Solid State Physics* vol 24 (Braunschweig: Vieweg) pp 163–189
[6] Stavola M, Levinson M, Benton J L and Kimerling L C 1984 *Phys. Rev.* B **30** 832–9
[7] Chadi D J and Chang K J 1988 *Phys. Rev. Lett.* **61** 873–6
[8] See other papers in this issue
[9] Huber A M, Linh N T, Valladon M, Debrun J L, Martin G M, Mitonneau A and Mircea A 1979 *J. Appl. Phys.* **50** 4022–6
[10] Akkerman Z L, Borisova L A and Kravchenko A F 1976 *Sov. Phys.–Semicond.* **10** 590–1
[11] Schnell R, Gisdakis S and Alt H Ch 1991 *Appl. Phys. Lett.* to be published
[12] Zhong X, Jiang D, Ge W and Song C 1988 *Appl. Phys. Lett.* **52** 628–30
[13] Watkins G D and Corbett J W 1961 *Phys. Rev.* **121** 1001–14
[14] Schneider J, Dischler B, Seelewind H, Mooney P M, Lagowski J, Matsui M, Beard D R and Newman R C 1989 *Appl. Phys. Lett.* **54** 1442–4
[15] Song C, Pajot B and Porte C 1990 *Phys. Rev.* B **41** 12330–3
[16] Silverberg P, Omling P and Samuelson L 1988 *Appl. Phys. Lett.* **52** 1689–91
[17] Mitonneau A, Mircea A, Martin G M and Pons D 1979 *Rev. Phys. Appl.* **14** 853–61
[18] The metastability of EL2 is reviewed, for example, by Martin G M and Makram-Ebeid S 1986 *Deep Centers in Semiconductors* ed S Pantelides (New York: Gordon and Breach) pp 399–487
[19] Skowronski M, Neild S T and Kremer R E 1990 *Appl. Phys. Lett.* **57** 902–4
[20] Look D C 1981 *Phys. Rev.* B **24** 5852–62
[21] Martin G M, Mitonneau A and Mircea A 1977 *Electron. Lett.* **13** 191–3
[22] Kaufmann U, Klausmann E, Schneider J and Alt H Ch 1991 *Phys. Rev.* B **43** 12106–9
[23] Herzberg G 1954 *Molecular Spectra and Molecular Structure* vol II (New York: Van Nostrand) p 168

Semicond. Sci. Technol. 6 (1991) B130–B133. Printed in the UK

A spectroscopic study of a metastable defect in silicon

J H Svensson, E Janzén and B Monemar

Department of Physics and Measurement Technology, Linköping University,
S-581 83 Linköping, Sweden

Abstract. A metastable defect in silicon is discussed. The defect appears after irradiation with electrons at room temperature with 2 MeV electrons, and is studied by means of infrared absorption spectroscopy. An absorption spectrum with a lowest no-phonon line at 615.0 meV is associated with a metastable configuration of the defect. This spectrum is initially not observable when the sample is cooled down to temperatures below ~65 K. It is created, however, after the sample is irradiated with light below this temperature. The spectrum disappears when the sample is heated in darkness at temperatures exceeding 70 K. It is concluded that the transformation to the infrared-active metastable configuration is induced by a capture or recombination process involving charge carriers. The observation of an isotope shift of the no-phonon line at 615.0 meV confirms that carbon is one of the constituents of the defect.

1. Introduction

A well known example of a carbon-related metastable defect in silicon for which a detailed microscopic model exists is the carbon interstitial–carbon substitutional (C_i–C_s) pair [1].

Recently, an infrared absorption spectrum with a lowest no-phonon line at 615.0 meV was interpreted as being due to electronic excitations of a metastable configuration of a carbon-related defect in electron-irradiated silicon [2, 3]. The increase of absorption, corresponding to the transformation from the stable to the metastable configuration of the defect, is induced by excitation of the samples with photons with energy exceeding, or in close resonance with, the band gap. The transformation occurs at temperatures below ~65 K. At temperatures higher than ~70 K the spectrum disappears, reflecting the thermally activated transformation to the stable configuration [2].

In this paper we present proof that carbon is one of the constituents of this complex defect. We also present evidence that the transformation of the defect from the stable to the metastable configuration is induced by a recombination or capture of a charge carrier(s) at the defect.

2. Experimental details

The silicon samples studied were crucible-grown and doped with phosphorus. Their initial resistivity was ~40 Ω cm at room temperature. After irradiation with 2 MeV electrons at room temperature the samples were highly resistive. One sample was isochronically annealed at a series of temperatures from 150 °C to 350 °C at increments of 20 °C for 20 min at each temperature.

The excitation of the samples was performed either with the 1.06 μm line from a Nd:YAG (yttrium aluminium garnet) laser, or with white light. The absorption measurements were performed with a BOMEM DA3.20 Fourier transform infrared (FTIR) spectrometer. The luminescence measurements were performed with a SPEX 1404 spectrometer and the samples were excited with the 647.1 nm line of a Kr$^+$ laser.

3. Experimental results and discussion

The absorption spectrum related to the 615 meV defect is initially weak after cooling the sample in darkness. After 3 h of excitation of the samples with a laser power of 50 mW mm^{-2} of the 1.06 μm line at a sample temperature of 10 K, the absorption is maximized in all samples studied. A typical example of the spectrum is shown in figure 1. This spectrum has been attributed to symmetry-allowed electronic transitions from a non-degenerate ground state of the metastable configuration of the neutral defect to excited electron states. The excited electron is bound in the Coulomb potential of a hole tightly bound to the defect (a 'pseudo-donor' state) [3], and is described by ns states derived from the effective-mass approximation (EMA), but perturbed by the central-

0268-1242/91/10B130+04 $03.50 © 1991 IOP Publishing Ltd

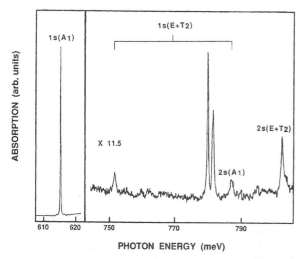

Figure 1. The absorption spectrum of the 615 meV defect after 3 h of excitation with the 1.06 μm line of the YAG laser at a sample temperature of 10 K. The notation of the lines refers to the different ns states of the excited electron (the symmetry notation corresponds to T_d symmetry) [3].

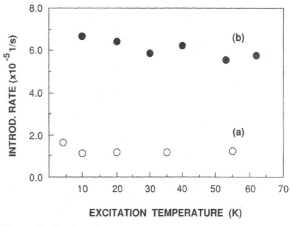

Figure 2. The introduction rate of the 615.0 meV line at different sample temperatures. The rate was measured in (a) a sample after electron irradiation at room temperature, and no heat-treatment exceeding $\sim$40 °C and (b) a sample that was isochronally annealed, as described in the text.

cell potential. Here, n corresponds to the main quantum number of the EMA wave functions. By assuming that the lines at 779.0–780.0 meV correspond to transitions to $1s(T_2 + E)$ pseudo-donor states that have a node at the defect site and that their binding energies are close to the 1s EMA value of 31.26 meV, the ionization edge of the bound electron is approximately at $780.4 + 31.26 = 811.7$ meV. The ground state of the metastable configuration is then approximately $E_v + (1169.5 - 811.7) \approx E_v + 358$ meV (measurement temperature 10 K).

Upon excitation of the sample with white light or with the 1.06 μm laser line the absorption increases monotonically versus time, a behaviour that has been explained as reflecting a transformation of the defect from the stable to a metastable configuration [2, 4]. Two irradiated samples have been studied. One sample was never exposed to temperatures exceeding $\sim$40 °C prior to the measurements. The other was isochronally annealed, as described in the previous section. If the temperature of the samples is varied while the excitation power is held constant it is found that the rate of increase of the absorption spectrum versus excitation time (the introduction rate) is independent of temperature between 4.2 and 60 K for both samples (figure 2). However, the introduction rate for the annealed sample is larger than for the unannealed sample. Furthermore, the introduction rate is found to increase linearly with excitation power [4]. If a germanium or a silicon wafer, held at room temperature, is used to filter the white light during excitation, the increase in absorption with excitation time is negligible within the time-scale of our experiments (hours).

The 1.06 μm line is energetically below the band gap of silicon at the lowest temperatures, and excites primarily free and bound excitons and not free charge carriers directly. When white light is used the absorption of photons in the sample occurs in a broad energy range. In

the heat-treated sample the concentration of other defects acting as competing capture or recombination channels has most likely decreased. For example, the divacancy (V_2) [5] and the C_i–C_s defect (figure 3) have disappeared and the vacancy–oxygen (V–O) defect has decreased substantially in concentration [6]. This means that more carriers are available for the transformation process and that the introduction rate should increase, as is indeed observed in figure 2. The experimental results thus show that the increase of absorption involves the

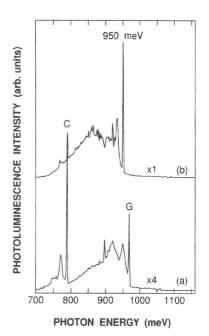

Figure 3. Photoluminescence spectra of (a) the sample after irradiation and no heat-treatment and (b) the isochronally annealed sample. The sample temperature was 2 K and the samples were excited with the 647.1 nm line of a Kr$^+$ laser. The power was 80 mW and the beam was unfocused ($\varnothing \approx 3$ mm).

capture or recombination of free or nearly free charge carrier(s). The transformation to the metastable configuration is then induced by a recombination or capture of a carrier(s) at the stable configuration of the defect. The question is then what mechanism governs this non-radiative capture or recombination.

The major proposals for non-radiative capture processes of charge carriers into defect-related states in the band gap have been: (i) capture via a cascade mechanism [7], (ii) capture through multiphonon emission [8], (iii) capture through different Auger mechanisms [9–15].

The absence of temperature dependence of the introduction rate observed in the present work is not in agreement with what is expected for a capture of carriers through a cascade process [7]. For capture of a carrier with multiphonon emission a thermally activated behaviour of the capture cross section is expected [8]. If the temperature is low enough to prevent this thermally activated capture, capture may still occur through a tunnelling process which would result in a temperature-independent capture cross section. Still, the capture rate, $C_{n,p}$, should be proportional to $T^{1/2}$, according to equation (1) below

$$C_{n,p} = \sigma n_{n,p}(3kT/m^*)^{1/2} \tag{1}$$

where $n_{n,p}$ is the free carrier concentration (n, electron; p, hole), σ is the capture cross section, k is Boltzmann's constant, T is the absolute temperature and m^* is the carrier effective mass. Thus a multiphonon capture mechanism does not account for the lack of dependence on temperature of the introduction rate.

An Auger-related recombination or capture mechanism is here proposed to be the process inducing the change in configuration. An Auger process involving two carriers of different kinds should yield both the linear dependence of the introduction rate on excitation power and the temperature independence of the introduction rate which were found. A recently proposed excitonic Auger capture mechanism [14, 15] is suggested to be the most likely [4].

The photoluminescence of the different samples was also investigated, as is shown in figure 3. As can be observed the spectra are dominated by luminescence from recombination of carriers at radiation-induced defects other than the 615 meV defect, and show that the concentration of deep defects is still high in the heat-treated sample. The defects corresponding to the no-phonon line at 0.789 eV (C line) [16, 17] and 0.950 eV [18–20] dominate the spectra of the annealed sample. The C line has been attributed to an interstitial carbon–interstitial oxygen complex defect [16, 17], and is already present before the heat treatment. The 0.950 eV line appears upon the annealing of the sample in the temperature region 250–475 °C [19], and the corresponding defect has been found to contain carbon [20].

Only after heating the sample in darkness at temperatures exceeding ~ 70 K does the spectrum disappear. The disappearance of the spectrum follows a simple exponential decay [2] of the form

$$A(t) = (A(0) - A(\infty))\exp(-Rt) + A(\infty)$$

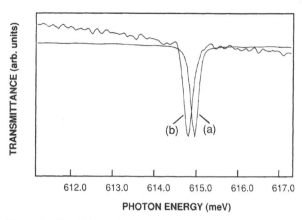

Figure 4. The 615.0 meV no-phonon line measured at a sample temperature of 10 K in (*a*) a sample containing carbon isotopes in the natural concentrations and (*b*) a sample enriched with the ^{13}C isotope.

where $A(t)$ is the absorption at time t and R is the rate of decrease in absorption of the spectrum. The rate, R, is consistent with the following relation

$$R = 1.8 \times 10^{11} \exp[-(0.21 \text{ eV})/k_B T].$$

This decrease is interpreted as being the thermal activation energy for changing the configuration of the defect from the metastable to the stable configuration. It is interesting to note that this activation energy of 0.21 eV is very close to the energy barrier for the carbon–silicon bond switching reported for the bistable C_s–C_i pair defect [1].

The effect on the 615.0 meV line caused by doping a crucible-grown sample with the ^{13}C isotope is shown in figure 4. The difference in energy between the two no-phonon lines due to the incorporation of the ^{12}C isotope and the ^{13}C isotope in the defect is

$$h\nu^{12} - h\nu^{13} = 0.11 \pm 0.01 \text{ meV}.$$

The relative absorption of the two no-phonon lines due to the ^{12}C and the ^{13}C isotopes in this ^{13}C-enriched sample is found to correspond to the ratio of the concentrations of the two isotopes. The concentrations of the two isotopes were estimated by the absorption of the local mode vibrations of the substitutional carbon atom.

4. Conclusion

The properties of a carbon-related complex defect ('615 meV') in electron-irradiated silicon studied by infrared spectroscopy have been summarized. It is confirmed that carbon is one of the constituents of the defect. The rate of increase of the absorption spectrum of the defect upon illumination at sample temperatures below ~ 65 K has been studied in detail. The temperature independence of the rate favours an Auger process as the initiating process for the change in configuration.

Acknowledgments

The experimental assistance of Anne Henry and Olaf Kordina is gratefully acknowledged.

References

[1] Song L W, Zhan X D, Benson B W and Watkins G D 1990 *Phys. Rev.* B **42** 5765

[2] Svensson J H and Monemar B 1989 *Phys. Rev.* B **40** 1410

[3] Svensson J H, Monemar B and Janzén E 1990 *Phys. Rev. Lett.* **65** 1796

[4] Svensson J H, Monemar B and Janzén E 1990 *Proc. 20th Int. Conf. on the Physics of Semiconductors* ed E M Anastassakis and J D Joannopoulos (Singapore: World Scientific) p 569

[5] Svensson B G, Johnsson K, Xu D-X, Svensson J H and Lindström J L 1989 *Radiat. Eff. Defects in Solids* **111–12** 439

[6] Svensson B G and Lindström J L 1986 *Phys. Rev.* B **34** 8709

[7] For a review see, for example, Abakumov V N, Perel V I and Yassievich I N 1978 *Fiz. Tekh. Poluprovod.* **12** 3 (Engl. transl. 1978 *Sov. Phys. Semicond.* **12** 1)

[8] Henry C H and Lang D V 1977 *Phys. Rev.* B **15** 989

[9] Landsberg P T, Rhys-Roberts C and Lal P 1964 *Proc. Phys. Soc.* **84** 915

[10] Haug A 1980 *Phys. Status Solidi* b **97** 481

[11] Haug A 1981 *Phys. Status Solidi* b **108** 443

[12] Riddoch F A and Jaros M 1980 *J. Phys. C: Solid State Phys.* **13** 6181

[13] Nelson D F, Cuthbert J D, Dean P J and Thomas D G 1966 *Phys. Rev. Lett.* **17** 1262

[14] Hangleiter A 1987 *Phys. Rev.* B **35** 9149

[15] Hangleiter A 1988 *Phys. Rev.* B **37** 2594

[16] Davies G 1989 *Phys. Rep.* **176** 83

[17] Trombetta J M and Watkins G D 1988 *Defects in Electronic Materials (Mater. Res. Soc. Symp. Proc.* **104**) ed M Stavola, S J Pearton and G Davies (Pittsburgh: Materials Research Society) p 93

[18] Awadelkarim O O, Weman H, Svensson B G and Lindström J L 1986 *J. Appl. Phys.* **60** 1974

[19] Tkachev V D and Mudryi A V 1977 *Radiation Effects in Semiconductors 1976 (Inst. Phys. Conf. Ser.* **31**) ed B Urlin and J W Corbett (Bristol: Institute of Physics) p 231

[20] Davies G, Lightowlers E C, Woolley R, Newman R C and Oates A S 1984 *J. Phys. C: Solid State Phys.* **17** L499

Semicond. Sci. Technol. **6** (1991) B134–B136. Printed in the UK

New mechanism for metastability of the 'red' luminescence in electron-irradiated CdS

S S Ostapenko

Institute of Semiconductors, Ukrainian Academy of Science, pr.Nauki 45, 252650, Kiev, USSR

Abstract. The effect of metastability of the 'red' photoluminescence band ($hv_{max} = 1.68$ eV) is studied in detail in electron-irradiated CdS. Spectral, temperature and polarization measurements are performed. A new mechanism of optical pumping and stimulation of luminescence intensity is proposed. This mechanism is attributed to the complex triple-luminescence centre and the recharging of its components under light illumination.

1. Introduction

The origin of the metastability of deep centres in semiconductors is still a point of current technological and physical interest. Some alternatives are usually available to explain the mechanism of the metastability of a centre, and a good example is given by the study of the DX centre in $Al_xGa_{1-x}As$ [1]. Only a few cases are known that provide well supported microscopical models of centres involved in metastable effects [2]. The majority of these models are based on the complex structure of the centre, which is demonstrated as its anisotropy in the ground state or in the metastable state of the defect. This is the case for Fe_i–Al_s and C_i–P_s pairs in Si, EL2 in GaAs and, presumably, DX centres in $Al_xGa_{1-x}As$. Complex centres with a symmetry lower than that of the local lattice usually give rise to the anisotropy of optical spectra and can be examined by polarized luminescence [3].

The purpose of this paper is to analyse the results of a comprehensive study of the effect attributed to the transient behaviour of the 'red' photoluminescence (PL) band ($hv_{max} = 1.68$ eV) in electron-irradiated CdS [4–6] and to show that the known mechanisms of centre metastability cannot be fitted to experiment. The new mechanism responsible for the PL metastability that has been proposed in [4] within the model of a multiparticle (triple) centre is supported by new data.

2. Experiment

Czochralski-grown bulk n-CdS single crystals were investigated. The crystals were irradiated at 200 K by 1.2 MeV electrons with a dose of 10^{16}–10^{18} cm^{-2}. Three

well known PL bands are observed in CdS crystals at 77 K before and after e-beam treatment: the 'green' band with hv_{max} of 2.42 eV, the 'orange' band (2.05 eV) and the 'red' band (1.72 eV). After electron irradiation a transient of PL intensity was observed within the spectral range of the 'red' PL band (figure 1). This transient behaviour was observed under PL excitation in a spectral region of stimulation light $hv_{st} = 2.5$ eV and unambiguously related to the preliminary illumination of the sample with pumping light of $hv_p = 2.0$–2.4 eV. The luminescence, pumping and stimulation spectra of ΔI value, i.e. the dependence of ΔI on the quantum energies of hv_{lum}, hv_p and hv_{st} respectively, are shown in figure 2. It is argued that the effect observed is caused by the specific centres

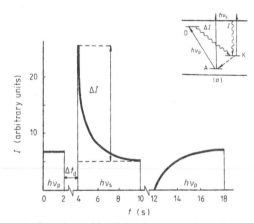

Figure 1. Transient of 'red' luminescence intensity ($hv_{lum} = 1.68$ eV) under pumping (hv_p) and stimulating (hv_s) light illumination of a sample; ΔI is the intensity of stimulated luminescence. (a) The scheme of electron transitions within the complex centre D–A–K giving rise to stimulated (ΔI) and stationary (I) luminescence.

0268-1242/91/10B134+03 $03.50 © 1991 IOP Publishing Ltd

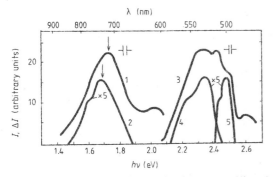

Figure 2. The spectra of stationary luminescence (1) and stimulated luminescence (2). Excitation spectra of stationary luminescence (3). Pumping (4) and stimulation (5) spectra of ΔI; $T = 77$ K [4].

which are labelled as the centres of stimulated 'red' luminescence (SRL). The concentration of SRL centres is gradually increased with the electron-irradiation dose and reaches a value of the order of 10^{13} cm^{-3}. Thus, the effect of PL stimulation is directly related to the e-beam treatment of CdS.

3. Discussion

The following data are particular reasons for evaluating the mechanism for the effect.

(i) SRL is not accompanied by the stimulation of photocurrent in n-type samples. This means that stimulated luminescence results from the recombination of localized electrons.

(ii) The kinetics of SRL measured with a pulse excitation is much slower ($\tau \simeq 5 \times 10^{-6}$ s) than that of free-hole capture in CdS. Thus, it can be concluded that stimulated PL originates from a bound-to-bound recombination.

(iii) The value of ΔI is quenched when the temperature is increased with an activation energy of $\Delta = 0.08 \pm 0.01$ eV. This energy is linked to the gap between the ground state ($E_c - 0.13$ eV) and first excited state ($E_c - 0.04$ eV) of the particular donor centre (D). The level of the D centre could be directly filled by electrons with pumping light and controlled by the measurement of thermostimulated current. These levels can be also measured independently as the shift of the stationary, I, and stimulated, ΔI, PL maxima, as shown by arrows in figure 2 (curves 1, 2). After switching off the pumping light the ΔI decays in the dark following the relation $\Delta I \sim \exp(-\Delta t_d / \tau_r)$ with $\tau_r = 5 \times 10^3$ s at 77 K, where the dark time interval, Δt_d, is shown in figure 1. This process results from the release of electrons from the D centre.

(iv) Besides the D centre, the level of a shallow acceptor (A) with energy position $E_v + (0.14 \pm 0.01)$ eV is found. This energy position is evaluated from the

threshold energy of the stimulation spectrum (curve 5, figure 2) attributed to an A → C-band transition and is given independently by the energy maximum in the pumping spectrum (curve 4, figure 2) which is attributed to an A → D transition (see below).

(v) The low symmetry of the SRL centre, being the consequence of anisotropy of the deep acceptor K with a level or $E_v + 0.68$ eV, is demonstrated by the PD method [5]. We emphasise the coincidence of the polarization characteristics of stationary, I, and stimulated, ΔI, 'red' PL. In particular, the degrees of polarization of PL and PL excitation as well as their polarization diagrams are identical for both bands [5]. The latter indicates the same orientation of the optical dipoles within both centres and identity of their symmetry.

(vi) SRL centres are annealed at $T_{an} > 160$ °C with an activation energy of $(0.45–0.50) \pm 0.05$ eV. The annealing mechanism is identified as the recombination of the D component enclosed in the SRL centre with mobile atoms of S_i due to their diffusion in the CdS crystal [6]. Thus, in experiment we observe the complex triple PL centre consisting of deep acceptor (K), deep donor (D) and shallow acceptor (A).

Some known mechanisms available to account for the metastability of the PL centre are given by models that suggest the strong change under illumination of the energy levels attributed to the centre. This is the case for the U centre [7] or a centre affected by large lattice relaxation [8]. On the contrary, the levels of K, D and A centres evaluated by optical and temperature experiments preserve their position before and after transient. Other mechanisms use the idea of an interband barrier for the carriers captured by shallow levels [9]. For direct-gap CdS and deep D and K components it can be ruled out. The small lattice relaxation model [10] seems to be adequate to explain our effect from the point of equality of the optical and thermal depth of the D centre. However, the light-stimulated D → K recombination does not fit this model.

The mechanism for the PL metastability adjusted to take into account all experiments with the SRL effect is proposed in [4] within the model of a triple D-A-K centre (figure 1(a)). According to this mechanism the components of a triple centre can be recharged as a result of pumping light illumination. This is a two-step process: (i) intercentre A → D optical electron transition, and (ii) hole tunnelling A → K. The total charge state of a complex is evidently unchanged after pumping, but recharging of D and K levels due to localization of electrons and holes takes place. We suggest that after pumping D → K recombination has a low probability as a result of two facts: (i) strongly localized electron and hole wavefunctions of the ground state for D and K centres (both deep) and separation of D and K in a lattice, and (ii) the existence of the repulsive Coulombic barrier from the negatively charged A centre which is a near neighbour to K. When (ii) dominates, the photoionization of the A centre by stimulated light removes the barrier and provides the D → K luminescent transition (figure 1(a)).

4. Summary

The model of a triple centre is developed in [5]. It is argued that A and D are the components of a Frenkel pair, S_i and V_S respectively, created by electron irradiation in the vicinity of the deep acceptor (K). The latter was tentatively identified with a substitutional Cu_{Cd} impurity. This model was justified by annealing experiments [6].

Acknowledgments

I thank Professor M K Sheinkman for helpful discussion.

References

[1] Bourgoin J C (ed) 1990 *Physics of DX Centers in GaAs Alloys (Solid State Phenomena 10)* (Vaduz: Sci. Tech. Publications) p 253
[2] Watkins G D 1989 *Mater. Sci. Forum* **38–41** 39
[3] Ostapenko S S and Sheinkman M K 1989 *Mater. Sci. Forum* **38–41** 809
[4] Bogdanyuk N S, Galushka A P, Ostapenko S S and Sheinkman M K 1984 *Sov. Phys.–Poluprovod.* **18** 189
[5] Bogdanyuk N S, Galushka A P, Ostapenko S S and Sheinkman M K 1985 *Sov. Phys.–Solid State* **27** 1155
[6] Bogdanyuk N S and Ostapenko S S 1986 *Phys. Status Solidi* a **96** 621
[7] Chadi D J and Chang K J 1989 *Phys. Rev.* B **39** 10063
[8] Mooney P M, Northrop G A, Morgan T N and Grimmeiss H G 1988 *Phys. Rev.* B **37** 8298
[9] Bourgoin J C and von Bardeleben H J 1989 *Phys. Rev.* B **40** 10006
[10] Henning J S M and Ansems J P M 1987 *Semicond. Sci. Technol.* **2** 1

Semicond. Sci. Technol. **6** (1991) B137–B142. Printed in the UK

The utilization of DX centres in high-pressure studies of low-dimensional doping structures in GaAs

R A Stradling†, E A Johnson†, A Mackinnon†, R Kumar†, E Skuras† and J J Harris‡§

† Physics Department and London University Interdisciplinary Research Centre for Semiconductor Materials, Imperial College, London WC1 2BZ, UK
‡ Philips Research Laboratories, Redhill RH1 5HA, UK

Abstract. High-pressure experiments with thin slabs of silicon donors in GaAs are employed to search for the correlation effects expected from the formation of D^+D^- pairs expected from the negative-U model of DX centres. The mobilities in the individual subbands are very dependent on subband energy but a preliminary analysis does not require the existence of such pairs.

1. Introduction

The ability of DX centres to trap out conduction electrons metastably can be exploited in quantum transport experiments with GaAs structures containing planes of silicon donors introduced during MBE growth. The results from earlier work involving the Imperial College group with such delta wells and thin doping slabs [1] are reviewed and discussed with respect to the different models for DX centres.

Transport measurements with bulk-doped samples by Maude *et al* [2] showed a large increase in mobility on increasing the pressure due to the change in charge state of the localized states. Controversy has arisen concerning the interpretation of this change in mobility. It was initially argued that the sign of the pressure-induced change in mobility provided conclusive qualitative evidence against the negative-U model of DX centres [2, 3]. However, it was shown later that a mobility increase of the required sign and magnitude could result from impurity correlation effects [4, 5]. Experiments with delta-doped samples provide an additional way to investigate this possibility as the extent of the electronic wavefunctions in the z-direction increases rapidly with increasing subband energy. Consequently, if the dipolar fields expected from correlation effects are present, the mobility changes should be greater for the higher-order subbands than for the $i = 0$ case (which should 'see' only monopolar fields). Furthermore the spacing of the individual subbands should be modified by the dipolar component

of the potential local to the doping plane. The electrons are expected to trap out onto donor sites which are favoured energetically and thus have an ionized (positively charged) donor close by. On a negative-U model, D^+D^- pairs will form and the dipolar component will be very strong. On the positive-U picture there will be no negative impurities present unless compensating acceptors are introduced during growth. Consequently the dipolar component of the local potential will be weak. The results of reference [1] are analysed in an attempt to detect correlation effects.

In reference [1] a method was developed for determining the individual subband mobilities quantitatively from the width of the peaks resulting from the Fourier analysis of the Shubnikov–de Haas oscillations. It was found that the mobility in the individual subbands can differ by as much as an order of magnitude but all the subband mobilities increase with increasing pressure by about the same amount. Consequently our initial interpretation appears to confirm the conclusions of Maude *et al* [2, 3] concerning the change in charge state of the Si donors. Another important result to emerge from this study is that Hall mobility measurements for delta samples must be analysed using a full multicarrier model involving estimates of the individual subband model. The Hall mobility as measured with the thinnest of the current samples can actually fall with increasing pressure, although all mobilities within the individual subbands are increasing. These apparently conflicting results can be reconciled on a multicarrier model, with the fall in the Hall mobility simply arising from the depopulation of higher-mobility upper subbands which are weighted more heavily in Hall experiments.

§ Now at the London University Interdisciplinary Research Centre, Imperial College.

0268-1242/91/10B137 + 06 $03.50 © 1991 IOP Publishing Ltd

2. Experimental techniques

Three GaAs samples with thin doping Si slabs were grown by MBE at Philips Research Laboratories at Redhill. The growth temperature was 400 °C. The value for the areal concentration was constant at a value of 1.1×10^{13} cm^{-2} and the slab thicknesses assuming no spreading of the Si donors were 2.0, 5.0 and 10 nm giving volume concentrations of 5.5×10^{19}, 2.2×10^{19} and 1.1×10^{19} cm^{-3} respectively. The thickness of the epilayer was 1 μm in each case and the doping slab was located 0.5 μm from the surface. The doping slabs were formed by sequential planar deposition of silicon at the appropriate doping level; e.g. the 5 nm slab was formed by depositing 18 planes of 5.56×10^{11} cm^{-2} of silicon separated by 2.8 Å. Local vibrational mode (LVM) studies of single-delta wells grown at 400 °C in the same reactor and having an areal concentration of 2×10^{13} cm^{-2} did not show any sign of silicon switching to the As site or of the formation of Si_{Ga}–Si_{As} pairs above the detection limit [6]. The volume concentration in these samples (estimated on the assumption that no spreading had taken place) would have been more than an order of magnitude higher than the concentrations employed in the present work. The temperature of growth and other growth conditions such as substrate orientation and growth rate are crucial in persuading high concentrations of silicon to become located on the gallium site and to act as simple substitutional donors. The experimental details of the electrical measurements and the techniques employed to analyse the Fourier data are described in greater detail in reference [1].

3. Carrier concentrations deduced from Fourier analysis of the Shubnikov–de Haas effect

Tables 1 and 2 show the carrier concentrations of the electric subbands deduced from Fourier analysis of the Shubnikov–de Haas data at ambient pressure and as a function of pressure. The total carrier density at zero pressure is estimated to be 7.2×10^{12} cm^{-2} for the 2 nm slab, 8.7×10^{12} cm^{-2} for the 5 nm slab and 1.1×10^{13} cm^{-2} for the 10 nm slab (in the case of the two thinnest slabs the population of the $i = 0$ subband has to be estimated from the theoretical fits, as the Shubnikov-de Haas peaks from this subband were only detectable after a few kilobars of pressure had been applied). These values suggest a loss of 3.8×10^{12} cm^{-2} and 2.3×10^{12} cm^{-2} carriers to localized states for the two thinnest slabs even at ambient pressure if no site switching had taken place. The value found with the thinnest slab agrees quite closely with the saturation value of 5.6×10^{12} cm^{-2} for the carrier concentration achievable with a truly delta-doped sample as estimated in reference [7]. There are two more items of evidence that at least a partial occupation of the localized centres occurs at atmospheric pressure with the two thinnest slabs. The first is that the free-electron concentration decreases immediately on increasing the hydrostatic pressure above atmospheric. The second concerns the persistent photoconductivity effect found after illuminating the sample with a red LED as described in reference [1]. All of these results confirm the LVM conclusion [6] that saturation of the carrier concentration in thin doping slabs for low growth temperatures at a value of 7×10^{12} cm^{-2} is occurring through capture into localized resonant states rather than by site switching of the silicon atoms.

The relative subband occupancies are sensitive to the Fermi level pinning remote to the well and hence to the background doping and to the nature of the defect centres present in the bulk. The sensitivity of the theoretical fit to the choice of depletion charge may provide the major limitation to the accuracy whereby such studies can be employed to study dopant diffusion.

It is worth commenting that 7×10^{12} cm^{-2} distributed over a monolayer would result in an effective volume concentration of 2.6×10^{20} cm^{-3} if no dopant spreading occurred. This is an order of magnitude greater than the limit for incorporation of silicon as a donor even

Table 1. Population of subbands at ambient pressure estimated from Fourier analysis and from self-consistent calculations (all values $\times 10^{12}$ cm^{-2}).

i	2 nm		5 nm			10 nm	
	Expt.	Theor.	Expt.	Theor. (no depletion)	Theor. (depletion)	Expt.	Theor. (depletion)
0	[4.6]	4.62	[5.2]	5.07	5.21	5.35	5.32
1	1.84	1.82	2.32	2.30	2.34	3.28	3.18
2	0.71	0.80	0.90	1.00	0.90	1.66	1.53
3		0.28	0.26	0.40	0.06	0.63	0.67
4		0.08		0.12			0.25
5				0.01			0.06
Total							
ΣN	7.15	7.60	8.68	8.90	8.51	10.92	11.01
E_{fermi} (meV)		222		219	223		206

[ ] $i = 0$ peak is not observed at ambient pressure with two thinnest slabs, so population is assumed to be the theoretical value.

Table 2. Pressure variation of subband occupancies (all concentrations $\times 10^{12}$ cm^{-2}).

Pressure (kbar)	Width (nm)	$i=0$	$i=1$	$i=2$	$i=3$	ΣN_i
	2	[4.6]	1.84	0.71		7.2
0	5 (expt)	[5.2]	2.32	0.90	0.26	8.7
	10	5.35	3.28	1.66	0.63	11.00
6.3	2	4.34	1.38	0.30		6.02
5.1	5(expt)	4.98	2.09	0.72		7.79
5.1	5(th-odp)	4.96	2.22	0.95	0.37	8.50
5.1	5(th-dep)	4.97	2.18	0.77		7.92
6.4	10	5.34	3.46	1.58	0.59	10.92
9	2	4.15	1.30			5.45
10.0	5(expt)	4.25	1.75	0.52		6.52
10.0	5(th-odp)	4.23	1.82	0.74	0.26	7.05
10.0	5(th-dep)	4.23	1.74	0.52		6.49
13.2	2	3.46	0.90			4.36
12.3	5(expt)	3.98	1.58	0.38		5.94
12.3	5(th-odp)	3.94	1.65	0.66	0.22	6.47
12.3	5(th-dep)	3.98	1.59	0.40		5.97
12.4	10	4.56	2.84	1.19	0.36	8.95
15.3	2	3.14	0.73			3.87
16.2	5(expt)	3.14	1.09			4.23
16.2	5(th-odp)	3.05	1.19	0.44	0.12	4.80
16.2	5(th-dep)	3.14	1.13	0.11		4.38
16.3	10	3.76	2.21	0.76		6.73
19	2	1.94				1.94
19.5	10	2.81	1.55	0.42		4.78

th-odp: theoretical value ignoring depletion effects.
th-dep: theoretical value including depletion effects.
expt: experimental values.

at the much reduced growth temperature of 400 °C. Comparisons of published data on Shubnikov–de Haas measurements [8, 9] with those for the sample of 2 nm slab thickness given in this paper and the results of SIMS measurements [10] demonstrate clearly that the silicon is diffusing distances at least of the order of 2–4 nm in supposedly delta-doped samples if the saturation limit is approached with growth temperatures much in excess of 400 °C. In a recent paper Koenraad et al [11] report that pressure-induced depopulation is not observed up to 9 kbar with a delta sample grown at 480 °C with a concentration of Si of 8×10^{12} cm^{-2} despite approximately a 50% loss in the free-electron concentration at ambient pressure. The carrier concentration reported for this sample is substantially less than the concentration measured for our thinnest slab at 9 kbar. Thus consistency between the two sets of experiments is achieved if it is assumed that a substantial proportion of the silicon has switched site or condensed into electrically inactive precipitates in the delta sample grown at 480 °C where the volume concentration assuming no spreading is equivalent to 2×10^{20} cm^{-3}. For a carrier concentration of 4×10^{12} cm^{-2} we would not expect to see any depopulation at pressures up to 9 kbar because the localized states will remain above the Fermi energy even for a very abrupt silicon profile.

4. Subband mobilities deduced from Fourier analysis of the Shubnikov–de Haas peaks

The half-width at half-height of each peak in the Fourier spectrum can readily be shown to be given by [1, 12]

$$\delta B_i = \sqrt{3}/2\,\mu_i$$

where μ_i is the mobility of the ith subband.

The results shown in table 3 are a compilation of a detailed analysis of the width and amplitudes of the Shubnikov–de Haas peaks and of a two-carrier analysis of the monotonic magnetoresistance. These show that the mobility in the $i=0$ subband is a factor of three lower than that of the $i=1$ subband and may be an order of magnitude lower compared with the higher-order subbands.

With the thinnest sample at the highest pressure (19 kbar) only a single subband is occupied as shown by a single Shubnikov–de Haas series. Under these conditions the carrier concentrations determined by the different experimental techniques and the Hall mobility should agree to within experimental error as should the Hall and Shubnikov–de Haas results for the carrier concentration. Agreement to within about 10% is indeed achieved (table 4).

Table 3. Subband mobilities deduced from Fourier analysis and galvanomagnetic measurements (all values in cm^2 V^{-1} s^{-1}).

	Pressure (kbar)	μ_0	μ_1	μ_2	μ_3
2 nm sample	0	600	1800	2900	
	6.3	950	2600	4000	
	9	1100	2900		
	13.2	1300	3100		
	15.3	1400	3500		
	19	1400			
5 nm sample	0	600	1100	2600	
	5.1	750	1500	3700	
	10	1000	1900	3900	
	12.3	1200	2300	5000	
	16.2	1500	2900		
10 nm sample	0	650	800	1400	2900
	6.4	900	1000	1700	3900
	12.4	1250	1350	2300	5800
	16.3	1500	1700	3900	

Table 4. Mobility and carrier concentration of the 2 nm sample at 19 kbar from different experimental techniques.

	Shubnikov–de Haas	Hall
Mobility (cm^2 V^{-1} s^{-1})	1400	1448
Carrier concentration (10^{12} cm^{-2})	1.94	1.83

Table 5. Hall and conductivity mobilities (measured and calculated values) (all mobilities in cm^2 V^{-1} s^{-1}).

	Pressure (kbar)	μ_{Hall} (meas.)	μ_{Hall} (calc.)	$\mu_{conduct}$ (meas.)	$\mu_{conduct}$ (calc.)
2 nm sample	0	2500	1700	1900	1200
	6.3	2500	2000	1700	1500
	9	2200	1900	1700	1500
	13.2	2200	2000	1800	1700
	15.3	1900	2200	1700	1800
	19	1450	1450		1450
5 nm sample	0				
	5.1	2400	1800	1900	1200
	10	2500	1900	2100	1500
	12.3	2400	2500	2100	1800
	16.2	2200	2000	2100	1900
10 nm sample	0	1900	1300	1700	900
	6.4	2000	1600	1700	1200
	12.4	2100	1700	1900	1600
	16.3	2400	1800	2200	1700

These values are used to estimate the pressure dependence of the Hall and conductivity mobilities and are compared with experiment in table 5. The qualitative agreement is good. For example it can be seen that with the thinnest slab the Hall and conductivity mobilities actually fall at the highest pressure although the mobilities in *all* the individual subbands increase monotonically with increasing pressure. This striking qualitative difference simply arises from the multicarrier nature of transport in the thin doping slabs. The depopulation of the high-mobility higher subbands with increasing pressure causes the relative occupancy of the low-mobility $i = 0$ subband to increase. The Hall mobility, which contains a weighted average of the mobility of all subbands, can therefore fall even if all the subband mobilities are increasing. It should also be noted that Yamada and Makimoto [13] report qualitatively very similar results with a rapidly increasing mobility with increasing subband energy. It should be stressed that the carrier concentrations from Hall measurements *without multiple carrier analysis* may be grossly in error because of the large differences in mobility in the different subbands. Because of the large differences in mobility for the individual subbands, the Shubnikov–de Haas peaks will always be superimposed on a large monotonic magnetoresistance background in delta-doped samples or thin doping slabs except in the case when only one subband is occupied.

The relative mobilities for different subbands in a two-dimensional electron gas (2DEG) have mostly been studied in various heterostructure systems [14–19] with the general conclusion that the mobility in the lowest subband is higher than that in the other bands, although the opposite case has also been reported [14]. In contrast, as mentioned above, work on a delta-plane doping sample has concluded that the mobility increases with increasing subband index [13]. These observations can be understood in terms of an interplay between (i) the physical extent and separation of the electron distribution from the ionized donors, which increases with increasing subband energy, giving reduced scattering, and (ii) the Fermi velocity of the carriers, which is lower in higher subbands, resulting in increased scattering. In single heterojunctions, in which all the electron distributions are remote from any ionized scatterers, it is the second factor which usually dominates, although as Mori and Ando [20] have calculated, it is possible for μ_1 to go from less than μ_0 (factor (ii) dominant) to above μ_0 (factor (i) dominant) as the carrier density in the 2DEG is increased by illumination. (It has been pointed out in [19] that the wavelength of the illumination used can affect the extent of the electron distribution In the delta wells, however, where the ionized donors are to be found in the centre of the well itself ($z = 0$), one notes that the electron probability of the lowest subband is centred in the well plane, while the next excited state has a node at this location. Higher excited states, whether or not they have a node at $z = 0$, have most of their weight even further from the region of dopants centred at $z = 0$.

Also, the electron concentrations in delta wells are generally higher than those in single heterostructures, so that the relative Fermi velocities in adjacent subbands are closer, making this a less important factor. The net result is thus a mobility which increases with subband index.

Since the lowest subband in delta wells is the one in which the electrons are confined to a region close to the ionized donors, a decrease in the donor density should have a strong effect on the ionized impurity scattering which limits the mobility. Electrons in the first excited state will be relatively unaffected by this change both because of their physical separation from the scatterers and because of the shielding of the donors by the electrons in the lowest level.

This change in the relative subband mobilities can be understood qualitatively in terms of a simple picture in which the effect of an increase in pressure is to allow condensation of some of the mobile electrons in the well onto localized states associated with the Si donors whose ionized states form the well itself. One is therefore comparing subband mobilities of two different systems, characterized by a higher (ambient pressure) or a lower (high pressure) net doping density. Furthermore when the dopants are spread into a slab, the donors located at the centre of the slab may be preferentially neutralized.

Another effect also operates to increase the relative mobility of the lowest subband when the donor density of the well is effectively reduced. Higher effective doping (a stronger well) acts to squeeze the wavefunctions of the various subbands into closer proximity to the doping plane. For a many-subband system, the lowest subband spread is relatively independent of the well strength and is limited instead by the effective Bohr radius, which decreases with pressure because of the increase in effective mass and the decrease in dielectric constant. When only one or two excited levels are present, however, a decrease in well strength can have an important effect on the wavefunction of the lowest level. In particular, if we consider a delta well with a uniform doping width of 2.2 nm, and two different doping densities typical of this experiment, namely 10^{13} cm^{-2} (the case for ambient pressure) and 4×10^{12} cm^{-2} (typical of the net doping of the same sample at high pressures), we find that the full width at half maximum of the lowest subband probability distribution is only 4.5 nm in the former case, compared with 6.2 nm in the latter. Thus, at these doping levels in GaAs, a decrease in the effective well strength not only reduces the number of scattering centres seen by the electrons in the lowest subband, but also reduces the probability that these will see the scatterers that remain. These factors are partially offset by the increased scattering probability due to the lower Fermi velocity.

The first calculations of the mobilities in different subbands have recently been reported by Mezrin and Shik [21] who developed a quasi-classical theory in the high-density limit. For 1×10^{13} cm^{-2} Si donors forming a delta spike in GaAs the mobilities in the first three subbands are predicted to be 1350, 3300 and 4800 cm^2 V^{-1} s^{-1} which are substantially larger than the experimental values given in table 4 for the thinnest sample. However, Mezrin and Shik [21] point out that the mobility derived from the Shubnikov-de Haas effect may be different from that measured in the limit of weak fields. Such a difference was found empirically for the Shubnikov-de Haas effect in bulk InSb [22] where the low-field mobility was a factor of 3 greater than that found from the Dingle temperature derived for the highest-field Shubnikov-de Haas peaks. Merzin and Shik also expect the mobility derived from the Shubnikov-de Haas amplitudes for delta-doped samples to be substantially less than the values quoted above, but only expect some 30% difference between the mobility in the ground state and those in the excited states.

Experimentally the pressure measurements show an increase in the mobility for all subbands and only a small relative increase for the lowest subband as the sample is subjected to high pressures. An increase in effective mass of 11.8% and decrease in dielectric constant of 3.3% are expected at 20 kbar from the results reported in reference [23]. From the results presented by Merzin and Shik the mobility in the $i = 0$ subband is expected to be 1120 cm^2 V^{-1} s^{-1} at 20 kbar, i.e. the theory predicts a small fall in the subband mobility on applying pressure in contrast to the greater than factor of two increase observed experimentally. It is too early to speculate whether this difference between theory and experiment arises from correlation effects or for other reasons. Qualitatively the behaviour observed experimentally is very similar to that reported by Maude et al [2, 3] for bulk samples. Thus, with a bulk sample doped with 1.2×10^{19} cm^{-3} silicon atoms, the electron concentration drops to 40% of the doping value at a pressure of 16 kbar and the mobility increases by about 1.5. For an Sn sample doped to 1.8×10^{19} cm^{-3} the carrier concentration drops to 40% of the initial figure and the mobility increases by 2.6. The figures for the 10, 5 and 2 nm samples are 60%, 40% and 32% for the carrier concentration and 2.6, 2.9 and 2.3 for the mobility changes respectively. Any changes in mobility are rather similar for different subbands. Furthermore the pressure dependence of the subband occupancies for the 5 nm slab sample can be fitted well by theory which assumes a uniform distribution of positive charge in the slab and a small amount of depletion charge, as can be seen from table 2.

5. Conclusion

We can see little evidence for correlation effects in our analysis of the Shubnikov-de Haas results and hence favour the interpretation that the final charge state of the silicon donors on capturing the electrons is neutral. However, as reported in references [1] and [24] we have evidence that a proportion of the localized states are also non-metastable and this could reduce the proportion of D^+D^- pairs formed. In addition the deliberate spreading of the donors over a finite slab to avoid site switching or the formation of precipitates rather than the deposition of the silicon onto a single plane may have acted to reduce the magnitude of correlation effects. It is intended to refine the analysis and to extend the measurements to thinner slabs in order to improve the detection limit for correlation effects.

Acknowledgments

The experimental measurements reported in reference [1] were performed at the High Field Magnet Laboratory at the University of Nijmegen with the assistance of C Skierbeszeswki, J Singleton, P J van der Wel and P Wisniewski. We gratefully acknowledge their assistance in carrying out the high-field measurements presented in reference [1] which are further interpreted in this paper. Conversations with Dr A Shik and access to his calculations prior to publication are also gratefully acknowledged.

References

[1] Skuras E *et al* 1991 *Semicond. Sci. Techol.* **6** 535–46

[2] Maude D K, Portal J C, Dmowski L, Foster T, Eaves L, Nathan M, Heiblum M, Harris J J and Beall R B 1987 *Phys. Rev. Lett.* **59** 815

[3] Maude D K, Eaves L, Foster T J and Portal J C 1989 *Phys. Rev. Lett.* **62** 1922

[4] O'Reilly E P 1990 *Appl. Phys. Lett.* **55** 1409–11

[5] Kossut J, Wilamowski Z, Dietl T and Swiatek K 1990 *Proc. 20th Int. Conf. on the Physics of Semiconductors* ed E M Anastassakis and J D Joannopoulos (Singapore: World Scientific) pp. 613–20

[6] Beall R B, Clegg J B, Castagne J, Harris J J, Murray R and Newman R C 1989 *Semicond. Sci. Technol.* **4** 1171–51

[7] Zrenner A, Koch F, Williams R L, Stradling R A, Ploog K and Weimann G 1988 *Semicond. Sci. Technol.* **3** 1203–9

[8] Zrenner A, Koch F and Ploog K 1988 *Surf. Sci.* **196** 671

[9] Santos M, Sajoto T, Lanzillotto A M, Zrenner A and Shayegan M 1990 *Sur. Sci.* **228** 225–9

[10] Lanzillotto A M, Santos M and Shayegan M 1990 *J. Vac. Sci. Technol.* A **8** 2009–11

[11] Koenraad P M, Voncken A P J, Singleton J, Blom F A P, Langerek C J G M, Leys M R, Perenboom J A A J, Spermon S J R M, van der Vleuten W C and Wolter J H 1990 *Surf. Sci.* **228** 538

[12] Williams R L 1988 *PhD Thesis* Imperial College, London

[13] Yamada S, Makimoto T 1990 *Appl. Phys. Lett.* **57** 1022–4

[14] Smith T P and Fang F F 1988 *Phys. Rev.*. B **37** 4303

[15] Nactwei G, Schulze D, Gobsch G, Paasch G, Kraak W, Kruger H and Hermann R 1988 *Phys. Status Solidi* b **148** 349

[16] Gobsch G, Schlze D and Paasch G 1988 *Phys. Rev.* B **38** 10943

[17] van Houten H, Williamson J G, Broekaart M E I, Foxon C T and Harris J J 1988 *Phys. Rev.* B **37** 2756

[18] Stormer H J, Gossard A C and Wiegmann W 1982 *Solid State Commun.* **41** 707

[19] Fletcher R, Zaremba E, D'Iorio M, Foxon C T and Harris J J 1990 *Phys. Rev.* B **41** 10649

[20] Mori S and Ando T 1980 *J. Phys. Soc. Japan* **48** 865

[21] Mezrin O and Shik A *Microstructures & Superlattices* submitted for publication

[22] Staromlynska J, Finlayson D M and Stradling R A 1983 *J. Phys. C.: Solid State Phys.* **16** 6373

[23] Wasilewski Z and Stradling R A 1986 *Semicond. Sci. Technol.* **1** 264

[24] Dmochowski J E, Wang P D and Stradling R A 1990 *Proc. 20th Int. Conf. on Physics of Semiconductors* ed E M Anastassakis and J D Joannopoulos (Singapore: World Scientific) pp 658–61

Semicond. Sci. Technol. **6** (1991) B143–B145. Printed in the UK

Shift of the DX level in narrow Si delta-doped GaAs

P M Koenraad†, W de Lange†, F A P Blom†, M R Leys†, J A A J Perenboom‡, J Singleton‡ and J H Wolter†

† Physics Department, Eindhoven University of Technology, PO 513, 5600 MB Eindhoven, The Netherlands
‡ High Field Magnet Laboratory and Research Institute for Materials, University of Nijmegen, Toernooiveld, 6525 ED Nijmegen, The Netherlands

Abstract. In this paper we present measurements under hydrostatic pressure on Si delta-doped GaAs. From the measurements we conclude that the energy position of the DX level is shifted away from the Γ conduction band minimum at high doping concentrations. This shift is consistent with measurements carried out on bulk GaAs heavily doped with silicon.

1. Introduction

The electronic properties of bulk doped $Al_x Ga_{1-x} As$ are controlled by the coexistence of a hydrogen-like shallow centre and a deep centre, the so-called DX centre [1]. This deep centre, which is dominant for $0.25 < x < 0.6$, is also responsible for the effect of persistent photoconductivity (PPC). From several experiments it has been concluded that the predominance of either the shallow level or the deep level depends on the relative positions of their energy levels [2]. Hydrostatic pressure experiments on GaAs by Mizuta *et al* [3] and Lifshitz *et al* [4] showed that the deep centre can be induced by pressure. At high doping concentrations in GaAs the Γ conduction band is filled to such an amount that the DX centre, which lies above the conduction-band edge, becomes populated and thus limits the maximum attainable electron concentration [5].

Recently there has been a strong interest in the physics of sharply confined (delta-doping) Si doping layers in GaAs [6]. It has been shown from secondary ion mass spectrometry (SIMS) [7] and subband population measurements [8, 9] that the spreading of the donors is strongly reduced when the growth temperature is lowered. Until now work on the DX centre in delta-doped GaAs and $Al_x Ga_{1-x} As$ is rather limited. Zrenner *et al* [8] have shown that in delta-doped GaAs a deep level can be induced by hydrostatic pressure. This level lies 200 meV above the Γ conduction band. In GaAs samples with a narrow delta-doping profile this level should limit the maximum attainable electron concentration. Following these arguments Santos *et al* [9] conclude that in samples they have grown at 400 °C the delta-doping profiles are only 5 Å wide. However, Beall *et al* [10] have shown from SIMS measurements that for a doping concentration of 10×10^{12} cm^{-2} and a growth

temperature of 520 °C clustering of the silicon atoms takes place in the delta-doping layer. Therefore the conclusion of Santos *et al* might be questionable.

Etienne and Thierry-Mieg [11] have grown $Al_{0.32} Ga_{0.68} As$ containing a superlattice of delta-doping layers. Van der Pauw measurements on these samples show that for high doping concentrations the number of DX centres decreases. At present it is not clear whether this is true also for delta-doped GaAs.

In this paper we resolve some of these controversies. We present measurements under hydrostatic pressure on the subband population in GaAs samples with doping concentrations of 2×10^{12} cm^{-2} and 8×10^{12} cm^{-2}, and a growth temperature of 480 °C or 620 °C.

2. Samples

The delta-doped samples were grown in a Varian modular MBE system. Before depositing the doping layer the growth was interrupted for 10 s. Then the Si furnace was opened either for 7.5 s or 30 s to deposit the doping layer. In this way we obtained a planar doping layer on a smooth surface. The samples were grown with a buffer layer between the doping layer and the substrate of 2.5 μm and a top layer of 1 μm on Si-GaAs substrates with orientation $(001) \pm 0.3°$. We have grown two sets of samples with doping concentrations of 2×10^{12} cm^{-2} and 8×10^{12} cm^{-2} at two different substrate temperatures, 480 °C and 620 °C. The background impurity concentration in the samples was less than 10^{15} cm^{-3}.

We used the van der Pauw method to determine the Hall mobility and electron density between 4.2 K and 300 K. The subband population measurements were performed on Hall-bar-shaped samples. Ohmic contacts were made by annealing small Sn balls. The samples were illuminated with GaAs outgap radiation of a red LED ($\lambda = 650$ nm) mounted inside the cryostat.

3. Experiments

Table 1 shows the measured Hall electron concentration at 4.2 K. In the samples grown at 620 °C, the Hall electron concentration at 4.2 K is close to the sheet doping concentration. On the other hand, for the samples grown at 480 °C the Hall electron concentration at 4.2 K appears to be smaller by almost a factor of 2. We note, however, that in delta-doped structures normally more than one subband is populated. Then Hall measurements in general do not give reliable information on the total electron density [12].

A method from which the electron concentration in the different subbands can be determined separately is based on the Shubnikov–de Haas effect. We carried out measurements on Hall-bar-shaped samples in magnetic fields up to 20 T. The results are shown in table 1. The subband population measured in the sample grown at 480 °C is in good agreement [12] with the calculations for a donor distribution with a width of 20 Å. This width is also in agreement with SIMS measurements. The true width of the donor distribution could be even smaller. However, it is difficult to give a very accurate value for the width below 20 Å because the subband population is nearly independent of the width of the donor distribution. The total electron concentration found in sample A is nearly equal to the intended donor concentration. In sample C, however, the total electron concentration is much smaller than the sheet doping concentration. For the same growth conditions Santos et al [9] have found a similar result. They argued that this is due to the DX centre at 200 meV above the Γ band.

If indeed DX centres were responsible for the discrepancy of the electron density as described above, one should expect to observe the PPC effect. Population of the DX centre could also be observed by applying hydrostatic pressure. When GaAs is put under hydrostatic pressure the L conduction band minimum and the DX level, which is coupled to the L minimum, are lowered relative to the Γ conduction band minimum. This implies that the electron concentration decreases with increasing pressure if the DX centre levels off the electron concen-

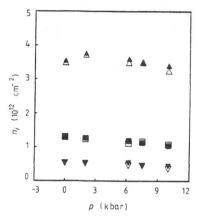

Figure 1. Population of the three lowest subbands in sample C before (open symbols: $\triangle$, $i = 0$; $\square$, $i = 1$; $\triangledown$, $i = 2$) and after illumination (full symbols: $\blacktriangle$, $i = 0$; $\blacksquare$, $i = 1$; $\blacktriangledown$, $i = 2$) as a function of hydrostatic pressure. $N_{don} = 8 \times 10^{12}$ cm^{-2}.

tration. In figure 1 the measured subband population in sample C is shown as a function of hydrostatic pressure. The measurements have been performed in darkness and after illumination with a red LED. Clearly the electron concentration has hardly changed after illumination of the sample or the application of hydrostatic pressure up to 10 kbar. This means that no DX centres are populated. Thus, contrary to the findings of Zrenner et al [8] for a delta-layer with the Si donors spread over 80 Å, we find that in GaAs samples with a very narrow delta-doping profile another mechanism rather than trapping by the DX level acts to limit the free electron concentration.

Recently Beall et al [10] presented SIMS and CV measurements on delta-doped GaAs structures in the same range of growth temperatures and doping concentrations as used in our experiments. They found evidence that at doping concentrations above 4×10^{12} cm^{-2} only a fraction of the silicon is deposited on electrically active sites. The remainder of silicon forms clusters. In view of their results we think that this mechanism is more likely to be responsible for the discrepancies between the electron concentration and the sheet doping concentrations found in our samples.

However, if a DX level is still present in sample C it must be shifted to an energy position above the position of the Fermi level at 10 kbar. Such a shift of the DX level can also be deduced from the hydrostatic pressure experiments of Zrenner et al [8] and Skuras et al [13]. The energy separation between the DX level and the Γ conduction band minimum at ambient pressure, as determined from these measurements, is given in Table 2. The results seem to indicate that the DX level shifts to a higher energy position when the doping profile becomes narrow. From hydrostatic pressure experiments on bulk GaAs with a high doping concentration Eaves et al [14] also found that the DX level shifts to a higher energy when the doping concentration increases. For a doping concentration of 10^{19} cm^{-3} Eaves et al found a distance between the Γ conduction band minimum and the DX

Table 1. The doping concentration (N_{don}), growth temperature (T_{growth}), Hall electron concentration (n_{Hall}) and total electron concentration ($n_{tot} = \Sigma\, n_i$) at 4.2 K.

	N_{don} (10^{12} cm^{-2})	T_{growth} (°C)	n_{Hall} (10^{12} cm^{-2})	n_{tot} (10^{12} cm^{-2})
A	2	480	1.1	1.79
B	2	620	1.9	1.76
C	8	480	3.8	5.24
D	8	620	7.5	≈7.5

Table 2. Energy of the DX level relative to the Γ conduction band minimum and width of the donor distribution, D_{don}, in three different samples. For comparison we also give the position of the DX level in bulk GaAs with a low doping concentration.

	$E_{DX} - E_{\Gamma}(x = 0)$ (meV)	D_{don} (Å)
Bulk [3]	175	
Zrenner et al [8]	200	80
Skuras et al [13]	270	≈ 30
This work	>260	20

level of ≈ 280 meV. Our measurements show a shift of the DX level of at least 260 meV for sample C which has a maximum Si concentration of 1.2×10^{19} cm^{-3} according to SIMS measurements. Thus our results are consistent with the work of Eaves et al.

The shift of the DX level as a function of doping concentration might be an apparent shift as has been pointed out by Wilamowksi et al [15]. They showed that the shift of the DX level in bulk GaAs can be removed if a broadening of the DX level due to potential fluctuations is taken into account.

Recently we have reported a reduction of the number of DX centres in $Al_{0.33}Ga_{0.67}As$ containing a single Si delta-doping layer [16]. Further analysis showed that this was just a GaAs sample. We just recently observed, in $Al_{0.33}Ga_{0.67}As$ samples containing a single delta-doping at low temperatures, a freeze-out of all the carriers and a strong PPC effect after illumination. Therefore we state that our previous conclusion concerning the reduction of the number of DX centres in delta-doped $Al_{0.33}Ga_{0.67}As$ is not correct for the samples we have grown up to now. The reason for the reduction of the number DX centres in $Al_{0.32}Ga_{0.68}As$ containing a superlattice of Si delta-doping layers with a high doping concentration as found by Etienne and Thierry-Mieg [11] is under investigation.

4. Conclusions

We have shown that in 20 Å delta-doped GaAs the maximum attainable electron density is not limited by the population of the DX level. Hydrostatic pressure experiments indicate that the DX level in delta-doped GaAs shifts away from the Γ conduction band minimum in samples with a narrow donor distribution.

Acknowledgments

We are grateful to W C van der Vleuten for growing the delta-doped samples, P A M Nouwens for preparing and contacting the samples, and J Harris and J Clegg from Philips Laboratories, Redhill, UK for performing the SIMS measurements. The research of one of us (PK) has been made possible by a fellowship of the Royal Netherlands Academy of Arts and Sciences.

References

[1] Lang D, Logan R and Jaros M 1979 *Phys. Rev.* B **19** 1015
[2] Blom, P W M, Koenraad P M, Blom F A P and Wolter J H 1989 *J. Appl. Phys.* **66** 4269
[3] Mizuta M, Tachikawa M, Kukimoto H and Minomura S 1985 *Japan. J. Appl. Phys.* **24** L143; Tachikawa M, Mizuta M, Kukimoto H and Minomura S 1985 *Japan. J. Appl. Phys.* **24** L821
[4] Lifshitz N, Jayaraman A and Logan R 1980 *Phys. Rev.* B **21** 670
[5] Theis T N, Mooney P M and Wright S L 1988 *Phys. Rev. Lett.* **60** 361
[6] Zrenner A, Reisinger H, Koch F and Ploog K 1985 *Proc. 17th Int. Conf. on the Physics of Semiconductors, San Francisco, 1984* ed J P Chadi and W A Harrison (New York: Springer) p 325
[7] Beall R B, Clegg J B and Harris J J 1988 *Semicond. Sci. Technol.* **3** 612
[8] Zrenner A, Koch F, Williams R L, Stradling R A, Ploog K and Weimann G 1988 *Semicond. Sci. Technol.* **3** 1203
[9] Santos M, Sajoto T, Zrenner A and Shayegan M 1988 *Appl. Phys. Lett.* **53** 2504
[10] Beall R B, Clegg J B, Castagné J, Harris J J, Murray R and Newman R C 1989 *Semicond. Sci. Technol.* **4** 1171
[11] Etienne B and Thierry-Mieg V 1988 *Appl. Phys. Lett.* **52** 1237
[12] Koenraad P M, Voncken A P J, Singleton J, Blom F A P, Langerak C J G M, Leys M R, Perenboom J A A J, Spermon S J R M, van der Vleuten W C and Wolter J H 1990 *Surf. Sci.* **228** 538
[13] Skuras E et al *Semicond. Sci. Technol.* to be published
[14] Eaves L et al 1989 *Shallow Impurities in Semiconductors 1988 (Inst. Phys. Conf. Ser 95)* ed B Monemar (Bristol: Institute of Physics) p 315
[15] Wilamowski Z, Kossut J, Jantsch W and Ostermayer G 1991 *Semicond. Sci. Technol.* **6** B38–46
[16] Koenraad P M, de Lange W, Blom F A P, Leys M R, Perenboom J A A J, Singleton J, van der Vleuten W C and Wolter J 1990 *Proc. Int. Conf. on Shallow Impurities in Semiconductors, London* ed G Davies (Zurich: Trans Tech Publications)

Semicond. Sci. Technol. **6** (1991) B146–B149. Printed in the UK

Te-related DX centre of GaAs and AlGaAs

**P Wiśniewski†, E Litwin-Staszewska†, T Suski†, L Kończewicz†,
R Piotrzkowski† and W Stankiewicz‡**

† UNIPRESS, High Pressure Research Center, Polish Academy of Sciences, ul. Sokołowska 29, 01–142 Warsaw, Poland
‡ Institute of Semiconductor Physics, Lithuanian Academy of Sciences, 232600 Vilnius, Lithuania, USSR

Abstract. The pressure dependences of the electrical conductivity and Hall coefficient have been studied for GaAs and $Al_xGa_{1-x}As$ ($x = 0.20$ and 0.25) heavily doped with Te. The results obtained show that for a GaAs crystal the resonant donor level related to Te is located approximately 0.45 eV above the conduction band minimum. The effect of persistent photoconductivity observed at high pressures proves the DX-like character of this donor state. Comparison of the data on the energetic position of the DX level illustrates its strong chemical energetic dependence (for Si, Sn and S dopants an E_{DX} value of around 0.3 eV has been reported previously). It appears that Te represents an impurity which is very suitable for testing the microscopic models of the DX centre. Studies of processes of the thermal recovery after high pressure freeze-out of electrons on the metastable states of the DX centres clearly demonstrate the multicomponent structure of the DX state in AlGaAs samples.

1. Introduction

Studies of donor-species-dependent properties of DX centres in AlGaAs have demonstrated similar behaviour for group VI elements (S, Se, Te) and different features of the DX states formed by Group IV dopants (Si, Ge, Sn) [1, 2]. The results obtained have been treated as supporting the microscopic model of the DX centre first proposed by Chadi and Chang [3]. Accordingly, a dopant replacing a cation in the AlGaAs crystal, e.g. an Si donor, is displaced significantly after the capture of two conducting electrons. This transformation from the substitutional to interstitial site (S → I) is accompanied by a rupture of one of four bonds which the impurity forms with neighbouring As ions. For donors substituting for As, such as Te, it is one of the surrounding cations which exhibits the S → I transformation to create the DX state. Since in the latter case it is not an impurity atom which undergoes the site-change rearrangement, one can expect that in this situation the nature of the donor plays a secondary role in determining the properties of the DX centre.

High-pressure studies of the energetics of the DX centre in GaAs crystal have strongly suggested that Te exhibits qualitatively different behaviour in comparison with other donors [4]. For example, the thermal ionization energy, E_{DX}, for the resonant DX state of Si, Sn and S is situated approximately 0.25–0.30 eV above the Γ conduction band minimum [4, 5]. For GaAs samples doped with these three impurities to the level $5 \times 10^{18} \, cm^{-3}$, increasing the pressure to about 10 kbar results in processes of carrier capture by the DX states.

On the other hand the Te-related DX centre remains electrically inactive for concentrations as high as $7 \times 10^{18} \, cm^{-3}$ and pressures up to 15 kbar [4, 6]. Suggestions of the occupation of the DX level by electrons in GaAs:Te were drawn from electron transport studies at higher pressures. The position of the DX level in the sample with the Hall concentration of electrons $n_H = 7 \times 10^{18} \, cm^{-3}$ was estimated to be about 0.5 eV above the Γ conduction band minimum [7].

Moreover, detailed examination of $Al_xGa_{1-x}As$:Te with $x = 0.15, 0.25, 0.35$ [8] makes it possible to estimate the value of E_{DX} for $x = 0$. The resulting $E_{DX} \sim 0.12 \, eV$ differs drastically from the result obtained in pressure studies of GaAs:Te. As has been demonstrated for AlGaAs:Si an effect of a so-called alloy splitting causes an appearance of the multicomponent structure of the DX centre [9–11]. Strong localization of the DX state and the possibly different Ga to Al ratios in the vicinity of the donor occupying the interstitial site, are responsible for the considered effect. Electron transport measurements give an 'effective' value of E_{DX} characterizing the average position of the level. In the case of the Si-related DX centre the described procedure of E_{DX} estimation for GaAs from data obtained in AlGaAs:Si leads to a value of about 0.17 eV [12] which is roughly 0.1 eV lower than the $E_{DX} \sim 0.3 \, eV$ determined in GaAs:Si [4, 5, 13]. Though the origin of the discrepancies in the magnitude of E_{DX} for GaAs:Te and $Al_xGa_{1-x}As$:Te ($x \to 0$) could be the same as for the Si-related DX centre, in the former case the effect of the alloy splitting should be more pronounced.

Inspection of deep-level transient spectroscopy

(DLTS) data obtained from AlGaAs:Te reveals that sometimes few peaks are observed (e.g. [8,14]). However, additionally observed components were usually treated as unrelated to the DX centre formed by the Te impurity.

The aims of this paper are:

(i) To determine the energy position of Te-related, localized and resonant donor level in GaAs crystals.

(ii) To prove the metastable behaviour of this state thus demonstrating the DX-like character of the Te dopant.

(iii) To perform an experimental test of the multicomponent structure of the DX centre in AlGaAs:Te showing whether the properties of donors replacing As in the AlGaAs exhibit a dependence on the local vicinity, qualitatively similar to that demonstrated by Si (group IV element).

2. Experimental details

The Hall coefficient R_H and resistivity as a function of pressure up to 22 kbar and temperature between 4.2 K and 300 K have been measured for Te-doped samples of GaAs ($n_H = 1.2 \times 10^{19}$ cm^{-3}), Al$_x$Ga$_{1-x}$As with $x = 0.20$ ($n_H = 0.26 \times 10^{18}$ cm^{-3}) and $x = 0.25$ ($n_H = 0.27 \times 10^{18}$ cm^{-3}). The samples were grown by the LPE technique on semi-insulating substrates of GaAs:Cr.

The majority of experiments were performed in a high-pressure clamp cell with light petroleum as a pressure-transmitting medium. In such cases pressure was applied at room temperature and for measurements at lower temperatures the cell was cooled while under pressure. To induce persistent photoconductivity (PPC) a red light-emitting diode (LED) was mounted inside the high-pressure cell. For studies of the multicomponent structure of Te-related DX centre in AlGaAs the high-pressure freeze-out (HPFO) procedure was employed. This uses an He gas compressor which enabled us to change pressure in a hydrostatic manner over a wide range of temperatures (77–300 K). The pressure was applied to the sample at room temperature and the changes of carrier concentration were induced. The temperature was then lowered to 77 K at which point the pressure was released. From the temperature dependence of n_H during the subsequent temperature increase we deduced that sequential electron emission occurs, caused by the multibarrier structure of the DX centre.

3. Results and discussion

3.1. GaAs:Te

It is well known that the application of hydrostatic pressure to GaAs raises the Γ point of the conduction band (Γ CB) with respect to the DX level. A transfer of electrons between the CB and the DX centre starts after the Fermi level coincides with E_{DX} and when the thermal energy of electrons exceeds the energy E_C, i.e. the poten-

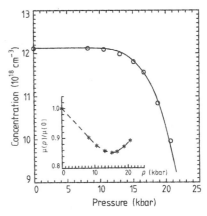

Figure 1. Pressure dependence of the Hall concentration of electrons for GaAs:Te ($T = 77$ K). The open circles are experimental points. The full curve is calculated assuming either the $U < 0$ model of DX centres or the $U > 0$ model with the following values of the fitting parameters: $E_{DX^0} - E_\Gamma = 0.47$ eV, $dE_{DX^0}/dp = -12.2$ meV kbar^{-1} ($U > 0$) and $E_{DX^-} - E_\Gamma = 0.41$ eV, $dE_{DX^-}/dp = -9.5$ meV kbar^{-1} ($U < 0$). The insert shows the relative changes of the electron mobility with pressure ($T = 77$ K). The asterisks are experimental points, the broken curve is drawn to guide the eye.

tial barrier for electron capture onto the localized state of the donor. If the DX level is very high above the the Γ CB minimum, a transfer of carriers to the subsidiary minima of the CB (L and X) may also occur. Figure 1 illustrates the variation of the Hall concentration of electrons, n_H, measured for a GaAs:Te sample at 77 K (pressure applied at 300 K). A significant decrease of n_H occurs for pressures higher than 15 kbar. A suggestion that this effect results mainly from the transfer of electrons to the deep state of the donor is supported by the observation that in this pressure range the electron mobility starts to increase (see insert in figure 1). Similar behaviour has been reported recently for GaAs samples doped with Si and Sn (see, for example, [3]). To prove whether the observed donor level exhibits features characteristic of the DX-like state, we performed a test consisting in inducing the PPC effect at low temperatures. The sample was cooled to 4.2 K at a pressure of 17.3 kbar and was illuminated with a red LED. Figure 2 illustrates the increase of the Hall concentration, n_H, after approximately 1 h of illumination. After the light was switched off the new value of n_H persisted, exhibiting a total recovery after the sample was heated up to temperature about 80 K at a rate of 1 K min^{-1}. The observed behaviour proves ultimately that this donor level represents the Te-related DX centre in GaAs. In addition it can be deduced that above ~ 80 K the electron distribution between the DX centres and the CB remains in equilibrium, whereas below this temperature the metastable population of the DX centres may occur. Therefore, it becomes possible to use the data on the pressure dependence of n_H detected at 77 K (see figure 1) for calculations of E_{DX} (77 K) and its pressure dependence.

Using the approach described in [4] and choosing the statistics appropriate for the neutral DX^0 centre ($U > 0$)

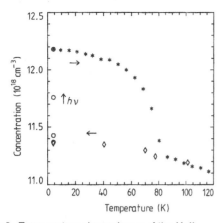

Figure 2. Temperature dependence of the Hall concentration of electrons for GaAs:Te. The sample was cooled under a pressure of 17.3 kbar ($\diamond$). At 4.2 K the sample was illuminated by a red light-emitting diode (LED) ($\bigcirc$), then the LED was switched off and temperature was raised at a constant rate of 1 K min^{-1} ($*$).

and for the negatively charged DX$^-$ centre ($U < 0$) we have determined for GaAs:Te at 77 K: (i) the pressure variation of E_F, (ii) E_{DX} and dE_{DX}/dp.

The theoretical curve seen in figure 1 was obtained for the following values of the fitting parameters:

$$E_{DX} - E_\Gamma = 0.47 \pm 0.04 \text{ eV}$$
$$\text{d}E_{DX}/\text{d}p = -12.2 \pm 2 \text{ meV kbar}^{-1} \qquad \text{for } U > 0$$

and

$$E_{DX} - E_\Gamma = 0.41 \pm 0.04 \text{ eV}$$
$$\text{d}E_{DX}/\text{d}p = -9.5 \pm 2 \text{ meV kbar}^{-1} \qquad \text{for } U < 0.$$

The satisfactory fit to the experimental dependence of n_H upon pressure was obtained after assuming that the minimum of the Γ CB is shifted to lower energies by approximately 55 meV.

At the highest pressures used in this work a significant number of electrons were transferred to the L CB. Uncertainty in the value of the electron mobility in the L CB as well as on the pressure dependence of the GaAs band structure led to the relatively low accuracy in the determination of E_{DX} and dE_{DX}/dp. In heavily doped GaAs the interdonor Coulomb interaction is thought to induce spatial ordering of the charged impurities in partially filled donor systems (e.g. [15, 16]) which may affect the measured value of E_{DX} and dE_{DX}/dp. This correlation assumes an involvement of positively charged centres, d$^+$, for the $U > 0$ model and dipoles created by d$^+$ and DX$^-$ pairs for the $U < 0$ model. Consequently a decrease of the total energy of the crystal and the renormalization of E_{DX} by about 20–30 meV are expected [15].

3.2. AlGaAs

For studies of the multicomponent barrier structure in AlGaAs we employed an HPFO procedure and the method proposed by Piotrzkowski *et al* [17]. After releasing the pressure at $T \sim 60$ K the sample temperature was

raised at a known rate (usually 1–2 K min^{-1}). After supplying electrons localized on the DX centres with sufficient thermal energy, the emission processes to the CB start. This behaviour is monitored by measurements of the increase of carrier density when electrons reach the CB (figure 3). One can easily see that the experimental curve in figure 3 consists of more than one component: the emission process occurs in various steps. Both samples studied exhibit qualitatively similar behaviour. More information on the dependence of n_H upon T is obtained assuming an energy barrier for electron emission, E_E, as a fitting parameter. We assume that in the situation examined no recapture of electrons by the DX states occurs.

In Al$_{0.25}$Ga$_{0.75}$As the experimental dependence of $n_H(T)$ can be fitted adequately with three equal barriers ($E_E \sim 85$ meV) for electron emission. Different emission rates are caused by three various pre-exponential factors [17]. Their magnitude is decisive in determining the temperature ranges where one of the three emission processes becomes more efficient (see the lower part of figure 3, showing the calculated components of the analysed emission).

4. Summary and conclusions

The most important result of this work is the observation of the localized and resonant state of the Te donor in GaAs. It exhibits DX-like character, i.e. at low temperatures the potential barriers for electron capture and emission to and from the metastable DX state, become relevant. Thus freezing of the non-equilibrium distribution of electrons between the CB and the donor state

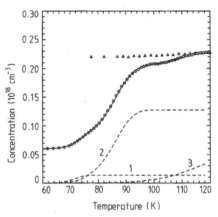

Figure 3. Hall concentration of electrons versus temperature. The open circles are experimental points measured after the sample of Al$_{0.25}$Ga$_{0.75}$As was cooled to $T \simeq 60$ K under a pressure of 7 kbar. Then the pressure was released and the temperature was raised at a constant rate of 1 K min^{-1}. The triangles represent the behaviour of the sample during cooling at ambient pressure. The full curve is calculated assuming three subsequent processes of electron emission from the DX centres to the conduction band. The broken curves represent calculations. The contributions from the subsequent emission processes (1, 2, 3) are shown.

occurs. However, since the DX level related to Te is situated much higher in the conduction band ($E_{DX} \sim 0.45$ eV) in comparison with Si, Sn and S ($E_{DX} \sim 0.3$ eV) the deleterious effects caused by the DX(Te) centre appear at much higher n_H and pressures.

The obtained result confirms the importance of the chemical nature of a donor in determining properties of the related DX centre. A similar trend in the donor-species-dependent character of the DX centre to the one reported here has been obtained theoretically by Yamaguchi *et al* [18] within the small lattice relaxation model ($U > 0$). To our knowledge there is a lack of this kind of calculation for the negative-U and large lattice relaxation model of Chadi and Chang [3]. Therefore our findings, which consist in the different behaviour of the Te-related DX centre in comparison with other donors, cannot be treated as pointing to one of the competing models of the DX centre to properly describe its microscopic configuration.

The multibarrier structure of the DX centre in AlGaAs:Te demonstrates that the properties of the group VI donor reveal qualitatively features studied intensively in the case of the Si-related donor [9-11]. More detailed studies of this effect, including its pressure dependence, can be used for further verification of different microscopic models of the DX centre.

References

[1] Lang D V 1986 *Deep Centers in Semiconductors* ed S T Pantelides (New York: Gordon and Breach) pp 489-539

[2] Kumagai O, Kawai H, Mori Y and Kaneko K 1984 *Appl. Phys. Lett.* **45** 1322

[3] Chadi D J and Chang K J 1988 *Phys. Rev. Lett.* **61** 873

[4] Suski T, Piotrzkowski R, Wiśniewski P, Litwin-Staszewska E and Dmowski L 1989 *Phys. Rev* B **40** 4012

[5] Maude D K, Portal J C, Dmowski L, Foster T, Eaves L, Nathan M, Heiblum M, Harris J J and Beall R B 1987 *Phys. Rev. Lett.* **59** 815

[6] Sallese J M, Lavielle D, Singleton J, Levcuras A, Grenet J C, Gibart P and Portal J C 1990 *Phys. Status Solidi* a **119** K41

[7] Suski T, Wiśniewski P, Skierbiszewski C, Dmowski L H, van der Wel P J, Singleton J, Gilling L J and Harris J J 1991 *J. Appl. Phys.* at press

[8] Shan W, Yu P Y, Li M F, Hansen W L and Bauser E 1989 *Phys. Rev.* B **40** 7831

[9] Calleja E, Gomez A, Munoz E and Camara P 1988 *Appl. Phys. Lett.* **52** 1877
Mooney P M, Theis T N and Wright S L 1988 *Appl. Phys. Lett.* **53** 2546

[10] Baba T, Mizuta M, Fujisawa T, Yoshino J and Kukimoto H 1989 *J. Appl. Phys.* **28** L891

[11] Azema S, Mosser V, Camassel J, Piotrzkowski R, Robert J L, Gibart P, Contour J P, Massie J and Marty A 1989 *Mater. Sci. Forum* **38-41** 857

[12] Chand N, Henderson T, Clem J, Masselink W T, Fischer R, Chang Y C and Morkoc H 1984 *Phys. Rev.* B **30** 4481

[13] Theis T N, Mooney P M and Wright S L 1988 *Phys. Rev. Lett.* **60** 361

[14] Lang D V and Logan R A 1979 *Phys. Rev.* B **19** 1015

[15] Kossut J, Wilamowski Z, Dietl T and Swiątek K 1990 *Proc. 20th Int. Conf. on the Physics of Semiconductors* ed E M Anastassakis and J D Joannopoulos (Singapore: World Scientific) p 613

[16] O'Reilly E P 1989 *Appl. Phys. Lett.* **55** 1409

[17] Piotrzkowski R, Suski T, Wiśniewski P, Ploog K and Knecht J 1990 *J. Appl. Phys.* **68** 3377

[18] Yamaguchi E, Shiraishi K and Ohno T 1990 *Proc. 20th Int. Conf. on the Physics of Semiconductors* ed E M Anastassakis and J D Joannopoulos (Singapore: World Scientific) p 501

Semicond. Sci. Technol. **6** (1991) B150–B153. Printed in the UK

The effect of high hydrostatic pressure on DX centres in GaAs and GaAsP

V Šmíd, J Krištofik, J Zeman and J J Mareš

Institute of Physics, Czechoslovak Academy of Sciences,
Cukrovarnická 10, 162 00 Praha 6, Czechoslovakia

Abstract. We report on results concerning the identification of DX centres obtained from high-pressure experiments. The sulphur-related DX centre in GaAsP alloys is characterized by two levels 40 meV apart which follow the X conduction band minimum. Properties of the DX centres in GaAs and GaAsP are compared.

1. Introduction

High hydrostatic pressure used as an external physical parameter can serve for the verification of different microscopic models, for the investigation of phase transitions, as well as for the substitution of changes in chemical composition.

In the past we have applied the hydrostatic pressure technique to amorphous–crystalline heterojunctions [1] and to glassy semiconductors [2] to evaluate transport mechanisms. This technique has also been extensively employed in the study of defect states in semiconductors [3–7]. Measurements of the influence of high hydrostatic pressure on the emission rate and the capture cross section of deep levels are described in [8]. Experimental results obtained on GaAs by Kukimoto's group [9], which demonstrated the possibility of inducing the DX centre in GaAs by hydrostatic pressure in excess of 20 kbar, give strong evidence of the fact that the DX centre is most probably associated with a single donor and not with a complex defect.

The aim of this paper is to present results of the influence of high hydrostatic pressure on the DX centre in GaAsP and to compare them with those obtained on GaAs. We will also discuss the properties of the DX centre in the context of recently proposed models.

2. Results and discussion

2.1. Sulphur-related DX levels in GaAs$_{1-x}$P$_x$

We have investigated samples with phosphorus content near to the Γ–X conduction band minima cross-over and observed two levels in samples with $x = 0.3$ and 0.4 which exhibit attributes of the DX centre, namely a

concentration comparable to that of the shallow dopant (sulphur), activated capture process, non-exponential shape of transients and low-temperature persistent behaviour. Results of our investigation are summarized in figure 1. For $x = 0.3$ the two levels at atmospheric pressure are resonant with the conduction band. The shallower level enters the gap at $x = 0.35$. For $x = 0.4$ the levels lie considerably deeper in the gap and the carrier freeze-out makes it difficult to evaluate the DLTS spectra. Therefore we employed, as a complementary method to DLTS, measurements of the frequency dependence of the complex impedance. From the Cole–Cole diagrams we obtained the values of the elements of the equivalent circuit and, from the measurements performed at different temperatures, the activation energy [10,11]. The full circles in figure 1 represent levels whose position was corrected with respect to the capture barrier. For the reasons mentioned above, in the case of level 2 in samples

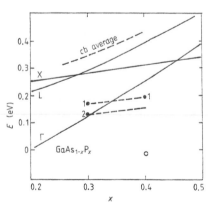

Figure 1. Position of DX levels in GaAsP:S. The shape of the conduction band average (in arbitrary units) is also shown.

with $x = 0.4$, the value of the capture barrier was not evaluated exactly. The experimental results, namely the pressure coefficients, made it possible to draw the conclusion that level 1 in both compositions belonged to the same defect. Its position approximately follows the X minimum. For $x = 0.4$ we determined the position to be 130 meV below the X minimum. The pressure coefficients with respect to the X minimum are $d(E_1 - E_{CX})/dp = -5$ meV GPa^{-1} and -20 meV GPa^{-1} for $x = 0.4$ and 0.3 respectively. These values are much smaller than those obtained with regard to the Γ minimum, $d(E_1 - E_{C\Gamma})/dp = 78$ meV GPa^{-1} and 110 meV GPa^{-1} for $x = 0.3$ and 0.4 respectively. That is why we observed an increase in the activation energy with increasing hydrostatic pressure. The capture barrier depends on composition in such a way that the sum of the thermal energy and the capture barrier (usually called the emission energy) remains almost independent of the composition. The capture barrier decreases with increasing amount of phosphorus from 0.25 eV at $x = 0.3$ to <0.07 eV at $x = 0.4$. The latter value is in good agreement with data published for $x = 0.38$ [12]. The capture barrier also decreases with increasing hydrostatic pressure: $dE_B/dp = -90$ meV GPa^{-1}. This prevailing tendency in the behaviour of the capture barrier and the thermal energy of this level is similar to that reported for the DX centre in GaAlAs [13]. This fact was also used to support the small lattice relaxation (SLR) model pointing out that the large lattice relaxation (LLR) model cannot account for the compositional dependence of both the capture barrier and the thermal energy [14]. More elaborate LLR models can, however, explain most of the published data [15,16]. It is also worth noting that small differences in the pressure coefficients of the energy levels observed for $x = 0.3$ and 0.4 are larger than the error of measurement. We believe that a plausible explanation originates from a difference in the band structure of both types of sample.

For $x = 0.3$ one can readily expect a stronger influence of the L minimum, whereas for larger amounts of phosphorus the DX level departs from the L conduction band minimum. This would also support the model of the DX centre as a localized defect with a wavefunction not derived from one conduction band minimum but composed of all the Bloch states in the Brillouin zone [17]. Nevertheless, it is evident from the results presented here that the deviation from the X-linked character of the DX centre is not very distinct. Our results show one important difference of the DX centre in GaAsP in comparison with that in GaAlAs where the data unambiguously indicate that the DX centre follows the L minimum or the conduction band average [15, 18] (see figure 1).

An important experimental fact presented here is that two levels in samples with $x = 0.3$ show all the attributes of the DX centre. At atmospheric pressure the second level (see figure 1) is resonant with the Γ conduction band minimum. The pressure coefficient with respect to the Γ minimum is indeed the same as that observed for level 1 in samples with the same composition. The capture barrier of level 2 is 0.24 eV at atmospheric

pressure. With increasing pressure the level is again pushed deeper into the gap and the capture barrier decreases with a pressure coefficient of 74 meV GPa^{-1}. For $x = 0.4$ samples the metastable behaviour was also proven but a direct measurement of the capture barrier for level 2 was not possible. Its value was estimated to be 0.16 eV, which is in good agreement with the level shown in figure 1 by the open circle. The second DX level is thus located about 40 meV below the first one. For $x = 0.4$ it means that at atmospheric pressure the second level is 105 meV below the edge of the Γ minimum. Hence the two DX levels in GaAsP are approximately 130 and 170 meV apart from the X minimum. These results differ from the classical work [19] where the authors reported on two levels being 60 and 160 meV below the X minimum. As concerns the theoretical calculation for the S-related DX centre in GaAsP, Yamaguchi's results [20] also show two levels. Their positions do not follow closely any conduction band minimum and, depending on the composition, their levels are 100 to 200 meV away from the lowest minimum.

2.2. Pressure coefficients used for distinction between LLR and SLR models

The above results, as well as data published for GaAlAs alloys, show that the fact that a level follows a certain band minimum cannot be taken as the definite proof of the validity of the effective mass theory for the defect under investigation. This can be strongly influenced by details of the band structure and properties of the defect. In the case of GaAsP we have shown that for $x = 0.3$ the influence of the L minimum is observed whereas for a higher phosphorus content the X minimum plays a predominant role. One of the reasons for this is that for higher x the DX levels depart considerably from the L minimum. In the case of GaAlAs near the transition into the indirect gap semiconductor a deviation from L-linked character was also reported. On the other hand, the depth of the level only cannot be taken as evidence of strong localization of the defect (see, for example, calculations of the effective mass theory with a modified potential [21,22]). Moreover, for the states derived from the X minimum one can obtain, as a consequence of higher effective mass in comparison with the Γ minimum, a set of hydrogenic levels appreciably deeper below the X minima than in the case of the Γ minimum. In fact, the simplest version of effective mass theory gives values which are not far from those experimentally observed for sulphur-related DX levels in GaAsP. This is another difference from GaAlAs where a typical separation of the DX level from the relevant L minimum is about 200 meV.

Another approach in solving the long-lasting controversy on microscopic models adequate for the DX centre is the evaluation of the configuration coordinate diagram with the help of experimentally obtained pressure coefficients of thermal energy and capture barrier, which was originally proposed by Li and Yu [23]. This analysis is based on the multiphonon theory [24]

$$\frac{dE_B}{dp} = \frac{\varepsilon_s^{-1} - 1}{2} \frac{dE_T}{dp} - \frac{\varepsilon_s^{-2} - 1}{4} \frac{dE_s}{dp}$$

with $\varepsilon_s = E_s/E_T$. E_s can be obtained from the equation $E_s^{\pm} = E_T + 2E_B \pm 2(E_B^2 + E_T/E_B)^{1/2}$, where E_s^+ and E_s^- refer to the LLR and SLR models as depicted in the inserts of figure 2. The quantity dE_s/dp may be obtained from the following equation:

$$\frac{dE_s}{dp} = \frac{dE_T}{dp} - \frac{E_T - E_s}{2E_B} \frac{dE_B}{dp}$$

which is based on the assumption (confirmed experimentally in our case) that the pre-exponential factor σ_∞ of the capture cross section $\sigma(T) = \sigma_\infty \exp(-E_B/kT)$ does not depend on pressure. This procedure was applied for both sulphur-related DX levels in $GaAs_{0.7}P_{0.3}$. We have obtained

level 1:

LLR: $dE_B/dp = -0.54(dE_T/dp) + 21$ (meV GPa^{-1})

SLR: $dE_B/dp = -6.6(dE_T/dp) + 2390$ (meV GPa^{-1})

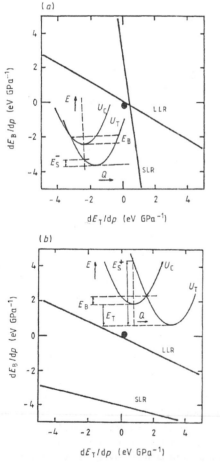

Figure 2. Relation of the pressure coefficients of E_B to E_T derived for the LLR and SLR models for (a) level 1 and (b) level 2 in samples with $x = 0.3$. The full circles represent experimental values. Configuration coordinate diagrams depicted in the inserts of (a) and (b) show the SLR and LLR models respectively.

level 2:

LLR: $dE_B/dp = -0.42(dE_T/dp) - 16$ (meV GPa^{-1})

SLR: $dE_B/dp = -0.23(dE_T/dp) - 4000$ (meV GPa^{-1}).

Both dependences are plotted for levels 1 and 2 in figure 2(a) and 2(b) respectively. Experimental values of dE_B/dp and dE_T/dp are shown in the figures as full circles. It is readily seen that this procedure is in favour of the LLR model in the GaAsP alloy.

2.3. Electric field dependence of emission rate

In addition, we would like to point out that the dependence of the emission rate on the electric field measured both at atmospheric and elevated pressures for the two levels in samples with $x = 0.3$ and 0.4 did not reveal any Poole–Frenkel effect. The electric field dependence of the emission rate observed in GaAlAs:Te [25] was used to support the SLR model by Bourgoin [14], who also reinterpreted the high hydrostatic pressure experiments on GaAs:Si [26]. The idea of these experiments consisted in the application of a sufficiently high pressure at room temperature to allow free carriers to be trapped at the DX centres pushed by pressure into the band gap. Cooling down the sample without relieving the pressure, the authors [26] simulated the situation which is often observed in $Ga_{1-x}Al_xAs$ ($x > 0.2$). This experiment is depicted in figure 3. As can be seen, at low temperatures the electrical conductivity does not change significantly even if the pressure is decreased. In other words, the carriers also remain trapped when the DX centre is resonant with the conduction band. At low temperatures the state with higher conductivity may be reached by illumination as shown in figure 3. This experiment has

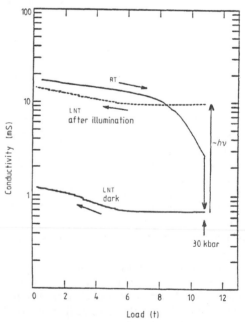

Figure 3. Conductivity of GaAs:Si against pressure at room temperature (RT) and at liquid nitrogen temperature (LNT).

been explained [26] within the framework of the LLR model because it proves the existence of a capture barrier high enough to prevent the transitions from the DX centre into the conduction band at low temperatures. The Poole–Frenkel dependence of the activation energy on the applied electric field can be considered as evidence of the existence of a long-range potential surrounding the DX centre and thus supports the SLR model.

The remaining question is whether the square-root dependence of the activation energy decrease induced by electric field holds in general for the DX centre. We have performed a series of measurements of the electric field dependence of the emission rate of the DX centre in GaAlAs:Si with different sample structures and Al contents without observing any well defined decrease of the activation energy. A very similar conclusion drawn also for GaAsP alloys can be used as support for the LLR model in both types of material.

3. Conclusions

In $GaAs_{1-x}P_x$ sulphur induces two levels exhibiting the attributes of the DX centre. Both levels seem to follow the X minimum of the conduction band. For $x = 0.3$ an influence of the L minimum has also been identified. The levels are located at 130 and 170 meV below the X minimum. The variation of thermal energy and capture barrier with composition and pressure shows that the emission energy (equal to the sum of these two quantities) is nearly independent of the composition, as observed in the GaAlAs alloys.

The fact that the DX centres are found in different semiconductor alloys allows one to distinguish between the individual and more general properties of the DX centre. The conduction band minimum which the DX centre follows may serve as an example of the former feature, whereas the applicability of the LLR model bears witness to a general property of the DX centre.

References

[1] Šmíd V, Mareš J J, Štourač L and Krištofik J 1985 *Tetrahedrally Bonded Amorphous Semiconductors* ed D Adler and H Fritzsche (New York: Plenum) pp 483–500

[2] Krištofik J, Mareš J J and Šmíd V 1985 *Phys. Status Solidi a* **89** 333

[3] Zylberstein A 1978 *Appl. Phys. Lett.* **33** 200

[4] Šmíd V, Zeman J, Krištofik J, Mareš J J, Vyzhigin Yu V, Kostylev V A, Sobolev N A, Eliseev V V and Likunova V M 1989 *Mater. Sci. Forum* **38–41** 231

[5] Kumagai O, Wünstel K and Jantsch W 1982 *Solid State Commun.* **41** 89

[6] Gilling L J 1988 *Proc. 19th Int. Conf. on the Physics of Semiconductors* ed W Zawadski (Warsaw: Institute of Physics, Polish Academy of Science) pp 1047–50

[7] Suski T and Baj M 1991 unpublished

[8] Zeman J, Šmíd V, Krištofik J, Hubík P, Mareš J J, Prinz V Ya and Rechkunov S N 1989 *Crystal Prop. Prep.* **19** 29

[9] Mizuta M, Tachikawa M, Kukimoto H and Minomura S 1984 *Japan. J. Appl. Phys.* **23** L734

[10] Zeman J, Šmíd V, Krištofik J, Hubík P. Mareš J J, Prinz V Ya and Rechkunov S N 1990 *Defect Control in Semiconductors* ed K Sumino (New York: Elsevier) pp 1067–71

[11] Hubík P, Šmíd V, Hulicius E, Krištofik J, Mareš J J, Hlinomaz P and Zeman J 1990 *Gallium Arsenide and Related Compounds* ed T Ikoma and H Watanabe (Bristol: Institute of Physics) pp 303–8

[12] Craven R A and Finn D 1979 *J. Appl. Phys.* **50** 6334

[13] Calleja E and Muñoz E 1988 *Solid State Phenomena* **10** 73

[14] Bourgoin J C 1989 *Solid State Phenomena* **10** 253

[15] Mizuta M 1990 *Defect Control in Semiconductors* ed K Sumino (New York: Elsevier) pp 1043–50

[16] Dmochowski J E, Dobaczewski L, Langer J M and Jantsch W 1990 *Defect Control in Semiconductors* ed K Sumino (New York: Elsevier) pp 1055–60; 1989 *Phys. Rev.* B **40** 9671

[17] Chadi D J and Chang K J 1989 *Phys. Rev.* B **39** 10063

[18] Mooney P M, Theis T N and Wright S L 1988 *Appl. Phys. Lett.* **53** 2546

[19] Craford M G, Stillman G E, Rossi J A and Holonyak N Jr 1968 *Phys. Rev.* **168** 867

[20] Yamaguchi E 1987 *J. Phys. Soc. Japan* **56** 2835

[21] Resca L and Resta R 1980 *Phys. Rev. Lett.* **44** 1340

[22] Lannoo M 1989 *Solid State Phenomena* **10** 209

[23] Li Ming-fu and Yu P Y 1987 *Solid State Commun.* **61** 13

[24] Huang K 1981 *Prog. Phys.* **1** 31

[25] Zazoui M, Feng S L and Bourgoin J C 1990 *Phys. Rev.* B **41** 8485

[26] Fujisawa T, Krištofik J, Yoshino J and Kukimoto H 1988 *Japan. J. Appl. Phys.* **27** L2373

Semicond. Sci. Technol. 6 (1991) B154–B155. Printed in the UK

Closing Address

Following on the after-dinner entertainment provided by the self-styled oldest participant at the banquet yesterday, I find myself, as the most elderly of the Organizers, providing the Closing Address. I should make it clear that my fellow organizer, Janusz Dmochowski, and I both agree that this was Wolfgang Jantsch's conference, both in conception and most certainly in execution. We are all indebted to him for the foresight in choice of subject matter, and for the magnificent planning and attention to detail that made this workshop such a success. What Wolfgang failed to mention in his Introductory Talk is that the first Mautendorf Conference, which incidentally was also concerned with defects, was not in Mautendorf at all but took place ten kilometres away at Mariapfarr and in somewhat less grand surroundings, i.e. in the local schoolhouse, as opposed to this magnificent medieval castle. Indeed I remember sitting in a most unrelaxed state in a desk designed, I am sure, for a ten-year-old. In all other respects the standards of these meetings have been uniformly high since that time.

Now, to some subjective and somewhat inaccurate statistics for our meeting. As far as the distribution of papers between the subject areas covered by 'DX and other metastable defects', DX centres accounted for 19.5 papers out of the 33; EL2 for 2.5, with the balance being made up of mainly unidentified levels. As far as materials are concerned, GaAs accounted for 10 papers, (AlGa)As for 14, other III–Vs for 1.5, Si for 2 and II–VIs for 3 papers.

The community has traditionally referred to 'The DX centre' and has thought only of the relaxed state of substitutional silicon donors in GaAs and (AlGa)As. One can see from the statistics quoted above how this tendency has arisen, and (AlGa)As:Si still remains the most popular system. We now know that four different DX levels occur for (AlGa)As:Si, depending on the local Al to Ga ratio. With GaAs, different energies are found for different chemical species, with extremes being formed for Te (shallow) and Ge (deep). Dmochowski, in particular, discussed experimentally the evidence for an intermediate A_1 state, and this was also the focus of much theoretical discussion. In the case of group VI donors it appears probable that the donor atom remains on the As site in the relaxed state in contrast to the Chadi and Chang mechanism for group IV donors.

Unfortunately, a number of the major theoretical groups were unable to attend, but we were treated to a masterly and, at times, almost scurrilous review by Baraff of the different approaches, and to an elegant exposition by Lannoo, which was comprehensible even to experimentalists. Suski brought experiment and theory together in an excellent review. It would be presumptuous of me to attempt to improve on these talks.

Watkins led the comparison of III–Vs with other materials with his very clear discussion of the metastable defects in Si. The off-centre distortions can be described as Jahn–Teller and pseudo Jahn–Teller in origin, which provides an interesting viewpoint with respect to DX and EL2 phenomena in III–V materials and alloys.

There was general agreement that this conference was timely and profitable. We are approaching a consensus on many of the significant features of metastability, but even such issues as positive or negative U and large versus small relaxation remain to some extent unresolved. As a consequence, the validity of various theoretical models has yet to obtain universal acceptance. Indeed, some of the basic experimental facts remain controversial, e.g. the magnetism of the DX relaxed state. Baraff warned us of 'band wagon' effects and the need to avoid forcing data to fit preconcieved notions. The statistics show that 7 papers supported a negative U model, with no support for positive U and the results of 3 papers giving no clear distinction of sign. This would appear to give a clear decision in favour of negative U except that 23 papers did not express an opinion. Baraff cited my papers as an example where the evidence was not clear-cut because of the non-random spatial distribution of changed impurities. These correlation effects, as discussed by Wilamowski and Kossut, certainly can very much enhance the mobility, and in the case of negative U can even change the energies expected in experiment. Thus, rather than accusing me of indecisiveness, Baraff was offering this as an example of the need to keep an open mind until the interpretation is clear.

Of all the papers presented, that from Morgan, proposing a double site-switch and a triple ground state, was the most extreme example of an alternative view challenging existing ideas. Baraff, in contrast, gave examples of disbelief leading to unwillingness to look for other levels which have subsequently been established experimentally.

0268-1242/91/10B154+02 $03.50 © 1991 IOP Publishing Ltd

Some 30 % of the papers employed hydrostatic pressure, but this should not be thought of as simply enabling the experimentalist to study AlGaAs without introducing aluminium. Rather, pressure assists the identification of alloy effects, which are now fairly well understood, and multicomponent features are becoming resolved.

With respect to other alloy systems, we heard something of GaAsP but nothing of AlInAs, where metastable centres are readily formed and are of importance in device operation. Are the metastable centres in the binaries InSb, GaSb and the II–VIs DX in character or are other mechanisms and features involved? There was one viewgraph showing Hall data for InSb under pressure, arguing that negative U centres were not involved. Dobaczewski emphasized the different features of GaSb.

Particularly evident at this meeting was the richness and precision of the experimental data that are now emerging. ESR, MDC, ODESR as discussed by Kaufmann, von Bardeleben, Kennedy and Spaeth are by now well established and give very precise and site-specific information, enabling configurations to be determined unambiguously. Added to resonance in telling us about the local environment, we have local vibrational mode (LVM) spectroscopy as presented in the beautiful high resolution data of Alt for oxygen in GaAs, which show clearly the three levels involved. Most excitingly, preliminary LVM results for the DX relaxed state from Wolk and the Berkeley group were reported. This latter contribution represented potentially the most significant advance reported at this workshop, although the shift of the new line from the main Si:LVM line is surprisingly small and remains to be explained.

One striking feature of this workshop in comparison with IPCS is the few papers that involved heterostructures and low-dimensional physics.

Several authors were concerned with barriers and transients. Of these, the Chairman's presentation provided some lively discussion on the question of the factor of two that occurs in the emission energy when two particles are involved.

It was a pleasure to see a high proportion of younger scientists making a contribution to the workshop and taking part in discussions in the castle and on the ski slope. The meeting was indeed timely, and an extraordinary variety of ideas was examined critically and new perspectives emerged. While the interpretation of DX phenomena in GaAs and AlGaAs is becoming clearer, how much of the interpretation can be carried over into other III–Vs and the II–VIs remains to be established. established.

Finally, we should express our thanks to Wolfgang and to Gerhard Brunthaler for organizing the meeting so superbly and to Alex Falk of the Apple Mac and 'organizational announcements'; Gerhard Brunthaler and Gernot Ostermayer, particularly, for assistance with the manuscripts and Ulla Hanneschtaeger who overcame any local problems so imperturbably. They all contributed magnificently to give us a superb meeting.

R A Stradling
Honorary Editor

T - #0344 - 101024 - C0 - 280/208/9 [11] - CB - 9780750301534 - Gloss Lamination